In Situ Testing in Geomechanics

T0199712

Demanding a thorough knowledge of material behaviour and numerical modelling, site characterization and in situ test interpretation are no longer just basic empirical recommendations.

Giving a critical appraisal in the understanding and assessment of the stress-strain-time and strength characteristics of geomaterials, this book explores new interpretation methods for measuring properties of a variety of soil formations.

Emphasis is given to the five most commonly encountered in situ test techniques:

- standard penetration tests
- cone penetration tests
- vane test
- pressuremeter tests
- dilatometer tests.

Ideal for practising engineers in the fields of geomechanics and environmental engineering, this book solves numerous common problems in site characterization. It is also a valuable companion for students coming to the end of their engineering courses and looking to work in this sector.

Fernando Schnaid is Associated Professor of Civil Engineering at Federal University of Rio Grande do Sul, Brazil.

This book is dedicated to my family and to people I have walked with along the way – mentors, teachers, colleagues, friends and students – irrespective of nationality, colour, religion and social background.

To Carmen, Gabby, Mano and
Lu with love

In Situ Testing in Geomechanics

The main tests

Fernando Schnaid

Taylor & Francis
Taylor & Francis Group

LONDON AND NEW YORK

First published 2009
by Taylor & Francis
2 Park Square, Milton Park, Abingdon, Oxon OX14 4RN

Simultaneously published in the USA and Canada
by Taylor & Francis
711 Third Avenue, New York, NY 10017, USA

Taylor & Francis is an imprint of the Taylor & Francis Group, an informa business

© 2009 Fernando Schnaid

Typeset in Sabon by
Book Now Ltd, London

British Library Cataloguing in Publication Data
A catalogue record for this book is available
from the British Library

Library of Congress Cataloging in Publication Data
Schnaid, Fernando.
In situ testing in geomechanics: the main tests/Fernando Schnaid.—1st ed.
    p. cm.
Includes bibliographical references and index.
1. Soils—Testing. 2. Engineering geology. I. Title. II. Title: In situ testing in
geomechanics.
TA710.5.S36 2008
624.1′51—dc22   2007047981

ISBN10: 0–415–43385–1 (hbk)
ISBN10: 0–415–43386–X (pbk)
ISBN10: 0–203–93133–5 (ebk)

ISBN13: 978–0–415–43385–3 (hbk)
ISBN13: 978–0–415–43386–0 (pbk)
ISBN13: 978–0–203–93133–2 (ebk)

Contents

Foreword

Professor Fernando Schnaid, author of *In Situ Testing in Geomechanics: The Main Tests*, has been deeply involved in this subject matter since completing his Ph.D. dissertation at the University of Oxford in the UK. A significant part of his research since then has been devoted to the use of in situ testing for geotechnical site characterization as evidenced in several journal papers and a number of state-of-the-art reports presented at important international conferences and symposia.

This valuable book is informed by Professor Schnaid's excellent theoretical engineering mechanics background and covers in depth both the execution and interpretation of the main in situ tests used in both in practice and research. The book covers the core in situ methods: the standard penetration test (SPT); the static cone penetration test with (CPTU) and without (CPT) penetration pore pressure measurement; pressuremeter tests (PMTs); the field vane test (FVT), and flat dilatometer tests (DMTs). Extensive and advanced guidance is given for the interpretation of each in situ testing technique covered by the book, comprising testing procedure, soil classification, assessment to design parameters and the direct use of results to foundation design.

As to the in situ tests dealt themselves, this excellent book is a commendable attempt at broadening the existing interpretation framework for bonded, aged and residual soils that benefits greatly from the author's research and his first-rate professional experience.

The most salient feature of *In Situ Testing in Geomechanics*, openly declared in the author's preface, is the attempt to improve, in accordance with the available knowledge, the interpretation of in situ tests, moving from purely empirical site-dependent correlations to more sound approaches based on simplified soil mechanics theory.

With this wide-ranging and updated book on in situ testing, the author offers an invaluable service to the geotechnical community, affording a new perspective for a more rational geotechnical site characterization and geotechnical design. *In Situ Testing in Geomechanic* will turn out to be an important reference text for all of those operating in the areas of Geotechnical Engineering and Engineering Geology. It is likewise recommended to practising engineers as well as to university researchers and students.

Michele Jamiolkowski
Emeritus Professor, Technical University, Torino, Italy
Past President, International Society of Soil Mechanics and
Geotechnical Engineering

Preface

In situ testing in geomechanics has been a familiar subject for engineers for about half a century. Some basic concepts and ideas in this field are so firmly rooted in intuition that professionals often tend to rely on experience rather than knowledge, on empiricism rather than theory. Although I fully recognize that engineering judgement and experience are key factors for safe and economical geotechnical design, there is a growing awareness that current design practice has evolved from early experimental approaches into a sophisticated subject area demanding a thorough knowledge of material behaviour and numerical modelling.

My own assessment is that our analytical and numerical knowledge has now reached a state of maturity where it is possible to give a reasonably coherent and logical account of the theory associated with the mechanical behaviour of soils and soil testing. If developments in the field of in situ test investigation have been for decades the subject matter of theoretical research and technological innovation, today the results of those developments have established themselves as standard textbook subjects, presented to senior students in most engineering schools. Surprisingly enough, a comprehensive review compiling the technical information and theoretical background necessary to equip professionals to employ sound engineering judgement when selecting appropriate tools for the geo-characterization of natural soils has not yet appeared in a single textbook form.

This book is an attempt to fill that gap. It provides engineers with the information and data necessary for the assessment of constitutive parameters to enable the proper design of geotechnical structures. It attempts to combine the principles of soil mechanics with the more advanced topics which are of prime importance to undergraduate and graduate students in the field of site investigation. In addition, it has been planned to provide a reference guide for the engineer who is interested in following the development of this field throughout the past decades. Going from theory to practice, the book is written with the intended purpose of:

- briefly describing the equipment, characteristics and procedures associated with the main commercial in situ testing techniques used worldwide;
- introducing the necessary concepts and background theory for interpretation of each testing technique;
- presenting current interpretation methods, developed to estimate the values of design parameters, such as strength and stiffness;
- discussing some useful empirical and semi-empirical approaches that help to predict the performance of geotechnical structures;
- reviewing the applicability of these methods to practical problems by summarizing the knowledge and experience acquired in geotechnical design, ranging from ordinary materials encountered in many conventional types of structures and foundations to unusually difficult ground conditions.

The book is intended to be viewed as a working reference by engineers in the civil engineering, geological, geotechnical and environmental fields.

Acknowledgements

As may be expected, I am indebted to a number of people. Writing a textbook involves a major commitment of time, which could only be accomplished with the collective effort of my colleagues and students. The final result has benefited immensely from these interactions, which have given me guidance in correcting and improving the original work. I should start by expressing my gratitude to the late Peter Worth, to Michele Jamiolkowski, William Van Impe, Guy Houlsby, Victor de Mello, Mark Randolph, Martin Fahey, Silvano Marchetti and Antonio Gens. I will be always indebted to Michele Jamiolkowski for his encouragement and support in the early stages of my career. When asking Silvano Marchetti for permission to use some of his photographs and images, he sanctioned their use, adding that 'our common interests really create a strong link and a sense of familiarity in our community'. His response is a measure of the privilege I have enjoyed in sharing views and experience with these remarkable scientists. I am also very grateful to colleagues with whom I have been collaborating throughout these years: Hai-Sui Yu, Barry Lehane, Luiz Guilherme de Mello, Leandro Moura Costa Filho, Sandro Sandroni, Marcio Almeida, Fernando Danzinger, Roberto Coutinho and Marcelo Rocha. Many thanks to my dearest colleagues from the ISSMGE Technical Committee TC 16: Peter Robertson, Paul Mayne, John Powell, Tom Lunne, Martin Fahey, Antonio Viana da Fonseca, An-Bin Huang and Zbigniew Mlynarek.

I wish to thank the members of staff and students from Federal University of Rio Grande do Sul. The group has grown around my dearest friend Prof. Jarbas Milititsky and has developed into a stimulating environment for working and researching. I am therefore indebted to my colleagues, and in particular to Nilo C. Consoli and to my PhD students José Mário Soares, Fernando Maria Mántaras, Pedro Domingos Marques Prietto, Edgar Odebrecht, Leandro Spinelli, Luiz Artur Kratz de Oliveira, Gabriela Maluf Medeiros, Maria Isabel Timm, Raymundo Carlos Machado Ferreira Filho, Bianca de Oliveira Lobo, Jucélia Bedin and Francisco Dalla Rosa.

Grateful acknowledgement is made to the following firms and organizations which were kind enough to provide me with information, photographs and drawings from which material for this book has been extracted: Fugro Ltd, A.P. van den Berg, Cambridge In Situ and Silvano Marchetti. Finally, I should say that I would not have been able to complete the work without the dedicated organization of each chapter by Ms Ana Luiza Oliveira and the tracing of most figures by Mr Rubens Renato Abreu.

List of symbols

a = area ratio $(= A_N/A_T)$
a_c = particle acceleration
a_i = pressuremeter cavity radius during expansion
a_{max} = maximum cavity displacement
a_{max} = maximum horizontal acceleration
a_0 = initial cavity radius
a_0 = particle acceleration
A = area
A_b = area of the pile base
A_r = strength anisotropy ratio
A_s = friction sleeve shaft area
A_{sb} = bottom end area of friction sleeve
A_{st} = top end area of friction sleeve
b = coefficient
b_c, b_q, b_γ = base inclination factors
B = width of the footing
B_q = pore pressure parameter ratio
c = soil cohesion intercept
c = velocity of wave propagation
c' = soil cohesion intercept (effective stress)
$c'_{(b)}$ = cohesion intercept at elastic-plastic boundary
c'_0 = Mohr-Coulomb triaxial peak cohesion intercept
C = grain characteristics
C_c = compression index
C_c^* = remoulded (or intrinsic) compression index
C_h = horizontal coefficient of consolidation
C_k = hydraulic conductivity change index
C_N = depth correction factor
C_v = vertical coefficient of consolidation
C_1, C_2, C_3, C_4 = coefficients
CRS = cyclic shear stress ratio
d = probe diameter

d_c, d_q, d_γ = depth factors
d_s = thickness of sublayer
D = foundation depth
D = diameter
D_r = relative density
D_{50} = 50% by weight of particles having smaller diameter
e = blade thickness
e = void ratio
e_0 = initial void ratio
e_L = void ratio at liquid limit
e_{max} = maximum void ratio
e_{min} = minimum void ratio
E = Young's modulus
$E_A(t)$ = energy delivered to the soil
E_c = isotropic stiffness
$E_C(t)$ = total kinetic energy
E_d = deviatoric stiffness
E_D = dilatometer modulus
$E_D(t)$ = energy dissipated in all nodes by the viscous damping criteria
E_m = Ménard modulus
$E_P(t)$ = total potential gravitational energy
$E_{PE}(t)$ = total potential elastic energy
E_s = reaction modulus
E_{si} = initial Young's modulus
$E_{sampler}$ = sampler potential energy
E_{spt} = energy on SPT
$E_T(t)$ = total potential elastic energy
E_{TOT} = instantaneous total energy
E_u = undrained Young's modulus
f = function
f' = derivative of f
f_p = pile unit side friction
f_s = sleeve friction
f_t = sleeve friction corrected for pore pressure effects
F = force
F_d = dynamic reaction force
$F(e)$ = void ratio function
F_r = normalized friction ratio
$F(t)$ = force
F_1, F_2 = coefficients
g = gravitational acceleration
g_c, g_q, g_γ = ground surface inclination factors
G = gravitational acceleration
G = shear modulus

G_0 = small strain shear modulus
G_{cor} = corrected shear modulus
G_{hh} = shear modulus measured from horizontally propagated, horizontally polarized shear waves
G_{hv} = shear modulus measured from horizontally propagated, vertically polarized shear waves
G_{max} = maximum shear modulus
G_{mea} = measured shear modulus
G_s = specific gravity
G_{sys} = pressuremeter system shear stiffness
G_{ur} = unload–reload pressuremeter shear modulus
G_{hh} = shear modulus measured from horizontally propagated, horizontally polarized shear waves
G_{vh} = shear modulus measured from vertically propagated, horizontally polarized shear waves
G_{hv} = shear modulus measured from horizontally propagated, vertically polarized shear waves
G_z = specific gravity
H = height of fall
H = layer thickness
i_c, i_q, i_γ = load inclination factors
$i_{\delta,\beta}$ = reduction factor
I_c = soil behaviour type index
I_D = material index
I_L = liquidity index
I_p = plasticity index
I_r = rigidity index
I_R = relative dilatancy index
J = empirical coefficient
J = additional resistance coefficient
k = hydraulic conductivity or coefficient of permeability
k_c, k_e, k_p = bearing capacity factors
k_h = coefficient of hydraulic conductivity in horizontal direction
k_v = coefficient of hydraulic conductivity in vertical direction
K = soil structure
K_C = correction factor
K_D = horizontal stress index
K_0 = coefficient of earth pressure at rest
l = total rod length
L = foundation length
LL = liquid limit
LP = plastic limit
m = rod mass
m_v = coefficient of volume compressibility

M = constrained deformation modulus
M = earthquake magnitude
M = slope of critical state line
M_{DMT} = dilatometer constrained modulus
M_h = hammer mass
M_r = total rod mass
M_0 = reference constrained deformation modulus
MSF = magnitude scaling factor
n = porosity
N = average value of blow count number
N_c, N_q, N_γ = bearing capacity factors
N_{ke} = cone factor
N_{kt} = cone factor
N_{SPT} = blow count number
$N_{\Delta u}$ = cone factor
N_1 = blow count number for a stress reference of 100 kPa
$(N_1)_{en}$ = blow count number for an energy reference value
$(N_1)_{60}$ = normalized blow count number
N_{60} = blow count number for 60% potential energy
N_{72} = blow count number for 72% potential energy
OCR = overconsolidation ratio
p = mean stress
p' = mean effective stress
p_a = atmospheric pressure
p_0 = effective overburden pressure of soil at foundation level
p_L = pressuremeter limit pressure
p_{LM} = Ménard pressuremeter parameter
p_0 = initial pressuremeter cavity pressure
p_0, p_1 = dilatometer pressures
P = pile perimeter
P_0 = total horizontal stress
P_u = ultimate lateral soil resistance
PE^*_{h+r} = theoretical potential energy
q = average net applied stress
q = shear strength
q_b = unit end-bearing resistance
q_c = measured cone tip resistance
$q_{c,avg}$ = average value of q_c
q_{cdr} = cone resistance under drained conditions
q_{cund} = cone resistance under undrained conditions
q_{c1} = normalized cone tip resistance
$(q_{c1N})_{cs}$ = stress-normalized cone tip resistance corrected for fines content
q_n = net cone resistance
q_s = unit skin friction

q_t = corrected cone resistance
q_u = uniaxial compression strength
Q = function of grain compressive strength
Q = applied vertical load
Q = maximum elastic deformation
Q_b = pile end-bearing capacity
Q_s = pile shaft friction capacity
Q_t = normalized cone resistance
r = radial distance
r_d = stress reduction factor
r_p = radius of plastic zone
r_0 = radius of cavity
R = empirical correlation coefficient
R_f = friction ratio
R_M = coefficient
R_u = ultimate static soil resistance
RR = recompression ratio
s = coefficient of settlement
s_c, s_q, s_γ = shape factors
s_{hh} = velocity of horizontally propagated, horizontally polarized shear waves
s_{hv} = velocity of horizontally propagated, vertically polarized shear waves
s_u = undrained shear strength
s_{uh} = undrained shear strength in horizontal direction
s_{ur} = remoulded undrained shear strength
s_{uv} = undrained shear strength in vertical direction
s_{vh} = velocity of vertically propagated, horizontally polarized shear waves
$(s_u)_{ps}$ = plane strain undrained shear strength
$(s_u)_{tc}$ = triaxial undrained shear strength
s_0 = dilatometer blade displacement
S = slope of pressuremeter loading data ($\ln p'$ versus $\ln \varepsilon_c$ plot)
S_d = slope of pressuremeter unloading data
S_r = degree of saturation
S_t = sensitivity
S_{1DMT} = settlement of shallow foundations from dilatometer tests
t = time
t_{flex} = time derived from DMT decay dilatometer test
t_{50} = time for 50% dissipation of excess porewater pressure
T = temperature
T = torque
T^* = modified time factor
T_c = torque at any rotation rate
T_{dr} = drained torque
T_H = torque mobilized on horizontal surfaces of vane blade

T_m = maximum value of measured torque corrected for apparatus and rod friction

T_{und} = undrained torque

T_V = torque mobilized on vertical surfaces of vane blade

$T(t)$ = kinetic energy

T_{50} = dimensionless time factor at $U = 50\%$

u = porewater pressure

u = longitudinal displacement

u_a = pore air pressure

u_i = in situ pore pressure $t = 0$

u_t = pore pressure at time $= t$

u_w = porewater pressure

u_0 = in situ pore pressure

u_1 = pore pressure measured on the cone

u_2 = pore pressure measured behind the cone

u_3 = pore pressure measured at friction sleeve

u_4 = pore pressure measured behind friction sleeve

U = normalized excess pore pressure

U = degree of drainage

U_c = uniformity coefficient

UU = unconsolidated undrained triaxial test

v = rate of rotation

v = penetration rate

v_p = peripheral velocity

v_s = velocity of shear wave

v_{s1} = normalized shear wave velocity

V = non-dimensional parameter

V_0 = initial cavity volume

$V(t)$ = velocity

$V(t)$ = potential energy

w = angular strain rate

w = pile settlement

w = water content

w_L = water content at liquid limit

w_p = water content at plastic limit

$W_s(t)$ = work done by non-conservative forces acting on the sampler–soil system

$W_{nc}(t)$ = work done by other non-conservative forces related to energy losses

y = radial displacement

y_c = pile deflection

z = depth

Z_h = rod impedance

Z_m = gauge zero offset when vented to atmospheric pressure

Greeks

α = constant
α = cone roughness
α_p = pile class factor
α_s = factor depending on the pile class and soil conditions
β = constant
β = ratio of plastic strain component
$\beta_{(b)}$ = dilation at elastic-plastic boundary
γ = soil density unit weight
γ_d = dry unit weight
γ_p = shear strain
γ_{nat} = unit weight
γ_w = unit weight of water
Γ = critical state parameter
δr = thickness of soil element
δy = change in thickness of soil element
Δ = change
ΔA = external pressure applied to dilatometer membrane in free air
ΔB = internal pressure which, in free air, lifts the dilatometer membrane
    centre 1.1 mm
ΔL = soil thickness
Δt = time interval
Δu = excess pore pressure
ΔV = change in cavity volume
$\Delta \rho$ = permanent penetration of the sampler
ε = strain or unit deformation
ε_c = cavity strain
ε_{cmax} = maximum cavity strain
ε_r = radial strain
ε_v = vertical strain
ε_z = axial strain
ε_θ = circumferential strain
$\dot{\varepsilon}_r^e$ = elastic component of radial strain increments

$\dot{\varepsilon}_\theta^e$ = elastic component of circumferential strain increments

$\dot{\varepsilon}_r^p$ = plastic component of radial strain increments

$\dot{\varepsilon}_\theta^p$ = plastic component of circumferential strain increments
η_1, η_2, η_3 = efficiency coefficients
λ = cone roughness
λ = slope of critical state line in $e - \ln p$ plot
Λ = critical state parameter
μ = longitudinal displacement

μ_r = coefficient
v = Poisson's ratio
ρ = mass per unit volume
ρ = density
ρ = settlement
ρ_s = mass per unit volume of the bar
ψ = soil dilation
Ψ = state parameter
σ = stress component
σ_{atm} = atmospheric pressure
σ_c = cylindrical cavity pressure
$\sigma'_{c,lim}$ = cylindrical cavity limit pressure
$\sigma'_{s,lim}$ = spherical cavity limit pressure
σ_h = total *in situ* horizontal stress
σ'_h = effective *in situ* horizontal stress
σ_{h0} = initial horizontal stress (total stresses)
σ'_{h0} = initial horizontal stress (effective stresses)
σ'_p = preconsolidation pressure
σ_r = total radial stress
σ'_r = effective radial stress
σ_s = spherical cavity pressure
σ_v = total vertical stress at foundation level
σ_{v0} = total overburden stress
σ'_{v0} = effective vertical stress
σ_z = total axial stress
σ_θ = total circumferential stress
σ'_θ = effective circumferential stress
σ_0 = in situ total stress
σ'_0 = in situ effective stress
σ'_1 = major effective principal stress
σ'_3 = minor effective principal stress
Σ = sum
τ_{av} = average applied shear stress
τ_H = value of shear in the vane horizontal failure surface
τ_{mH} = maximum value of shear in the vane horizontal failure surface τ_{mV}
ϕ' = effective angle of internal friction
ϕ'_{cv} = critical state angle of internal friction
ϕ'_p = effective peak angle of internal friction
ϕ'_{ps} = plane strain friction angle
ϕ'_{tc} = triaxial friction angle

Chapter 1

Introduction

The Earth is a complex three-dimensional object with numerous structures and hydro-geomorphologies that are difficult to visualize and characterize. It is therefore essential to develop concepts and techniques that will support the identification and description of the Earth's spatial arrangements in manipulable two-dimensional representations. Although this challenge is one to be approached by a coalition of hybrid disciplines, it has prompted the development of the scientific field of soil mechanics, and in particular the area of site characterization.

(Schnaid, 2005)

General considerations

The basic objectives of site characterization are the following (modified from Jamiolkowski *et al.*, 1985):

- to acquire topographical, hydro-geological, geotechnical and geo-environmental information that is relevant to the requirements of a project;
- to produce detailed and representative soil profiles, including determining the position of the groundwater table and soil index properties;
- to provide suitable geotechnical data to support safe and reliable design, including:
  o assessment of the initial geostatic stresses and the stress history of the soil prior to any construction activity;
  o prediction of the stress-strain-time and strength characteristics of the soil layers encountered; and
- to assess environmental changes in groundwater and drainage conditions of the site and the surrounding ground and structures.

Several publications provide surveys of many of the technical developments within the field and give a good overview of their immediate applications to site characterization (e.g. Clayton and Hababa, 1985; Rowe, 2001; US Army Corps of Engineers, 2001). It is now recognized that the work of geo-engineers has expanded enormously over the past decades and professionals have become more aware of, and are better equipped to comprehend, the interdisciplinary issues covered by the planning and organization of site characterization programmes. Complex projects, such as dams, large infrastructure projects and electricity generating plants, necessitate a comprehensive field investigation led by a professional team consisting of geologists, engineering geologists, geotechnical engineers and civil engineers with expertise in geotechnical investigation.

Any project or related activity requires a fundamental understanding of its geotechnical and environmental constraints and, for that reason, each professional involved should comprehend the interplay of processes that leads to site characterization. Figure 1.1 illustrates the stages associated with characterization in a flowchart that identifies the relevance of desk studies, site reconnaissance and the stages of preliminary and main ground investigation. This process consists of a desk study in which the relevant information relating to the site is collected and organized. Site reconnaissance is a vital stage and should cover, in addition to the whole site, any areas in the vicinity that might influence the hydro-geological aspects of the project. From project conception to the early stages of design, initial geotechnical investigations will be general and will provide broad information relevant to a particular project. As project development continues, ground investigation becomes more detailed and appropriate exploration methods are selected according to the ground conditions of the site. At this stage, mechanical testing is shown to be an integral element of site investigation, determining the basic soil classification, supporting the conceptual model adopted in design and establishing the representative strength and stiffness parameters required for engineering design calculations. A firm understanding of laboratory and in situ tests and the constitutive relationships that link material behaviour is therefore essential to optimize engineering geotechnical design.

Important interrelationships exist between the field and laboratory which, given the numerous techniques currently available, offer different strategies for characterizing soils. A general recommendation is to examine the characteristics of the soil from a macro to a micro perspective. Surface wave methods provide a spatial three- or two-dimensional subsurface representation of large areas. Geophysical methods give a qualitative picture of the site which, however, does not remove the need for direct measurements attained by in situ tests. Standard penetration tests (SPTs), cone penetration tests (CPTs), dilatometer tests (DMTs) and pressuremeter tests (Ménard pressuremeters [MPTs] or self-boring pressuremeters [SBPs]) are designed to gather one-dimensional information about the ground that, with the application of some simplified assumptions, can be interpreted to assess average properties of soil profiles. Finally, the laboratory deals with close examination of the elemental material properties. Since each testing technique responds to different physical properties and gives information about the

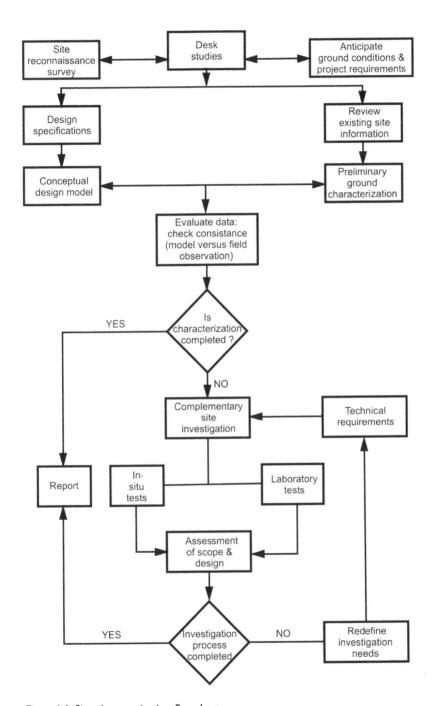

Figure 1.1 Site characterization flowchart.

varying nature of different soils, a successful and cost-effective site characterization programme should consist of an appropriate combination of field
tests (three-, two- and one-dimensional representations) and laboratory
tests, so that the relevant information can be synthesized and applied with
confidence.

The approach of this book is to give a general picture of the field of geo-
characterization with an emphasis on in situ testing soil mechanics. Every
effort has been made to produce a broad view of the technical information
and theoretical backgrounds that are necessary to equip professionals
to employ sound engineering judgement when selecting appropriate tools
for the geo-characterization of natural soils. The aim is to develop an
understanding of the ways in which the key experimental and theoretical
elements are correlated in the domain of geotechnical characterization,
acknowledging the potential and the limitations of different tests and measurements, and recognizing hypotheses and assumptions associated with
mechanical models developed to describe the behaviour of natural soils.

A variety of in situ tests is commercially available to meet the needs of
geotechnical engineers. These existing field techniques can be broadly
divided into two main groups:

- *non-destructive tests* that are carried out with minimal overall disturbance of soil structure and little modification of the initial mean effective stress during the installation process. The non-destructive group
  comprises seismic techniques, pressuremeter probes and plate loading
  tests: a set of tools that is generally suitable for rigorous interpretation
  of test data under a number of simplified assumptions;
- *invasive, destructive tests* where inherent disturbance is introduced by
  the installation of the probe into the ground. Invasive-destructive techniques comprise SPTs, CPTs and DMTs. These penetration tools are
  robust, easy to use and relatively inexpensive; however, the mechanism
  associated with the installation process is often complex and therefore
  a rigorous interpretation is only possible in a limited number of cases.

A geophysical survey is, for example, a truly non-destructive, non-invasive
subsurface exploration technique. In recent years there has been a steady
increase in the perceived value of geophysics in representing complicated
subsurface conditions which display extensive spatial variability. Cross-
and downhole methods, including the introduction of sensors within the
seismic cone, are defined in geophysics as intrusive methods, since they are
generally performed within boreholes. However, since shear waves propagate in a soil mass that has not been disturbed by installation in the field of
geomechanics these are regarded as a *non-destructive* type of test is required,
rather than an *invasive* penetration technique in which interpretation is sensitive to the shear zone created around a penetrating probe. Since the in situ

behaviour of geomaterials is complex, current research efforts are placing emphasis on correlations with mechanical properties that are based on the combination of different sensors in a single test device, usually combining a non-destructive with an invasive technique, such as the seismic cone and cone pressuremeter. A summary of the key information relating to the commonly used in situ tests is given in Table 1.1, in which measurements of each testing technique are described and common applications are identified.

Commercial in situ tests can be deployed into the ground using a wide range of pushing equipment, such as manual lifting of the SPT weight, very lightweight rigs, trucks and vessels that support remote-controlled operations. Examples of available technologies for both onshore and offshore operations are given in Figures 1.2 to 1.8. Marine geotechnical offshore and near-shore investigations can be performed from dedicated drillships equipped for geological survey, geotechnical stratigraphy and reservoir characterization (Figures 1.2 and 1.3). These ships are often equipped with seabed units that can perform tests mechanically, with the unit submerged and placed onto the seabed where the test is performed automatically according to prescribed procedures (Figure 1.4). Contact between the surface and the seabed unit is by means of acoustic data modem or via a bypass cable.

A variety of driving systems is available worldwide for onshore operations. Trucks, such as those illustrated in Figures 1.5 and 1.6, are ideal vehicles for providing quick, cost-effective mobile platforms capable of high production rates. The main obstacle to performing site investigation operation tests in these robust, off-road 100 to 200 kN vehicles is mobility on soft ground. This can be overcome by the integrated track systems shown in Figure 1.6, which combine hydraulically driven tracks with the truck's wheels (i.e. by retracting the tracks the vehicle can run as an ordinary truck), giving the required mobility in all ground conditions.

Operations in locations with restricted access can be performed using lightweight rigs that are mounted on trailers or mini-crawlers with compact dimensions, supporting single or double hydraulic cylinder units (see Figures 1.7 and 1.8). Screw anchors can be used to provide the required reaction force of up to 150 kN.

Codes of practice

The field of site investigation requires a significant degree of expertise because the extent of an investigation plan is shaped by a large number of factors, such as the variability of soil conditions, financial constraints, time restrictions, risk assessment and size of the project. The individuals carrying out the ground investigation are among the first to assess the site and therefore senior-level, experienced personnel should be involved in selecting, planning and supervising the exploration methods for the foundation work.

Table 1.1 Commercial in situ testing techniques (modified from Schnaid et al., 2004)

Category	Test	Designation	Measurements	Common applications
Non-destructive or semi-destructive tests	Geophysical tests: seismic refraction surface waves crosshole test downhole test	SR SASW CHT DHT	P waves from surface R waves from surface P and S waves in boreholes P and S waves with depth	Ground characterization Small strain stiffness, G_0
	Pressuremeter test pre-bored self-boring	PMT SBPM	G, $(\psi \times \varepsilon)$ curve G, $(\psi \times \varepsilon)$ curve	Shear modulus, G Shear strength In situ horizontal stress Consolidation properties
	Plate loading test	PLT	$(L \times \delta)$ curve	Stiffness and strength
Invasive penetration tests	Cone penetration test electric piezocone	CPT CPTU	q_c, f_s q_c, f_s, u	Soil profiling Shear strength Relative density Consolidation properties
	Standard penetration test (energy control)	SPT	Penetration (N value)	Soil profiling Internal friction angle, ϕ'
	Flat dilatometer test	DMT	p_0, p_1	Stiffness Shear strength
	Vane shear test	VST	Torque	Undrained shear strength, s_u
Combined tests (invasive + non-destructive)	Cone pressuremeter	CPMT	$q_c, f_s, (+u)$, G, $(\psi \times \varepsilon)$	Soil profiling Shear modulus, G Shear strength Consolidation properties
	Seismic cone	SCPT	$q_c, f_s, v_p, v_s, (+u)$	Soil profiling Shear strength Small strain stiffness, G_0 Consolidation properties
	Resistivity cone	RCPT	q_c, f_s, ρ	Soil profiling Shear strength Soil porosity
	Seismic dilatometer		p_0, p_1, v_p, v_s	Stiffness (G and G_0) Shear strength

Figure 1.2 Geotechnical vessel *Markab*, Panama (courtesy of Fugro Ltd).

Figure 1.3 Geotechnical vessel *Fugro Explorer* (courtesy of Fugro Ltd).

Figure 1.4 Roson seabed rig (courtesy of A.P. van den Berg).

Figure 1.5 CPT-truck at the Kwai Chung Container Port, Hong Kong (courtesy of Fugro Ltd).

Figure 1.6 Tracked truck (courtesy of A.P. van den Berg).

Figure 1.7 Lightweight rig (courtesy of A.P. van den Berg).

Figure 1.8 Mini-crawler (courtesy of A.P. van den Berg).

Although the design of a site investigation programme should ideally be carried out by a soil mechanics specialist, this is not always the case in practice. The standard of practice ranges from services offered by reputable companies with considerable expertise and experience to companies that do not appreciate the risks and economic implications that arise from poor or inaccurate ground evaluation.

In an area where expertise is of ultimate importance, codes of practice are intended to provide guidance on the practical aspects of the field work. Codes and recommendations should be viewed as the sum of experience and consensus of opinion of a group of experts with respect to techniques and procedures in a given subject. These recommendations help in defining the content and number of tests necessary to establish the compatibility of the site with the proposed construction, bearing in mind that expenditure generally increases with larger projects and more difficult site conditions.

Engineers are always struggling with the balance between efficiency and economy. This balance, recognized early on by Peck (1969), is now incorporated into codes of practice such as the Eurocode. Peck stated that investigation methods form three broad groups:

- *Method I*: carry out limited investigation and adopt an excessive factor of safety during the design.
- *Method II*: carry out limited investigation and make design assumptions in accordance with general average experience.
- *Method III*: carry out very detailed investigation.

As recognized by Clayton *et al.* (1995), in the first method the engineer will make conservative estimates of the constitutive material properties and the design for the given conditions. The second method requires considered estimates of parameters and of the ranges within which they could deviate from the expected values. Observation of performance is recommended so that appropriate corrective action can be taken in case of unexpected construction response. Comprehensive site investigation, as prescribed in the third method, increases design reliability – in this case the engineer will have to prove that the extra cost of detailed investigation is worth expending in terms of a particular project.

Eurocode 7 (1997) extended these early concepts by dividing investigation, design and implementation of geo-constructions into three categories:

- *Category I*: includes small and relatively simple structures for which it is possible to guarantee that the fundamental requirements will be satisfied on the basis of experience and qualitative geotechnical investigation. Geotechnical Category I procedures will only be sufficient in ground conditions which are known, from comparable experience, to be sufficiently straightforward to permit routine methods to be used for foundation design and construction.
- *Category II*: includes conventional types of structures and foundations with no abnormal risks or unusual or exceptionally difficult ground or loading conditions. Structures in Geotechnical Category II require quantitative geotechnical data and analysis to ensure that the fundamental requirements will be satisfied. Routine procedures for field and laboratory testing and for the design and execution of construction may be used.
- *Category III*: includes structures or parts of structures which do not fall within the limits of Categories I and II. Geotechnical Category III includes very large or unusual structures, structures involving abnormal risks or unusual or exceptionally difficult ground or loading conditions and highly seismic areas.

The extent of ground investigation is related to the classification of the structure or part of the structure, which in turn is a function of factors such as the nature and size of the project, local conditions (traffic, utilities, hydrology, subsidence, etc.), ground and groundwater conditions, and regional seismicity, among others. Since the sequence of stages and the extent of the investigation will vary from site to site, costs of the site investigation cannot be determined a priori. Weltman and Head (1983) argued that:

> Sufficient finance should be allocated for a thorough investigation to facilitate economic and safe geotechnical design and to reduce the possibility of unexpected ground conditions being encountered during the construction of the works which frequently lead to costly delays in a contract. Such delay can cost many times more than would a properly conducted ground investigation.

According to the US Army Corps of Engineers (2001):

> Insufficient geotechnical investigations, faulty interpretation of results and failure to portray results in a clear, understandable manner may contribute to inappropriate designs, delays in construction schedules, costly construction modifications, use of substantial borrow material, environmental damage to the site, post-construction remedial work, and even failure of a structure and subsequent litigation.

Interpretation methods

The frontiers of a scientific field are defined as much by the tools available for observation as by the breakthroughs in theoretical developments. Several important concepts have been introduced or adapted for soil mechanics in past decades, bringing new perspectives for the characterization of geomaterials. Interpretation of in situ tests has become a highly specialized subject, continually evolving by employing formal analytical or numerical solutions in addition to previously accepted semi-empirical approaches. The following paragraphs provide guidance on how to link theory to practice by establishing four classes within which the interpretation of in situ tests can be broadly grouped (Schnaid, 2005).

Class I

These are analytical solutions capable of idealizing the field test into an equivalent realistic form. The ability to extract a plausible solution depends on the accuracy of constitutive models in representing soil behaviour and on the degree of accuracy of the imposed boundary conditions. This is achieved in a limited number of cases. Exact closed-form solutions are only applicable in tests that contain sufficient geometric symmetries to reduce the problem to a simple form, such as the expansion of spherical and infinite long cylindrical cavities in a semi-infinite elastic-plastic continuum.

Class II

In cases in which closed-form solutions cannot be obtained, the accuracy of numerical solutions largely depends on the constitutive model adopted to represent soil behaviour. There are solutions that offer a very close approximation of the physical mechanism of a test. These solutions, which are regarded as rigorous and defined as Class II, comprise the family of numerical analyses applied to describe the penetration mechanism of a cone or flow penetrometer in normally consolidated highly plastic clays (e.g. Randolph, 2004; Yu, 2004).

Class III

These are approximate analytical solutions, developed on the basis of simplified assumptions imposed to reduce the requirements of a more rigorous approach. For penetration tools, the concepts of bearing capacity (e.g. de Mello, 1971; Durgunoglu and Mitchell, 1975) and cavity expansion (e.g. Vesic, 1972; Salgado et al., 1997) form the background for this category of analysis. Since bearing capacity theories carried out using limit analysis are unable to account for soil stiffness and volume change, they are regarded only as approximate solutions to model penetration problems. Calibration chamber tests and centrifuge tests are recommended to validate Class III type correlations.

Simple solutions, such as Class III, should preferably be achieved from a combination of measurements derived from independent tests. Since, in the interpretation of in situ tests, the number of controlling variables (soil parameters) far exceeds the number of measured variables, combining the results with independent measurements reduces the degree of uncertainty. The author foresees the increasing use of interpretation methods under this third category, with a growing trend in favour of the use of a range of sensors incorporated within a single penetration probe. This book explores the ratio of the elastic stiffness to ultimate strength (G_0/q_c, G_0/N_{60}), the ratio of cone resistance and pressuremeter limit pressure (q_c/Ψ) and the association of strength and energy measurements (N_{60} and energy).

Class IV

This class of solution comprises empirical analysis based on direct comparisons with structure performance and correlations with laboratory test results. Complexities in the interpretation of in situ tests in unusual geomaterials and

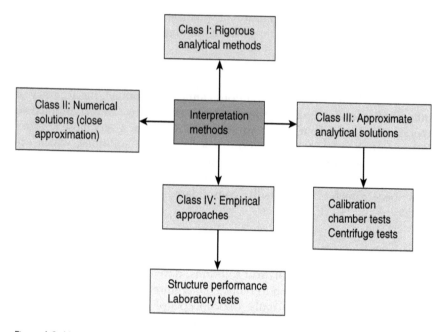

Figure 1.9 Alternative interpretation methods to *in situ* tests (after Schnaid, 2005).

poorly defined ground conditions still prompt the use of empirical approaches in geotechnical engineering practice (e.g. Schnaid *et al.*, 2004).

A schematic representation of alternative interpretation methods of in situ testing is presented in Figure 1.9. The difficulty of isolating the independent factors that control the test mechanism limits the adoption of Class I and II types of solutions and prompts the development and use of *laboratory physical modelling* to support Class III solutions. Two techniques are now recognized as reference: centrifuge and large calibration chamber tests. Well-planned series of laboratory physical modelling tests can be used to establish correlations for field tests that, although empirical, are set against controlled variables (such as density, stress history, horizontal and vertical stresses) and reproduce some of the environmental requirements for calibration against field data.

A calibration chamber consists of a large cylindrical homogeneous sample of known density (see Figure 1.10). Chamber walls can either support flexible rubber membranes, from which uniform stresses are applied, or can be rigid in the lateral direction when imposing zero lateral strain conditions (K_0) on the sample. Stress-controlled boundary conditions are more generally adopted, in order to investigate the independent effects of vertical and horizontal stress on penetration tests. For tests carried out in sand, where a

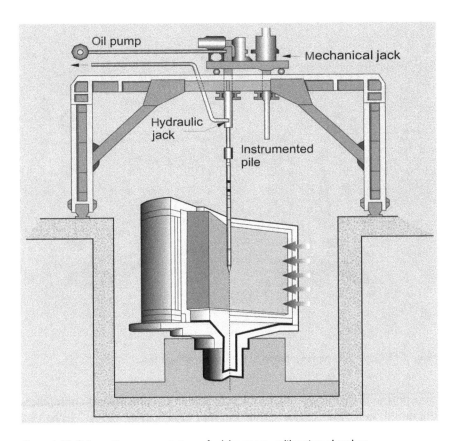

Figure 1.10 Schematic representation of a laboratory calibration chamber.

rigorous theoretical analysis is difficult, this technique represents the best way of establishing well-defined calibrations of in situ testing devices, providing databases that support empirical interpretation procedures between tip cone resistance (q_c) and relative density (D_r) or friction angle (ϕ'). Several examples of correlations developed in laboratory calibration chamber tests have been incorporated into geotechnical engineering practice. Note that this is in contrast with the approach required for clays, for which calibration chamber tests are practically unfeasible due to time constraints, and calibration of analytical procedures at well-documented test sites is a more reliable approach.

Centrifuge tests have been increasing in popularity over the past 20 years as a physical modelling technique in all types of soils. The technique illustrated in Figure 1.11 has been successfully applied to model classical problems such as foundations and slopes, environmental aspects of contaminant

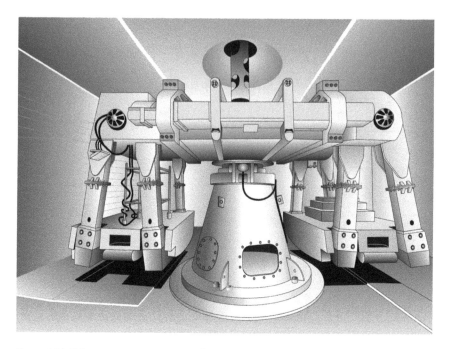

Figure 1.11 Schematic representation of a beam centrifuge test.

transport and seismic events. Results have been documented in conferences organized under the auspices of the International Society of Soil Mechanics. The fact that scale models can be prepared with prescribed soil property profiles in clay, sand or silt soils, and shaken in the simulated gravity environment under controlled input motion (e.g. Schofield and Steedman, 1988; Taylor, 1995), makes this approach particularly attractive as a means of producing controlled sets of in situ test data.

This brief review highlights the fact that geo-characterization of the properties of natural soils can be achieved using different testing techniques and can rely on various methods of interpretation (Classes I, II, III and IV). These approaches are extensively covered in this book and are applied to geomaterials such as clay, sand, intermediate permeability silt and bonded soils.

Scope and purposes

To summarize, various in situ testing methods are introduced and discussed in this book. Two approaches are covered, in which the results of in situ tests are used directly for design purposes, by means of empirical correlations, or, alternatively, where they are used to determine the values of design parameters such as strength and stiffness. Preference is given to procedures developed to estimate engineering properties of soils by adopting some form of analysis or correlations based on laboratory physical models. These methods can be used with confidence, provided they are established on the basis of three fundamental recommendations (Wroth and Houlsby, 1985):

- they are based on a physical appreciation of why the properties can be expected to be related,
- they are set against a background theory, however idealized this may be, and
- they are expressed in terms of dimensionless variables so that advantage can be taken of the scaling laws of continuum mechanics.

Engineers should ideally relate measurements made from in situ tests to fundamental geotechnical properties. For instance, it is more useful to relate SPT blow count number and CPT tip resistance to shear strength and shear stiffness rather than relating them directly to bearing capacity or settlement of piles. Once basic soil parameters and soil properties are quantified, engineers will have the necessary information to facilitate the proper design of geotechnical structures.

Emphasis is given to the five more commonly encountered in situ test techniques. Chapter 2 provides guidance for *standard penetration tests* and introduces both empirical and energy-based analytical methods designed for the interpretation of dynamic penetration. Chapter 3 describes *cone penetration tests*, the theoretical and experimental knowledge on the piezo-cone and the combined applications of seismic and environmental measurements. Chapter 4 reviews the experience gained over decades with the *vane test*. Chapter 5 provides guidance on the background of cavity expansion theory and its application to *pressuremeter tests*. Chapter 6 details the experimental experience and interpretation methods associated with *dilatometer tests*. Finally, a number of case studies and typical geotechnical properties are summarized in Chapter 7.

Only certain selected aspects of the in situ tests are covered in this book. Equipment and test procedures are briefly described, since the main focus of the text is on the interpretation methods conceived to predict horizontal stress, stress history, shear strength, soil stiffness and consolidation characteristics of geomaterials. It will be seen that site characterization and in situ test interpretation have evolved from basic empirical recommendations into a sophisticated subject area demanding a thorough knowledge of material behaviour and numerical modelling. Through these interpretation methods it is now possible to measure soil properties in clay, sand, silt, hard soils and soft rocks, among other soil formations.

Chapter 2

Standard penetration test (SPT)

Engineering involves action based on decision despite uncertainty. Prediction is one necessary vehicle for adequate decision, but most often an engineer's capacity at predicting is disparately poor regarding what will happen, and quite competent, in comparison, at predicting what will not happen; in other words, we inherently know that there are upper and lower bounds, and that we may place ourselves beyond such bounds. ... As civil engineers, and especially as geotechnical engineers, we cannot presume to simplify our problem, but must needs simplify the solution; which frequently comprises resorting to upper and lower bounds, even if for lumped parameters only roughly estimated.

(V. F. B. de Mello, 1977)

General considerations

Of the various methods of ground investigation currently practised worldwide, the standard penetration test (SPT) remains the most popular and widely used in situ testing technique. Primarily recommended for granular soils and other ground conditions in which it is difficult to sample and test in the laboratory or to perform other in situ tests, the SPT is being used for preliminary exploration in virtually any ground condition. Additionally, it provides assessment of soil properties, foundation design parameters and liquefaction potential. The simplicity, low cost and robustness of its method of driving a split-barrel sampler into the ground from the bottom of a pre-bored hole partially explains its wide acceptance.

The SPT provides a measure of the soil resistance to penetration through computation of the number of blows required to drive the sampler 300 mm into the ground, after it has already been advanced 150 mm, except in cases where 50 blows are insufficient to advance it through a 150 mm interval. Interpretation of test results rely on the measured N-value, which in recent years has been subjected to various 'corrections' to account for the lack of standardization in test procedures, effects of overburden pressure and influence of rod length. Scatter produced by the highly variable and unknown values of energy delivered to the SPT rod system can now be properly accounted for by standardizing the measured N-value to a reference value of 60% of the potential energy of the SPT hammer, as suggested by Seed *et al.* (1985) and Skempton (1986).

State-of-the-art reviews of the SPT are presented by de Mello (1971), Nixon (1982), Decourt (1989) and Clayton (1995).

Standards and procedures

The SPT is an in situ dynamic penetration test designed to provide information on the geotechnical properties of soils. The test originated in the late 1930s in the USA and was first standardized in 1958 under ASTM Designation D 1586–58T. It is currently covered by several national and regional standards (i.e. American ASTM D 1586-99: 1999, British BS 1377-9: 1990, Brazilian NBR 6484: 2001, Eurocode 7: 1994) that fully comply with the ISSMGE International Reference Test Procedure (IRTP) published in 1988 at the 1st International Symposium on Penetration Testing (Decourt *et al.*, 1988).

The equipment shown in Figure 2.1 uses a thick-walled sample tube that is driven into the ground at the bottom of a borehole by blows from a slide hammer that weighs 63.5 kg (140 lb) falling through a distance of 760 mm (30 in.), yielding a maximum energy of 474 J.[1] The range of acceptable dimensions for the SPT split-spoon sampler is given in Figure 2.2, following IRTP (1988) specifications of an average outside diameter of 51 mm, an inside diameter of 35 mm and a length greater than 457 mm. Appropriate tolerances are described in ASTM D 1586 specifications.

There are still significant differences between the drilling techniques, equipment and test procedures used in different countries. Hammer design varies considerably (Figure 2.3) and includes donut, safety and trip release hammers. The donut hammer consists of a simple cylindrical mass which falls down a guide rod. The safety hammer was developed to protect the rig operator from injury by internalizing the point of impact between the falling mass and the anvil rod. These hammers can be lifted manually by pulling on the free end of the rope or by wrapping the rope around a rotating cathead. In trip release hammers, the drop height is mechanically set to improve the repeatability of the SPT hammer and the hammer is often lifted automatically by a chain drive mechanism, which is hydraulically powered by the rig itself. Automatic trip hammers are standard in Australia, Japan, the UK and the USA.

Rod stiffness varies widely even within countries (Clayton, 1993). The Brazilian Standard recommends 1 in. *Schedule 80* (3.23 kg/m) rods to be used, whereas ASTM and British Standards require that the drill rod must have a stiffness greater than A rod (> 4.57 kg/m).

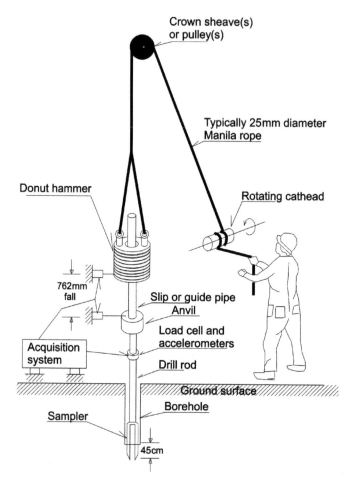

Figure 2.1 Schematic view of STP using donut hammer, rope and cathead method (after Kovacs and Salomone, 1982).

An additional procedure has recently been incorporated into the SPT in Brazil. This new procedure consists of attaching a torque wrench to the drill rods and measuring the maximum torque required to rotate the sampler at the end of the test drive (Ranzini, 1988). This operation produces another measurement which, in combination with the blow count number, is useful for soil characterization.

The SPT is often considered unreliable for gravel deposits, primarily because gravel particles can be larger than the opening of the SPT sampler.

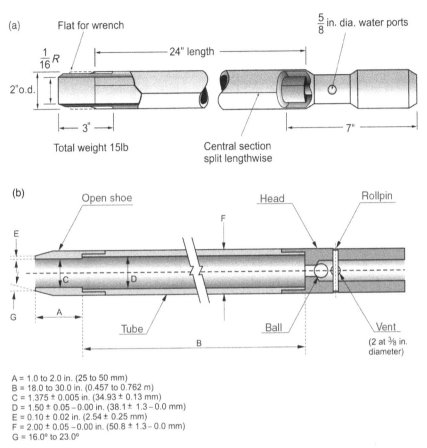

(a) Flat for wrench

$\frac{5}{8}$ in. dia. water ports

$\frac{1}{16}R$

24" length

2" o.d.

3"

7"

Total weight 15lb

Central section
split lengthwise

(b) Open shoe   Head   Rollpin

E   F

C   D   Ball   Vent

G

A

Tube

B

(2 at ⅜ in.
diameter)

A = 1.0 to 2.0 in. (25 to 50 mm)
B = 18.0 to 30.0 in. (0.457 to 0.762 m)
C = 1.375 ± 0.005 in. (34.93 ± 0.13 mm)
D = 1.50 ± 0.05 – 0.00 in. (38.1 ± 1.3 – 0.0 mm)
E = 0.10 ± 0.02 in. (2.54 ± 0.25 mm)
F = 2.00 ± 0.05 – 0.00 in. (50.8 ± 1.3 – 0.0 mm)
G = 16.0° to 23.0°

The 1½ in. (38 mm) inside diameter split barrel may be used with a 16-gauge wall thickness split liner.
The penetrating end of the drive shoe may be slightly rounded. Metal or plastic retainers may be
used to retain soil samples.

Figure 2.2 SPT split-spoon sampler: (a) IRTP (1988) and (b) ASTM D 1586 (1999).

For gravelly materials, it is becoming routine to use non-standardized over-
sized split-spoon samplers with outer diameters as large as 140 mm.
Geometry and material properties of these different, larger penetrometers
(LPTs) are given in Table 2.1, in which it is seen that each device delivers
different input energies. As a consequence, the penetration process and the
actual sampler penetration reflect the energy delivered to the sample in
each test configuration. Although the similarity of the SPT and LPT suggests that nothing more than a scaling factor is required to correlate the

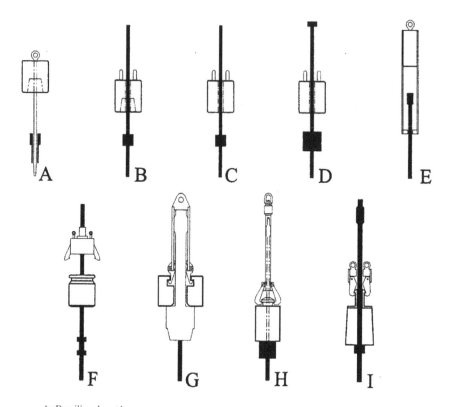

A: Brazilian donut hammer
B: Brazilian donut hammer
C: USA donut hammer
D: USA donut hammer
E: USA safety hammer
F: Booros Co Ltd darp hammer
G: USA Pilcon trip hammer
H: British automatic hammer
I: Japanese automatic hammer

Figure 2.3 SPT hammers.

blow counts in these two test configurations, this is done on the basis of correlations that empirically relate LPTs to the SPT N-values in granular and gravelly soils (e.g. Burmister, 1948; Tokimatsu, 1988; Crova *et al.*, 1993; Daniel *et al.*, 2004).

It is clear from the preceding discussion that, despite its definition as a standard penetration test, the SPT is far from standardized. Although

Table 2.1 Details of the SPT and LPT (modified from Daniel, 2000)

Detail	SPT[a]	SPT[b]	JLPT[c]	NALPT[d]
Hammer weight (kg)	65.0	63.5	98.1	133.5
Hammer drop height (m)	0.75	0.762	1.50	0.762
Hammer length (m)	0.23	0.533	0.368	0.699
Hammer diameter (m)	0.20	14.0	21.1	17.8
Rod area (cm^2)	4.1	8.0	10.1	9.3
Rod weight (kg/m)	3.230	6.304	7.959	7.328
Hammer velocity at impact (m/s)	3.836	3.866	5.425	3.866
Sampler area[e] (cm^2)	10.807	8.800	18.800	13.900
Sampler length (m)	0.45	0.45	0.45	0.45
Sampler weight (kg/m)	8.516	6.934	14.814	10.953
Potential energy (J)	478	474	1143	998

Notes
a   Brazilian Standard NBR 6484.
b   ASTM D 1586.
c   Kaito *et al.* (1971);Yoshida *et al.* (1988).
d   Daniel (2000); Koester *et al.* (2000).
e   Steel cross-sectional area.

equipment and procedures follow general requirements from IRTP, different equipment and techniques are used worldwide. As recognized by Ireland *et al.* (1970):

> one of the main problems of the Standard Penetration Tests remains that neither the test equipment nor the associated drilling techniques have been fully standardized on an international basis. Differences in equipment and technique exist partially because of the level of capital investment current in different countries, but more importantly because of the adaptation of drilling techniques to local ground conditions.

Wave propagation analysis

Although dynamic tests in geomaterials are primarily interpreted by correlations of an empirical nature, solutions for the motion produced in an elastic body by suddenly applied forces have been recognized for almost a century (e.g. Timoshenko and Goodier, 1970; Skov, 1982). The action of a hammer blow, not transmitted at once to all parts of the body, can be analysed by wave theory and can provide the necessary framework for interpretation of STPs.

The analysis considers that each hammer blow propagates as elastic shear waves moving down the length of the rod stem, so that the time taken for the waves to traverse the body becomes of practical importance. The method of solution of these propagation waves in elastic solid media to obtain the energy delivered to the soil is discussed in this section and the boundary conditions embedded into the representation of dynamic tests are presented.

The wave equation

The idealized representation of the action of a hammer having a mass M_h dropping from a height h towards the top end of a bar is shown in Figure 2.4. Let us assume the bar to be perfectly elastic with a modulus E, cross-section A remaining plane during deformation and energy losses stemming from irreversible deformation of the rod to be negligible. Under these conditions, the relationship between components of stress and strain is deduced from Hooke's law:

$$\sigma = E\varepsilon \tag{2.1}$$

where σ is the uniformly distributed stress component and ε the unit deformation. The value of ε due to a longitudinal displacement u caused by a propagated wave is $\Delta u/\Delta x$, and the corresponding force in the bar at a given time interval Δt is:

$$\frac{F}{A} = E\frac{\Delta u}{\Delta x} \quad \text{or} \quad \frac{F}{A} = E\frac{\Delta u}{\Delta t}\frac{\Delta t}{\Delta x} \tag{2.2}$$

The velocity of the wave propagation c is defined as $\Delta x/\Delta t$ and should be distinguished from the velocity V of the particles in the compressed zone of the bar produced by compressive forces and defined as $\Delta u/\Delta t$. Hence, the force in equation (2.2) can be written as:

$$F = \frac{EA}{c}V \tag{2.3}$$

The proportionality between force and velocity expressed in equation (2.3) is defined as the rod impedance and represented as Z_h:

$$Z_h = \frac{EA}{c} \tag{2.4}$$

Introducing Newton's law for motion of the same section Δx:

$$F = ma_c \tag{2.5}$$

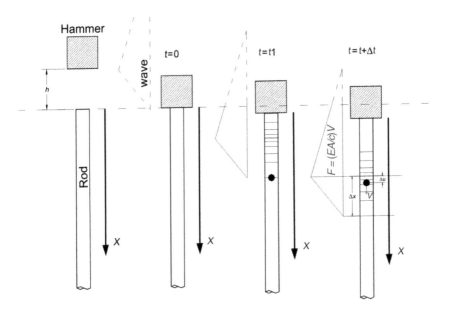

Figure 2.4 Idealized representation of the action of a hammer.

where a_c is the particle acceleration (= $V/\Delta t$), m the rod mass (= $\rho A \Delta x$) and ρ the mass per unit volume of the bar (in the case of steel it can be assumed to be 7,810 kg/m^3). The force can now be expressed as:

$$F = \rho A \frac{\Delta x}{\Delta t} V = \rho A c V \qquad (2.6)$$

From equations (2.4) and (2.6), the velocity of the wave propagation c is obtained:

$$c = \sqrt{\frac{E}{\rho}} \qquad (2.7)$$

An equation for wave propagation can now be derived by combining Hooke's law with Newton's law. The difference between the forces acting on the sides of an element of the bar is:

$$F_1 = F_2 + m a_c \qquad (2.8)$$

which leads to:

$$EA\left(\frac{\partial u}{\partial x} + \frac{\partial^2 u}{\partial x^2}dx\right) - EA\frac{\partial u}{\partial x} - \rho A dx\frac{\partial^2 u}{\partial t^2} = 0 \tag{2.9}$$

Finally, the equation of motion of the element can be written as:

$$\frac{\partial^2 u}{\partial t^2} = c^2\frac{\partial^2 u}{\partial x^2} \tag{2.10}$$

The general solution for equation (2.10) can be represented in the form:

$$u(x,t) = f(x - ct) + g(x + ct) \tag{2.11}$$

The physical interpretation of functions f and g can be easily explained: the second term on the right-hand side represents a wave travelling in the direction of the x-axis with a constant speed c, whereas the first term represents a wave travelling in the opposite direction. The velocity depends only on the modulus E and the density of the material bar.

The mechanics of dynamic penetration tests

Wave equations introduced in the previous section can be solved numerically to simulate the propagation in an elastic medium in order to illustrate the mechanisms of in situ dynamic penetration test devices. In the model, dynamic equilibrium equations can be solved by finite difference analysis in the time domain to produce the discretization of a penetration system – including hammer, rod, sampler and soil. Results of a numerical analysis give the reader a better understanding of the penetration mechanism, which helps in evaluating the consistency of recommended methods of interpretation.

The numerical model is a one-dimensional representation of the complete SPT system, as illustrated in Figure 2.5. Nodal equilibrium equations are solved by means of a finite-difference scheme with explicit integration in the time domain. Special boundary conditions are applied to the contact surface between hammer and rod top in order to ensure displacement compatibility during impact. The analysis takes into account the fact that the hammer blow induces strains on both hammer and rod and consequently produces pressure waves (P-waves) that propagate in opposite directions, moving down the length of the rod and up the height of the hammer. Hammer, rod and sampler are modelled as linear elastic, whereas the assumed load–deformation relationship for the soil during sampler

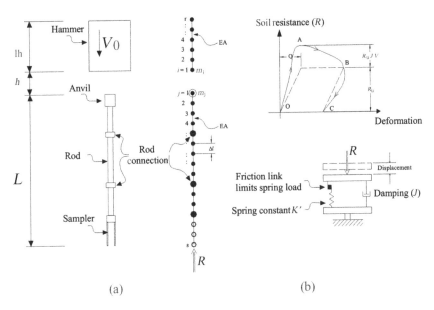

Figure 2.5 Schematic representation of the model: (a) hammer and rod discretization and (b) Smith soil model.

penetration is represented by the original Smith model (1960) illustrated in Figure 2.5(b), expressed by the ultimate static soil resistance, R_u, the maximum elastic deformation (quake), Q, and an additional resistance coefficient, J, representing a dynamic resistance proportional to the velocity of loading. Although fairly simple, the Smith model has frequently been adopted for characterizing the soil in pile-driving analysis (e.g. Poulos, 1989).

In performing the computations, it is possible to keep track of the energy balance within the complete hammer, rod, sampler and soil system. Assuming that no further external forces are applied to the system after releasing the hammer, the total initial energy must remain constant with time. The instantaneous total energy, $E_{TOT}(t)$, is calculated as the sum of:

$$E_{TOT}(t) = E_{TOT}(t=0) = E_P(t) + E_C(t) + E_{PE}(t) + E_D(t) + E_A(t) \quad (2.12)$$

where $E_P(t)$ is the total potential gravitational energy, $E_C(t)$ is the total kinetic energy, $E_{PE}(t)$ is the total potential elastic energy, $E_D(t)$ is the energy dissipated in all nodes by the viscous damping criteria and $E_A(t)$ is the energy delivered to the soil.

Examples of numerical simulations are presented to describe the penetration mechanism of an SPT sampler in an attempt to demonstrate the influence of the energy employed to drive the sampler into the soil. Geometrical conditions comply with recommendations given by the Brazilian Standard (NBR 6484/2001), which is also in accordance with the IRTP (ISSMGE, 1988). A split barrel connected to a *Schedule 80* rod stem (3.23 kg/m) is driven by a 65 kg hammer falling 750 mm. Young's Modulus E is adopted as 205 GPa for both hammer and rod. In these simulations, Newton's law of impact was used without incorporating energy losses during impact and wave propagation. Despite the fact that energy losses can be easily implemented in the numerical analysis, these losses would deviate from its main purpose of gaining a better understanding of the mechanism involved in driving a penetrometer into the soil.

Results for the numerical simulation of 35.8 m and 12.6 m long rods are presented in Figure 2.6, in which force, rod displacement and sampler penetration are plotted against time. In this simulation, it is possible to identify the several consecutive impacts produced by a single SPT blow, as well as the displacement paths of both the bottom of the hammer and the top of the rod stem. The time interval between points 1 and 2 represented in Figure 2.6 corresponds to the period of the first impact, 3 and 4 the second impact, 5 and 6 the third impact, and 7 and 8 the fourth and last impact (observed only on the 12.6 m long rod). The signals of force and velocity (also expressed in units of force) make it possible to visualize the moments of impact and the time of contact between the hammer and rod, represented by points A and B and G and H for the first and second impacts respectively on the 35.8 m long rod and points A and B, E and G and K and L for the first, second and third impacts respectively for the 12.6 m long rod.

The first impact of the falling hammer in the time interval from A to B produces a first compression wave that propagates downwards to the sampler. The contact between hammer and rod is interrupted at B when a tension wave arrives at the hammer at a time corresponding to $2l/c$. At this time, the rod temporarily pulls away from the hammer, which produces a second longitudinal compression wave that still reflects the first impact and is represented by the interval between B and C. The same mechanism is then reproduced in the rods between points C and D and D and E, giving rise to the third and fourth compression waves respectively. In the 12.6 m rod, the fourth compression is disturbed by the second impact of the hammer. The mechanism, observed for the second and third impacts at points E and K, is similar to the one described for the first impact.

The penetration produced by the hammer blow starts when the first compression wave reaches the sampler after a time l/c from the hammer impact at the top of the rod. The arrival of the second and third compression waves at the sampler at points B and C is also perceptible. In the

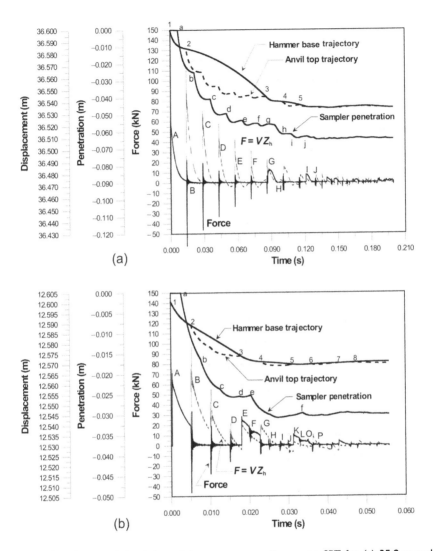

Figure 2.6 Numerical simulation of the wave propagation on an SPT for (a) 35.8 m and (b) 12.6 m long rods.

35.8 m long rod penetration is almost entirely produced by the sequence of compression waves produced by the first impact. In the 12.6 m rod the second impact contributes to a further and more significant penetration of the sampler, whereas the third impact will not have any significance with respect to the penetration in this particular test.

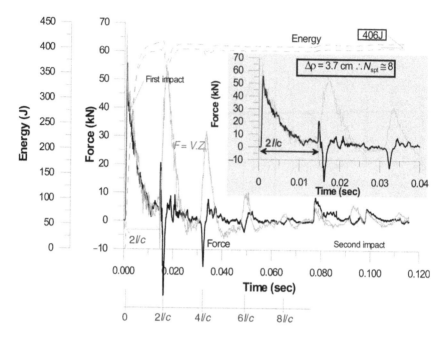

Figure 2.7 Typical measured force–time relationships measured below the anvil for a 35.8 m long rod stem – an extended plot for the first hammer impact is shown on the right-hand side.

Dynamic penetration force

The maximum energy transmitted to the rod stem computed by the F–V method is known as the Enthru energy and can be calculated as:

$$E = \int_{0}^{\infty} F(t)V(t)dt \qquad (2.13)$$

where $F(t)$ and $V(t)$ are force and velocity respectively, and integration limits (from zero to infinity) are set to account for all compression waves produced by a single hammer blow. From carefully monitored measurements of force and acceleration it is therefore possible to calculate the energy effectively delivered to the sampler during a test and to compare these measurements to those predicted numerically.

Typical records of the force measured by a load cell and the force calculated from the readings of accelerometers are presented in Figures 2.7 and 2.8. The difference between these two figures is that, in the first case,

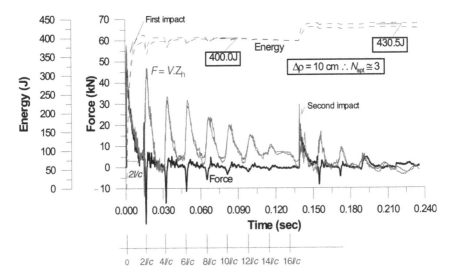

Figure 2.8 Typical measured force–time relationships measured below the anvil for a 35.8 m long rod stem.

a condition that represents a sample with an average N of about 8 is shown, whereas the second case represents a looser sample where the blow produces a large permanent penetration ($N = 3$). Figure 2.7 illustrates the characteristics of a standard result that displays features already discussed in the previous section and that is in accordance with previously reported data (Kovacs and Salomone, 1982; Schmertmann and Palacios, 1979; Skempton, 1986). The energy is fully transmitted to the rods during a time interval of $2l/c$ and, in this case, the energy transmitted to the rod can be calculated by equation (2.14):

$$E = \int_0^{2l/c} F(t)V(t)dt \qquad (2.14)$$

However, a slightly different pattern is observed in Figure 2.8. The impact of the falling hammer produces a first compression wave that propagates downwards to the sampler and reflects upwards as a tension wave, arriving at the hammer at a time corresponding to $2l/c$. At this time, the rod temporarily pulls away from the hammer until a second impact is produced. For the large penetration induced by the hammer, this second and late impact ($\Delta t > 100$ ms $\gg 2l/c$) produces a further increase in energy that eventually

contributes to the penetration of the sampler. Note that the energy delivered
to the 35.8 m long composition of rods is different in these two reported
cases (406 J and 430 J, respectively), suggesting that energy effectively transmitted to the rod stem is not only a function of the so-called nominal potential energy E^* (474 J) but is also affected by the permanent penetration of
the sampler.

The energy delivered by the hammer impact, defined in equation (2.13),
can be conveniently expressed as a function of the height of fall, H, increased
by the permanent penetration of the sampler, $\Delta\rho$. In this framework, the
energy actually associated with penetration becomes a function of three
groups of parameters: (1) soil properties, (2) hammer mass M_h and height of
fall, and (3) rod geometry (length and cross-sectional area), which yields the
total rod mass M_r. Hence, the *theoretical potential energy* (PE^*_{h+r}) of the
system (considering simultaneously hammer, rods and soil) is expressed as
(Odebrecht *et al.*, 2004):

$$PE^*_{h+r} = (H + \Delta\rho)M_h g + \Delta\rho M_r g \qquad (2.15)$$

where g is the gravitational acceleration. Equation (2.15) represents an
ideal condition, where energy losses related to the penetration process
are not computed. These losses depend on test characteristics and type
of equipment (hammer design, rod size, sampler geometry, automatic
or hand-controlled hammer, etc.), whose influence on penetration is
internationally recognized as the *equipment efficiency* (e.g. Skempton,
1986). Three efficiency coefficients have to be defined (Odebrecht *et al.*,
2004), called η_1, η_2 and η_3, which are used to account for losses in
equation (2.15). Hence, equation (2.16) now defines the *potential
energy* $(E_{sampler})$:

$$E_{sampler} = \eta_3[\eta_1(H + \Delta\rho)M_h g + \eta_2(M_r g \Delta\rho)] \qquad (2.16)$$

As recently postulated by Aoki and Cintra (2000), the work effectively
delivered to the soil can be derived from the principles of energy conservation, known as Hamilton's principle. The principle states that the time
variation of kinetic and potential energies, summed with the work done by
non-conservative forces during a given time interval, is equal to zero. Using
the same notation adopted by Clough and Penzien (1975), the principle can
be written as:

$$\int_{t_1}^{t_2} \delta[T(t) - V(t)]dt + \int_{t_1}^{t_2} \delta[W_s(t) + W_{nc}(t)]dt = 0 \qquad (2.17)$$

where:

T(t) is kinetic energy
V(t) is potential energy
$W_s(t)$ is work done by non-conservative forces acting on the sampler–soil system
$W_{nc}(t)$ is work done by other non-conservative forces related to energy losses
t_1, t_2 is time interval.

Defining t_1 as the time at the liberation of the hammer in free fall and t_2 as the time after the complete penetration of the sampler, when the kinetic energy has been completely dissipated, and considering that all losses have occurred during this time interval, leads to:

$$\int_{t_1}^{t_2} \delta V(t)dt = \int_{t_1}^{t_2} \delta\left[W_{s(t)} + W_{nc(t)}\right]dt = 0 \qquad (2.18)$$

Hence, the variation of the potential energy of the hammer–rod system equals the work done by all non-conservative forces:

$$E_{sampler} = W_s + W_{nc} \qquad (2.19)$$

Since equation (2.19) gives the energy effectively delivered to the soil, with all losses accounted for by means of the efficiency coefficients, it is now possible to calculate the mean dynamic reaction force (F_d) on the soil.

$$E_{sampler} = W_s = F_d \Delta\rho \qquad (2.20)$$

$$F_d = E_{sampler}/\Delta\rho \qquad (2.21)$$

Substituting equation (2.16) for equation (2.21) gives:

$$F_d = \frac{\eta_3\eta_1(0{,}75M_hg) + \eta_3\eta_1(\Delta\rho M_hg) + \eta_3\eta_2(\Delta\rho M_rg)}{\Delta\rho} \qquad (2.22)$$

Equation (2.22) requires a previous calibration of coefficients, η_1, η_2 and η_3. The preliminary estimation for the SPT is proposed by Odebrecht et al. (2005): $\eta_1 = 0.76$, $\eta_2 = 1$ and $\eta_3 = (1 - 0.0042l)$, where l is the total rod length. Penetration values ($\Delta\rho$) recorded during a dynamic test can then be used to compute the dynamic force, presenting new opportunities for the interpretation of SPT results.

Measurements and corrections

The SPT does not satisfy the requirements of full standardization. It follows that the interpretation of test results, which rely on the blow count number N, has in recent years been subjected to various 'corrections' to account for the lack of standardization in test procedures, effects of overburden pressure and the influence of rod length. Normalization of the blow count number with respect to a reference effective overburden pressure in granular soils has become internationally accepted. Scatter produced by the highly variable and unknown values of energy delivered to the SPT rod system can now be properly accounted for by standardizing the measured N-value to a reference value of 60% of the potential energy of the SPT hammer. These mandatory corrections are detailed in the following section.

Overburden pressure

Since the pioneering work by Gibbs and Holtz (1957), there has been recognition that the geostatic mean stress affects the magnitude of the blow count number in granular soils. This effect can be removed by normalizing N_{SPT} values with respect to a reference effective stress. This recommended correction is essential because penetration resistance is known to increase approximately linearly with depth and, at a constant vertical effective stress, penetration increases approximately as the square of relative density D_r (Meyerhof, 1957):

$$N_{SPT} = D_r^2(a + bp')$$

(2.23)

where a and b are material dependent factors and p' is the mean effective stress. Equation (2.23) is often expressed as a function of the effective vertical stress σ'_v, since σ'_v can be estimated with reasonable accuracy at any given site. The influence of overburden pressure can then be expressed by a depth correction factor C_N (e.g. Liao and Whitman, 1985; Skempton, 1986; Clayton, 1993):

$$C_N = \frac{N_1}{N_{SPT}} = \frac{D_r^2(a + b100kPa)}{D_r^2(a + bp')} = \frac{a/b + 100}{a/b + p'}$$

(2.24)

where N_1 is the equivalent SPT resistance under a vertical stress of 100 kPa. Based on this concept, several expressions have been developed to represent C_N supported by both in situ and laboratory tests (Gibbs and Holtz, 1957; Marcuson and Bieganousky, 1977; Seed, 1979; Seed et al., 1983; Skempton, 1986) as summarized in Table 2.2, notably the correlations proposed by Peck et al. (1974) and Skempton (1986). Typical values of C_N for both normally consolidated and overconsolidated sands are shown in Figure 2.9.

Table 2.2 Typical calculations of C_N values

Reference	Depth correction factor C_N	σ'_{v0}	Observation
Skempton (1986)	$C_N = \dfrac{200}{100 + \sigma'_{v0}}$	kPa	Seed et al. (1983) $D_r = 40–60\%$ NC sand
Skempton (1986)	$C_N = \dfrac{300}{200 + \sigma'_{v0}}$	kPa	Seed et al. (1983) $D_r = 60–80\%$ NC sand
Peck et al. (1974)	$C_N = 0.77\log\left(\dfrac{2000}{\sigma'_{v0}}\right)$	kPa	NC sand
Liao and Whitman (1985)	$C_N = \sqrt{\dfrac{100}{\sigma'_{v0}}}$	kPa	NC sand
Liao and Whitman (1985)	$C_N = \left[\dfrac{(\sigma'_{v0})_{ref}}{\sigma'_{v0}}\right]^k$	–	$k = 0.4–0.6$
Skempton (1986)	$C_N = \dfrac{170}{70 + \sigma'_{v0}}$	kPa	OC sand OCR = 3
Clayton (1993)	$C_N = \dfrac{143}{43 + \sigma'_{v0}}$	kPa	OC sand OCR = 10
Robertson et al. (2000)	$C_N = \left(\dfrac{\sigma'_{v0}}{\sigma_{atm}}\right)^{-0.5}$	kPa	NC sand

Notes
NC = normally consolidated.
OC = overconsolidated.

Table 2.3 Correction for overburden pressure

Depth (m)	γ_{nat} (kN/m³)	σ'_{v0} (kN/m²)	N_{SPT}	C_N	N_1
2	18	36	5	1.667	~8
20	18	360	17	0.527	~9

Consider the simple case of a site with a homogenous thick layer of a dry, normally consolidated sand (low water table), unit weight γ_{nat} = 18 kN/m³, with SPT blow count numbers recorded as N_{SPT} = 5 at 2 m depth and N_{SPT} = 17 at 20 m depth, as represented in Table 2.3. Adopting the correction factor

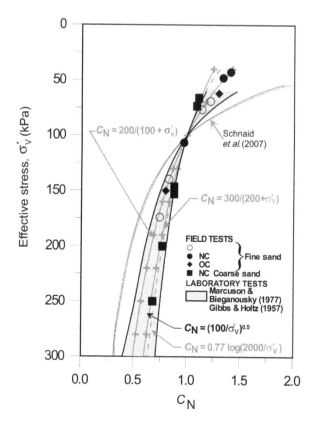

Figure 2.9 Values of depth correction factor C_N (modified from Stroud, 1988).

proposed in equation (2.25), it can be concluded that, although penetration resistance at 20 m ($N_{SPT} = 17$) is far greater than at 2 m depth ($N_{SPT} = 5$), the equivalent SPT resistance N_1 is similar and so is the relative density.

$$C_N = \left(\frac{\sigma'_{v0}}{\sigma_{atm}}\right)^{-0.5} \tag{2.25}$$

Energy measurements

The need for standardization of the measured blow count number N_{SPT} into a normalized reference energy value is now fully recognized. Scatter produced by the highly variable and unknown values of energy delivered to the SPT rod system can now be properly accounted for by standardizing the measured N-value to a reference value of 60% of the potential energy of the SPT hammer, as suggested by Seed *et al.* (1985) and Skempton (1986). This correction is carried out through a simple linear relationship in which:

$$N_{60}E_{60} = N_{SPT}E_{SPT} \qquad (2.26)$$

or

$$N_{60} = \frac{N_{SPT}E_{SPT}}{E_{60}} \qquad (2.27)$$

A test which was carried out in Brazil, for example, had a calibrated potential energy of around 72% of the free-fall energy (e.g. Decourt *et al.*, 1996) for which the measured N-value of 20 would be converted into an equivalent $N_{60} = 24$, i.e. $N_{60} = (20*0.72/0.60) = 24$.

This correction requires signals of both force and acceleration to be recorded in order to produce a reliable and accurate database of input energy. As illustrated in Figures 2.10 and 2.11, this can be done with a notebook computer provided with *A/D* conversion cards, an oscilloscope for real time signal monitoring, a load cell and accelerometers. To measure accelerations, which will be integrated to velocities, a pair of piezoelectric accelerometers symmetrically mounted on the load cell is recommended. A dynamic range from 100 µG to 10 kG is suitable for the 5 kG accelerations produced by the hammer impact.

Summaries of energy efficiency representative of the American and Brazilian practices are given in Tables 2.4 and 2.5 respectively. The type of hammer, anvil mass and release method all have considerable influence on the energy delivered to the rod. The lower efficiency of the donut hammer (Table 2.4) is attributable to the heavy anvil associated with this type of hammer in the USA, compared with the small anvil of the safety hammer, as postulated by Skempton (1986).

Strain rate effect

The influence of the rate of penetration on the soil shear response may account, at least in part, for the discrepancy between static and dynamic measurements. Typical rates of laboratory testing are of 1%/hour (or 0.3×10^{-3}%/s) whereas field tests are associated with strain rates that may reach 10^3%/s to 10^5%/s, this highest rate corresponding to SPT tests. In principle, the shear strength mobilized by a field test can be 1.5 to 2.5 times a reference shear strength mobilized by a laboratory test (e.g. Massarsch, 2004; Randolph, 2004). For the high rate of penetration of an SPT, excess pore pressures are generated even in uniform clean sands when the ground is relatively permeable (e.g. Clayton, 1993). This cannot be taken into account

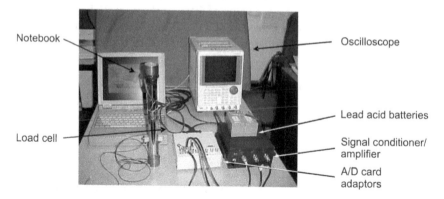

Notebook

Oscilloscope

Lead acid batteries

Load cell

Signal conditioner/
amplifier

A/D card
adaptors

Figure 2.10 Force and acceleration measurement system.

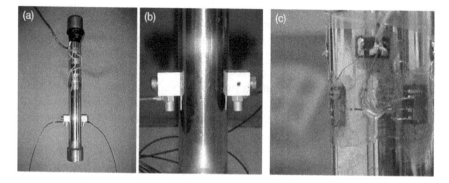

Figure 2.11 Detailed view of instrumentation: (a) complete load cell, (b) mounting of accelerometers and (c) strain gauge placement.

and engineers take the simplified route of assuming in every soil type either a drained penetration representative of a cohesionless soil or an undrained penetration representative of a cohesive geomaterial.

Characterization and classification

Characterization and classification using the SPT is possible because at every depth the test combines a sample with a measurement of penetration resistance. The SPT sampler is designed to retrieve disturbed specimens for a visual inspection which, despite being of very poor quality, enables soil mineralogy, particle shape, grading, colour and layering to be described. As demonstrated by Clayton (1993) and summarized in Table 2.6, SPT data shows that soils and rocks can be grouped in simple classification systems, supported by the

Table 2.4 American average rod energy ratios for two types of hammer and anvil, with two-turn slip-rope release method (adapted from Skempton, 1986)

Donut		Safety		Notes	Reference
Energy (%)	Number of tests	Energy (%)	Number of tests		
53	4	72	9	Laboratory tests	Kovacs and Salomone (1982)
48	8	52	9	Various field rigs	Kovacs and Salomone (1982)
		55	24		Schmertmann and Palacios (1979)
		52	5		Schmertmann and Palacios (1979)
48	23				Robertson *et al.* (1992b)
43	8	62	8		Robertson *et al.* (1992b)
–	–	58	12		Daniel (2000)

Table 2.5 Brazilian rod energy ratios for manual release method

Anvil mass (kg)	Depth (m)	Energy efficiency of manual release method			References
		Average (%)	Number of tests	Standard deviation (%)	
1.2 3.6 14.0	14	66.7	51	2.73	Belicanta (1998)
3.6	6–36	79.7	56	7.94	Cavalcante (2002)
3.6	6–36	81.0	40	10.8	Odebrecht (2003)

penetration resistance of the stratum. Residual soils that have not been considered in original classification systems are now included in Table 2.6. These formations are created by a different process, which results from the in situ weathering of parent rocks and gives rise to profiles containing materials ranging from intact rocks to completely weathered soils.[2]

An important feature in geotechnical characterization is the identification of aging and cementation in both sedimentary and residual soils. Since granular materials are difficult to sample, it is necessary to develop a methodology capable of identifying the existence of distinctive behaviour emerging from soil structure in field tests. Schnaid (1999) and Schnaid *et al.* (2004) proposed that SPT N-values be combined with seismic measurements of G_0 to assist in the

Table 2.6 SPT-based soil and rock classification system (Clayton, 1993)

Sand	$(N_1)_{60}$	0–3	Very loose
		3–8	Loose
		8–25	Medium
		25–42	Dense
		42–58	Very dense
Clay	$(N)_{60}$	0–4	Very soft
		4–8	Soft
		8–15	Firm
		15–30	Stiff
		30–60	Very stiff
		>60	Hard
Residual soils[*]	$(N)_{60}$	0–5	Completely weathered
		5–10	Very weathered (lateritic)
		10–15	Weathered
		>15	Moderately weathered (saprolitic)
Weak rock	$(N)_{60}$	0–80	Very weak
		80–200	Weak
		>200	Moderately weak to very strong

Note

[*]Author's classification (not included in the original classification system proposed by Clayton, 1993).

assessment of the presence of bonding structure, primarily because the effect of bonding on G_0 is stronger than on N_{SPT}. This system can be evaluated on the G_0/N_{60} versus $(N_1)_{60}$ space shown in Figure 2.12, where $(N_1)_{60}$ is the energy-corrected N-value normalized to a reference stress:

$$\frac{(G_0/p_a)}{N_{60}} = \alpha N_{60} \sqrt{\frac{p_a}{\sigma'_{v0}}} \quad \text{or} \quad \frac{(G_0/p_a)}{N_{60}} = \alpha (N_1)_{60} \tag{2.28}$$

and α is a dimensionless number that depends on the level of cementation, age, compressibility and suction. Values of α should be validated by site-specific correlations. Fresh, uncemented sand characterizes a defined region in this space, and soils that exhibit some bond structure fall outside and above the band proposed for uncemented sand. This produces values of normalized stiffness (G_0/N_{60}) that are considerably higher than those observed in fresh, cohesionless materials.

Soil properties in granular materials

When carrying out dynamic tests in granular soils it is assumed that no excess pore pressures are generated. Under fully drained conditions, measurements

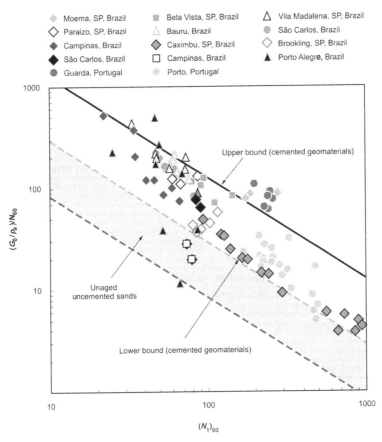

Figure 2.12 G_0/N_{60} versus $(N_1)_{60}$ space to identify cementation.

of blow count number are traditionally used as an indicator of strength and stiffness for sands and gravels.

Shear strength

In granular soils, shear strength can be estimated by two slightly different approaches:

- estimating the relative density D_r from penetration resistance, and then assessing the effective internal friction angle ϕ' by using density as an intermediate parameter; and
- estimating the effective internal friction angle ϕ' directly from the N_{SPT} blow count number.

Relative density, D_r is defined as:

$$D_r = \frac{e_{max} - e}{e_{max} - e_{min}} \qquad (2.29)$$

where e_{max} and e_{min} are the maximum and minimum void ratios. Difficulties in assessing e_{max} and e_{min} from laboratory tests are well recognized and field tests are viewed as a preferable option to estimate density in routine engineering design. Cubrinovski and Ishihara (1999) have related $(N_1)_{60}$ values calculated from equation (2.29) to the relative density and void ratio potential $(e_{max} - e_{min})$:

$$\frac{(N_1)_{60}}{D_r^2} = \frac{11.7}{(e_{max} - e_{min})^{1.7}} \qquad (2.30)$$

This work follows the evidence that, at a constant vertical effective stress, N-values are known to increase approximately as the square of relative density, a finding which has previously been explored by Gibbs and Holtz (1957) and Skempton (1986). According to Skempton (1986):

$$D_r = \left(\frac{(N_1)_{60}}{0.28\sigma'_{v0} + 27} \right)^{1/2} \qquad (2.31)$$

In equation (2.31), σ'_{v0} is expressed in kN/m^2 and N-values are corrected for both overburden pressure and energy levels, adopting 60% of the free-fall energy as reference (Skempton, 1986). For a given density, penetration resistance depends primarily on the in situ horizontal stress and, therefore, σ'_{h0} or mean stress p' must be accounted for in a rational interpretation of field tests (e.g. Clayton et al., 1995; Houlsby and Hitchman, 1988; Schnaid and Houlsby, 1992). Departing from this concept, it is possible to demonstrate that the relationship between D_r and $(N_1)_{60}$ varies with overconsolidation ratio (OCR), as indicated in Figure 2.13. For a given $(N_1)_{60}$ value it is evident that the effect of OCR on ϕ' is more significant for dense materials than for loose.

Values of D_r estimated from the above equations can be used to produce an estimate for the effective internal friction angle ϕ' that is expressed as a function of the critical state frictional component ϕ'_{cv} and the dilatancy component reflecting structural arrangements of particles (Rowe, 1962, 1963; Jamiolkowski et al., 1985; Bolton, 1986). The generalized form has been proposed by Bolton (1986):

$$\phi'_p - \phi'_{cv} = m[D_r(Q - \ln p') - R] \qquad (2.32)$$

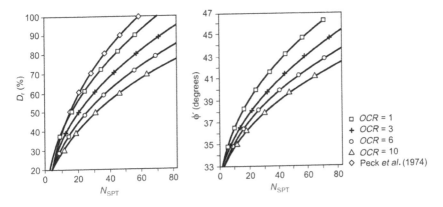

Figure 2.13 Relationship between D_r and $(N_1)_{60}$.

Table 2.7 Typical angles of friction

Material	Bolton (1979)	Bolton (1979)	Robertson and Hughes (1986)
	ϕ_p' (degrees)	ϕ_{cv}' (degrees)	ϕ_{cv}' (degrees)
Dense, well-graded sand or gravel	55	35	40
Uniform, medium-dense/ coarse sand	40	32	34–37
Dense, sandy silt with some clay	47	32	
Fine sand and sandy, silty clay	35	30	30–34
Clay-shale or partings	35	25	
Clay (London)	25	15	

where D_r is in decimals, p' is the mean effective stress (in kPa), R is an empirical factor found equal to 1 as a first approximation and Q is a logarithm function of grain compressive strength, known to range from about 10 for silica sand to 7 for calcareous sand (Jamiolkowski *et al.*, 1985; Randolph and Hope, 2004). Parameter m is stress path dependent and is taken as 5 for plane strain and as 3 for triaxial strain.[3] Typical values of the internal friction angle at critical state ϕ_{cv}' are presented in Table 2.7.

Alternatively, the effective internal friction angle can be assessed from charts such as those illustrated in Figure 2.14 (Peck *et al.*, 1974; Mitchell and Lunne, 1978). Although the chart proposed by Peck *et al.* (1974) underestimates the friction angle, it is still recommended in the UK as a

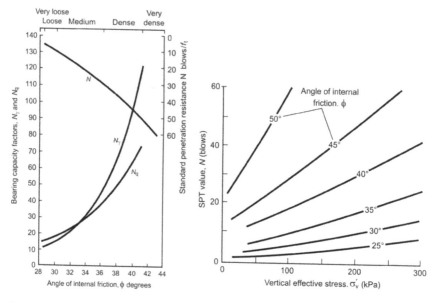

Figure 2.14 Recommended relationships between angle of friction and blow count number (Peck *et al.*, 1974 and Mitchell and Lunne, 1978).

basis for routine design (Clayton, 1993). Other proposed empirical relationships are:

$$\phi'_p \approx 15° + \sqrt{24(N_1)_{60}} \qquad \text{Teixeira (1996)} \qquad (2.33)$$

and

$$\phi'_p \approx 20° + \sqrt{15.4(N_1)_{60}} \qquad \text{Hatanaka and Uchida (1996)} \qquad (2.34)$$

The author has recently advocated that SPT results should be interpreted as a dynamic force F_d (equation (2.22)) from which soil properties can be assessed (Odebrecht *et al.*, 2004; Schnaid *et al.*, 2004; Schnaid, 2005). From a combination of bearing capacity and cavity expansion theory (Vesic, 1972) it is possible to estimate ϕ' as a function of the rigidity index, $I_v = (G / p' \tan \phi')$, and mean stress (Figure 2.15). The blow count number is defined as $(N_1)_{en}$ because the interpretation relies on both the energy transferred to the soil (equation (2.22)) and the applied mean stress. The relationship between $(N_1)_{en}$ and ϕ' is shown to be fairly sensitive to the variations in the rigidity index, an effect that is known to be more significant for dense materials than for loose.

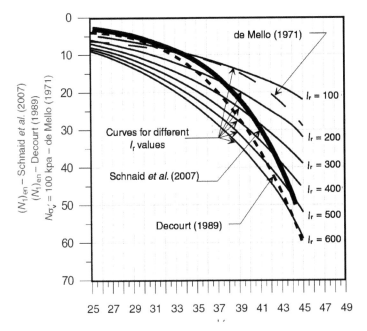

Figure 2.15 Friction angle estimated from the dynamic force F_d.

Alternatively, a rough estimate of ϕ' can be obtained without having to account directly for the effects of soil stiffness, which might help practitioners to select appropriate values of ϕ' without having to estimate operational values of I_r. Computations using the solution given by Berezantzev (1961) are also plotted in Figure 2.15. The friction angle is thus estimated directly from the equality between dynamic force F_d (equation 2.22) and the bearing capacity theory:

$$F_e = A_p(p'N_q + 0.5_\gamma dN_\gamma) + A_l \ (\gamma L \ K_s \tan \delta) \tag{2.35}$$

where N_c, N_q and N_γ are bearing capacity factors, A_p is the area of the sampler base, A_l the area of the sampler shaft, L the test depth, γ the unit weight of the soil, K_s the coefficient of the lateral pressure (taken as 0.8) and δ the soil-sampler angle of friction (adopted as 20°).

It is worth emphasizing that in the proposed approach (adopting either Vesic or Berezantzev) there is no need to rely on statistical analysis in order to estimate the friction angle because the influence of each variable controlling SPT penetration can be properly isolated. This takes into consideration the effects of the height of fall, sampler permanent penetration, and weight of both hammer and rods, as well as the dominant influence of soil compressibility. Strength values quoted from equation (2.35) yield friction angles that are not unreasonable for granular materials.

Soil stiffness

Geotechnical design requires the simulation of soil–structure interaction mechanisms for which it is necessary to know the stiffness of the soil. It is now recognized that soil stiffness is strain dependent, increases with soil structure and reduces with increasing OCR (e.g. Tatsuoka *et al.*, 1995). At very small strains, soils are believed to behave as elastic materials, represented by the initial elastic stiffness G_0. The magnitude of G_0 can be measured in the laboratory by using bender elements or resonant column tests (e.g. Seed and Idriss, 1971; Jardine *et al.*, 1984; Tatsuoka *et al.*, 1995; Lo Presti *et al.*, 2001) and in the field by seismic techniques (e.g. Santamarina *et al.*, 2001; Massarsch, 2004; Stokoe *et al.*, 2004). With increasing shear strains, the modulus decreases markedly. In addition, given the sensitivity to the magnitude of strains, it is not straightforward to use the elastic modulus determined by laboratory or field tests to obtain a serviceability modulus which can be used in routine calculations to estimate settlements.

The SPT is a large displacement test, associated with very high shear strain amplitudes, and insensitive to the soil-stress history. Correlations between N-values and stiffness are therefore very sensitive to a large number of factors and are likely to produce misleading predictions. Therefore, SPT results should be used with great caution.[4] Despite these limitations, there are reported attempts to correlate G_0 and N_{60} by a simple power law (e.g. Ohta and Goto, 1978; Imai and Yokota, 1982; Sykora and Stoke, 1983):

$$G_0(MPa) = a(N_{60})^b$$

(2.36)

Typical constant values are given in Table 2.8. The exponent b is relatively unaffected by the soil type and is consistent with laboratory data shown in Figure 2.16. In this figure, the variation of G_0 values are expressed as a function of void ratio e for a variety of soils and it is demonstrated that the slope that reflects the void ratio function $F(e)$ does not seem to be significantly influenced by the type of soil (e.g. Lo Presti, 1989; Jamiolkowski *et al.*, 1991).

Note that the penetration resistance is corrected to 60% of the free-fall energy, but it is not corrected for the stress level because it is argued that both stiffness and penetration resistance increase with increasing mean normal effective stress.

Another methodology has been proposed by Schnaid (1999) and Schnaid *et al.* (2004). As a first approximation, the variation of G_0 with N is expressed by upper and lower boundaries:

$$G_0 = 450 \sqrt[3]{N_{60}\sigma'_{v0}p_a^2} \text{ upper bound: fresh uncemented sand}$$

$$G_0 = 200 \sqrt[3]{N_{60}\sigma'_{v0}p_a^2} \text{ lower bound: fresh uncemented sand}$$

(2.37)

Table 2.8 Constant values for the correlation between G_0 and N_{60} (adapted from Crespellani and Vannucchi, 1991)

Soil type		a	b	R
Clay	Alluvial	10.4	1.070	0.500
	Alluvial	17.3	0.607	0.715
	Glacial	24.6	0.555	0.712
	Alluvial	16.6	0.719	0.921
Sand	Alluvial	12.3	0.611	0.671
	Glacial	17.4	0.631	0.728
Gravel	Alluvial	8.1	0.777	0.798
	Glacial	31.3	0.526	0.522
Residual soils*	Lower bound	20.0	0.40	> 0.500

Notes

R = correlation coefficient.

*Lower bound: author's classification (not included in the original classification system proposed by Crespellani and Vannucchi, 1991).

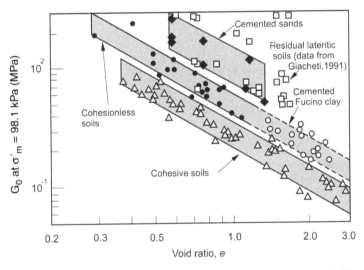

Figure 2.16 Normalized maximum shear modulus versus void ratio (modified from Jamiolkowski et al., 1991).

Equation (2.37) can match the range of recorded G_0 values and, despite the fact that this equation has originally been proposed to distinguish cemented and uncemented soils, it is likely that practitioners may be tempted to employ it to estimate G_0. For preliminary design, in the absence of direct measurements of shear wave velocities, the proposed lower bound is recommended for a conservative evaluation of the small strain stiffness from N_{60}.

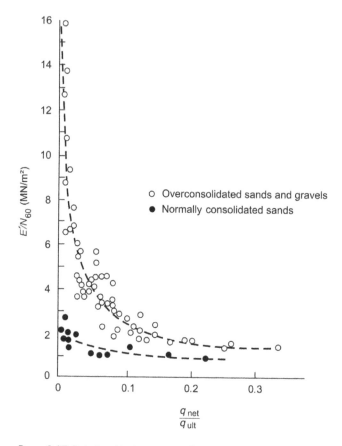

Figure 2.17 Relationship between stiffness, penetration resistance and degree of loading (after Stroud, 1988).

A logical approach to obtaining serviceability stiffness of granular soils from the SPT has been proposed by Stroud (1988). The proposed approach recognizes the importance of strain level by plotting the ratio E'/N_{60} as a function of degree of loading q/q_u, where q is the bearing pressure and q_u is the ultimate bearing pressure at the point of local failure. This suggests that q/q_u is an indirect measurement of shear strain. Figure 2.17 shows data from a large number of case records (Burbidge, 1982) for both normally consolidated and overconsolidated soils. The relationship between E' and N_{60} is strongly strain dependent, with the ratio E'/N_{60} ranging from 1 to 16. Adopting a factor of safety of 3 on bearing capacity (i.e. $q/q_u = 1/3$), a reasonable approximation is:

$$E'/N_{60} = 1\,(MPa) \tag{2.38}$$

Table 2.9 Typical range of Young's modulus derived from SPT data

Penetration resistance N_{60}	$E'/N_{60}(MPa)$			Residual soils[†]
	Sedimentary soils: Clayton (1993)*			Lower limit
	Mean	Lower limit	Upper limit	
4	1.6–2.4	0.4–0.6	3.5–5.3	1–2
10	2.2–3.4	0.7–1.1	4.6–7.0	2–3
30	3.7–5.6	1.5–2.2	6.6–10.0	3–4
60	4.6–7.0	2.3–3.5	8.9–13.5	6–7

Notes
*Data from Burland and Burbidge (1985).
[†]Author's experience.

A similar methodology was adopted by Clayton (1993). Results are summarized in Table 2.9, in which the likely range of E'/N_{60} ratios is given as a direct function of penetration resistance.

Soil properties in cohesive materials

SPTs in cohesive soils are undrained, which implies that porewater pressures are generated during penetration. Since these pore pressures are not recorded, interpretation – which is mainly empirical – is carried out against total stress parameters, such as the undrained shear strength and undrained stiffness.

At this stage, it is worth recalling that significant disturbance is produced in the soil immediately below the borehole due to the technique used to excavate and wash the soil prior to an SPT. Disturbance is particularly severe in highly compressible clays with blow counts lower than 5. Under this condition, the test does not give reliable measurements of penetration needed for an accurate estimate of soil parameters so complementary investigation by laboratory and other in situ test techniques is strongly recommended.

Shear strength

The undrained shear strength of clays, common in all soils, depends on the mode of failure, rate of shearing, soil anisotropy and stress history (e.g. Ladd et al., 1977; Jamiolkowski et al., 1985; Ladd, 1991). Short-term undrained shear strength is therefore not a unique soil parameter, being strongly dependent on the method by which it is determined and by factors that cannot be accounted for in dynamic penetration. In addition, the disturbance inherently associated with the test procedure impacts the strength characteristics of the soil and, for this reason, the SPT gives no more than a rough estimation of the magnitude of s_u values.

Yet, dynamic tests are often used as a preliminary tool for site investigation and, despite the fact that no accurate predictions of undrained shear

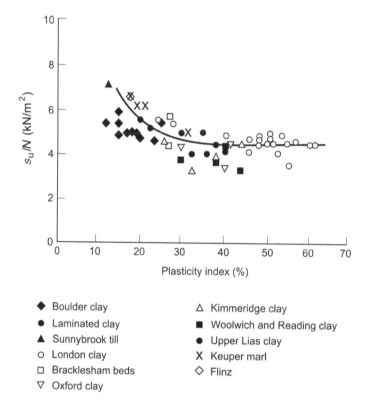

Figure 2.18 Undrained shear strength for insensitive clays (after Stroud, 1974).

strength can be expected, the first approximation of strength values is obtained from s_u/N ratios (e.g. Stroud, 1988; Decourt, 1989). De Mello (1971) has reported that comparisons for the s_u/N ratio should be viewed with caution, since they give a range of measured values varying from 0.4 to 20. Stroud (1974) has suggested that a correlation between plasticity index, N-values and s_u can be obtained for insensitive overconsolidated clays, as illustrated in Figure 2.18. These data yield a lower bound to s_u/N ratios and can be roughly expressed as:

$$s_u = 4.5N_{60}(\text{kN/m}^2)$$

(2.39)

From laboratory triaxial unconsolidated undrained tests, Decourt (1989) has proposed a less conservative $N-s_u$ relationship:

$$s_u = 10.5N_{60}(\text{kN/m}^2)$$

(2.40)

In an analogy with sand, the author has recently proposed a different approach for estimating the soil shear strength in cohesive soils, based on the dynamic force that represents the reaction of the soil to the penetration of the sampler. This is achieved by combining equation (2.22) with bearing capacity theory and cavity expansion (Vesic, 1972):

$$s_u = \frac{(F_d/1.5) - \gamma L A_b}{(N_c A_b + \alpha A_l)} \tag{2.41}$$

where L is the test depth, A_b is the area of the sampler base, A_l is the area of the sampler shaft and α is the sampler–soil adhesion. N_c values are in the range of 7 to 9 for normally consolidated clay (Skempton, 1951; Caquot and Kerisel, 1956; de Beer, 1977), but for stiff clays N_c is much larger and may approach the empirical suggested value of 30 (e.g. Lunne et al., 1997). The influence of shearing rate on undrained behaviour is accounted for by reducing the SPT dynamic force F_d by a factor of 1.5 (Tavenas and Leroueil, 1977; Vaid et al., 1979; Kulhawy and Mayne, 1990; Sheahan et al., 1996; Biscontin and Pestana, 2001; Einav and Randolph, 2005). Sampler–soil adhesion α can be estimated from any of the correlations proposed in the literature for pile bearing capacity analysis (e.g. Flaate, 1968; Tomlinson, 1969; McClelland, 1974).

Two well-documented case studies on the Brazilian experience for both normally consolidated and overconsolidated clays illustrate the proposed procedure. Figure 2.19 shows typical SPT and piezocone test (CPTU) profiles and soil characterization parameters for the Porto Alegre soft clay deposit. The profile reveals an approximately 5 m thick soft clay layer with an overlying sand deposit. Near the surface there is an overconsolidated crust within the depth affected by seasonal variations in the water table. Table 2.10 illustrates the step-by-step process adopted for estimating the undrained shear strength and its variation with depth. A value of I_r of 135 was adopted in the analysis (Schnaid et al., 1997) whereas the value of α was estimated as unity (Tomlinson, 1969). In this case study, estimated values of s_u from the SPT are comparable to values derived from the other test techniques (being slightly on the conservative side). Of particular interest are the values where the rod stem penetrates the soil by its own self-weight, corresponding to an N_{SPT} resistance equal to zero. In this limiting condition, undrained shear strength cannot be interpreted from s_u/N ratios. In contrast, equation (2.16) can be simply expressed as:

$$E_{sampler} = (M_h + M_r)g\Delta\rho \tag{2.42}$$

where η_1, η_2 and η_3 are unity. Values of s_u derived from equation (2.42) are in the order of 15 kPa, which is within the range of measured values.

Table 2.10 Determination of s_u from the dynamic force F_d

Depth	γ	σ'_{v0}	N_{60}	Permanent penetration	Penetration per blow	Energy	Force	s_u
(m)	(kN/m³)	(kN/m²)	(SPT)	(cm)	(m)	(J)	(kN)	(kN/m²)
1	17	17	2	30	0.15	438.87	1.76	44.07
2	15	22.19	0	45	0.45	315.33	0.70	14.06
3	15	27.38	0	45	0.45	329.58	0.73	14.70
4	15	32.57	0	45	0.45	343.84	0.76	15.33
5	15	37.76	0	45	0.45	358.09	0.80	15.97
6	15	42.95	0	45	0.45	372.34	0.83	16.61
7	15	48.14	2	30	0.15	455.44	1.82	45.73

Estimating the in situ undrained shear strength in overconsolidated clays is not a straightforward procedure. Heavily overconsolidated clay cannot be successfully modelled by the constitutive models of analysis currently in use and therefore theoretical values of bearing factors produce unrealistic predictions of undrained shear strength. Engineers should rely solely on empirical correlations, even when interpreting results from piezocone tests (see Chapter 3). Given the fact that penetration in overconsolidated geomaterials may be difficult and that there is a lack of realistic theoretical solutions, the SPT still offers considerable appeal for qualitative and quantitative evaluation of soil properties in situ.

The undrained shear strength calculated from F_d for the Guabirotuba overconsolidated clay in southern Brazil is presented in Figure 2.20. For an N_c (or $N_{k,SPT}$) equal to 30, adopted in the analysis, comparisons between calculated and measured shear strength show that values of s_u estimated for the SPT are similar to those obtained from unconsolidated, undrained (UU) triaxial tests, vane tests and self-boring pressuremeter tests.

Soil stiffness

Soil stiffness depends on interactions of state (bonding, fabric, degree of cementation, stress level), strain level (and effects of destructuration), stress history and stress path, time-dependent effects (aging and creep) and type of loading (monotonic or dynamic). Given these complex interactions, the characteristic response of clay with respect to small strain stiffness and stiffness non-linearity should preferably be assessed from either in situ seismic techniques or laboratory tests. However, in many countries the SPT is a routine site investigation tool and is used to assess soil stiffness.

Several empirical correlations have been proposed and are currently being adopted by engineering practice to estimate the coefficient of volume compressibility, m_v, and the undrained Young's modulus of stiff overconsolidated

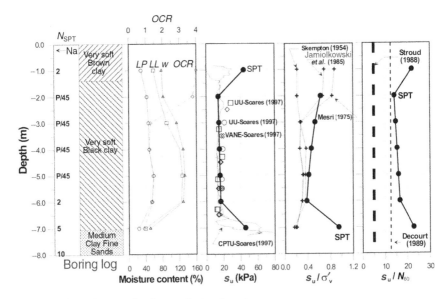

Figure 2.19 Prediction of s_u for a soft clay deposit.

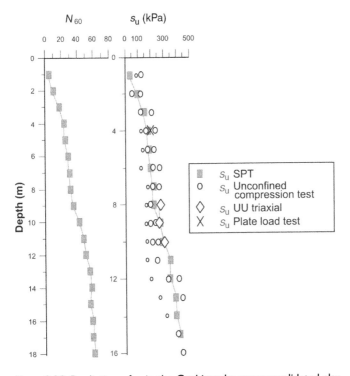

Figure 2.20 Prediction of s_u in the Guabirotuba overconsolidated clay.

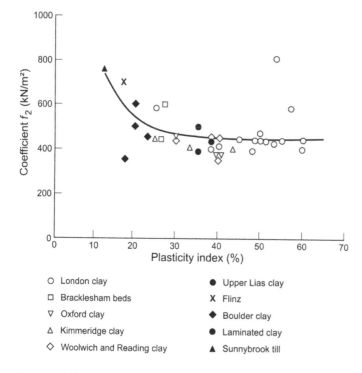

Figure 2.21 Correlation between stiffness coefficient f and plasticity index (after Stroud and Butler, 1975).

clays, E_u. The British experience shows that the coefficient of volume compressibility m_v can be obtained from the following equation (Butler, 1975):

$$m_v = fN(\text{m}^2/\text{MN}) \tag{2.43}$$

where f is obtained from Figure 2.21. The undrained Young's modulus E_u from case histories is (Butler, 1975):

$$E_u/N_{60} = 1.0 \text{ to } 1.2 \text{ (MPa)} \tag{2.44}$$

This ratio is close to values reported by Stroud (1988) for a degree of loading q/q_u greater than 0.2. It is interesting to note that for a soft clay (N_{60} < 3) the combination of equations (2.41) and (2.44) gives a stiffness to strength ratio E_u/s_u lower than 100, which is conservative when compared to the range of measurements reported by Simons (1975): $E_u/s_u \sim 100$ (large deformations) and $E_u/s_u \sim 500$ (small deformations).

For lower shear strains (i.e. for q/q_u lower than 0.1), Stroud (1988) reported E_u/s_u ratios within the range of 5 to 10 MPa.

Soil properties in bonded soils

Although most geomaterials are recognized as being 'structured', the natural structure of bonded soils has a dominant effect on their mechanical response (e.g. Vaughan, 1985, 1997; Blight, 1997). Soil improvement, especially with the addition of other materials such as fibres, cement or lime, can also produce marked structural changes and a cohesion/cementation component that dominates the resultant material's shear strength. Geotechnical problems involving slope stability, excavations, road pavements and other applications involving low stress levels in overconsolidated and weathered soils cannot be addressed without accounting for a cohesion/cementation component in the maintenance of long-term shear strength. In these geomaterials, the difficulties in sampling cause severe restrictions on the characterization of stress–strain behaviour through laboratory tests. Site investigation is often based on in situ tests and the SPT is the most commonly used testing technique.

The debate on an appropriate level of site investigation for any particular project is most pertinent in bonded soil formations. A limited ground investigation based on penetration tests (SPT, CPT or CPTU) will not produce the necessary database for any rational assessment of soil properties, for the simple reason that two strength parameters (ϕ', c') cannot be derived from a single measurement (q_c or N_{60}). Limited investigations are, however, often the preferred option. In such cases involving cohesive frictional soil, engineers tend to (conservatively) ignore the c' component of strength and correlate the in situ test parameters with the internal friction angle ϕ' (see above under *Shear strength* in the section on Soil properties in granular materials). Average c' values may be assessed later from previous experience and back-analysis of field performance.

As for settlement calculations, the magnitude of the small strain stiffness in bonded geomaterials is better understood in comparison with values derived from natural sands. A reference equation adopted in this comparison is:

$$\frac{G_0}{F(e)} = S(p')^n \tag{2.45}$$

with units in MPa and a void ratio function expressed as:

$$F(e) = \frac{(2.17 - e)^2}{1 + e} \tag{2.46}$$

Values of parameter S and n are given in Table 2.11 and a direct comparison is shown in Figure 2.22, having the data for alluvial sands from Ishihara (1982) as reference. Values of G_0 diverge significantly from those established for transported soils and exhibit the same granulometry but are uncemented. Parameter S is much higher than the value adopted for

Table 2.11 S and n parameters

Soil	S	n	Reference
Alluvial sands	7.9–14.3	0.40	Ishihara (1982)[*]
Porto saprolite granite	65–110	0.02–0.07	Viana da Fonseca (1996)[*]
Guarda saprolite granite	35–60	0.30–0.35	Rodrigues and Lemos (2004)[*]
Caxingui gneiss saprolite	60–100	0.30 (p' < 100 kPa)	Barros (1997)[†]

Notes
[*]$N = N_{60}$ likely to reflect Japanese and European experience.
[†]$N = N_{72}$ according to Brazilian practice.

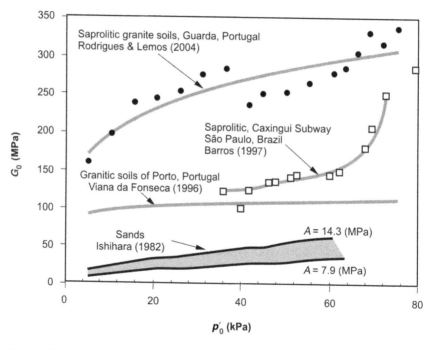

Figure 2.22 Relation between G_0 and p'_0 for residual soils (modified from Gomes Correia et al., 2004).

cohesionless soils, whereas n varies significantly because of local weathering conditions. Given the variations in both S and n, the need for site-specific correlations becomes evident.

Considering the variation observed in natural bonded soils, it is preferable to express correlations in terms of lower and upper boundaries designed to match the range of recorded G_0 values. The variation of G_0 with N_{60} can be expressed as (Schnaid *et al.*, 2004):

$$G_0 = 1{,}200 \sqrt[3]{N_{60}\sigma'_{v0}p_a^2} \quad \text{upper bound: cemented}$$

$$G_0 = 450 \sqrt[3]{N_{60}\sigma'_{v0}p_a^2} \quad \begin{array}{l} \text{lower bound: cemented} \\ \text{upper bound: uncemented} \end{array} \qquad (2.47)$$

$$G_0 = 200 \sqrt[3]{N_{60}\sigma'_{v0}p_a^2} \quad \text{lower bound: uncemented}$$

Once again it is emphasized that, given the considerable scatter observed for different soils, correlations such as those given in equation (2.47) are only approximate indicators of G_0 and do not replace the need for in situ shear wave velocity measurements.

With numerical modelling able to predict ground movements qualitatively, the assessment of operational stiffness has become the key element in the design of structures such as foundations and retaining walls. Very few comparisons have attempted to establish a direct relationship between an operational stiffness and N_{SPT} in cemented soils. Sandroni (1991) compiled a number of plate loading tests carried out in gneissic residual soils using the Theory of Elasticity to derive an operational Young's modulus E representative of ground movements under shallow foundations. Jones and Rust (1989) compiled data for a saprolitic weathered diabase. Despite the scatter, results are presented in Figure 2.23 and are used to support empirical correlations between E and N-values such as:

$$E = aN_{60} (\text{MPa}) \qquad (2.48)$$

with a ranging from 1 to 1.6 (Barksdale and Blight, 1997)[5] or

$$E = bN_{72}^c (\text{MPa}) \qquad (2.49)$$

with coefficients $b = 0.6$ and $c = 1.4$ (Sandroni, 1991).

Soil properties in weak rock

In weak rock, the difficulties involved in sampling and in driving limit the application of both in situ and laboratory tests. Given its robustness, the SPT can be a very useful tool to obtain qualitative measurements of weak, weathered rock properties, as well as estimates of the uniaxial unconfined compressive strength (σ_c). Stroud (1988) reports an attempt to correlate σ_c to N_{60} based on pile tests and pressuremeter tests (Figure 2.24). For $N_{60} < 200$ an extrapolation of the relationship found for clays is appropriate for weak rocks, expressed as:

$$\sigma_c = 10N_{60} (\text{kPa}) \qquad (2.50)$$

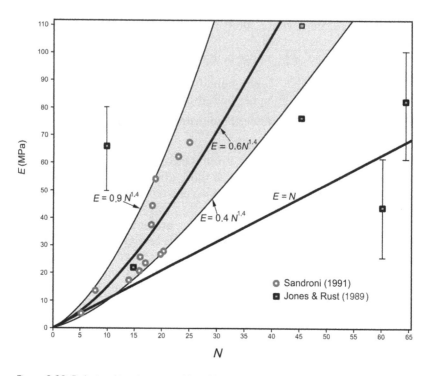

Figure 2.23 Relationships between N and Young's modulus for residual soils: (a) N_{72} for Sandroni (1991); (b) N_{60} for Jones and Rust (1989).

This correlation looks rather conservative for materials exhibiting an unconfined compressive strength greater than about 4 MPa, as seen in Figure 2.25. Clayton (1993) confirms the fact that correlations from N-values for weak rock cannot be expected to be particularly accurate because in rocks the SPT will typically be terminated after 50 to 100 blows and the value used in design must be extrapolated. Moreover, Figure 2.25 is plotted to a double-logarithmic scale and for $N_{60} < 200$ there is scatter of about 20% of the average predicted value.

Prediction of stiffness from SPT in weak rocks is expected to produce a result that could potentially be inaccurate by an order of magnitude and therefore seismic techniques and plate loading tests are always recommended for more accurate predictions. Field evidence collected by Stroud (1988) shows that for a wide range of weak rocks:

$$E' = 0.5 \text{ to } 2.0 N_{60} (\text{MPa}) \tag{2.51}$$

Values of stiffness increase with reducing shear strains, and therefore the coefficient in equation (2.51) increases with increasing factors of safety.

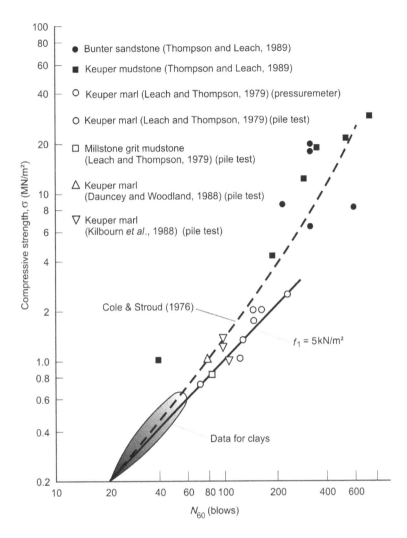

Figure 2.24 Unconfined compression strength versus penetration resistance in weak rocks (after Stroud, 1988).

Direct design methods

Since the early stages of soil investigation by penetration tests, a great amount of field experience has been published comparing the penetration database to empirical performance of foundations. Despite the rapid advance in theoretical methods applied to foundation design, this large volume of empirical knowledge is still used today for estimates of performance. This experience is summarized in this section for the design of both shallow and deep foundations, as well as for the assessment of soil liquefaction.

Generally, no reference is made to a standardized value of potential energy in the original publications. Methods that reflect American and British practices should be read as N_{60}, following recommendations from Harder and Seed (1986) and Skempton (1986). Brazilian practice correlates to N_{72}.

Shallow foundations

The design of shallow foundations is based on considerations of stability and tolerable settlements of the structure. Various analytical methods have been established for calculating the unit ultimate bearing capacity q_n for foundations. These methods, which can take into account the shape and depth of the footing, and the inclination of the loading, the ground level and the base of the footing (Brinch Hansen, 1961; Meyerhof, 1963), are generally accepted in practice and have recently been incorporated into Eurocode 7. The basic equation for calculating the bearing capacity of a footing is (Brinch Hansen, 1961):

$$Q/A = q_n = cN_cs_cd_ci_cb_cg_c + p_0N_qs_qd_qi_qb_qg_q + \frac{1}{2}BN_\gamma s_\gamma d_\gamma i_\gamma b_\gamma g_\gamma \quad (2.52)$$

where

Q	= applied vertical load
A	= footing area
B	= width of the footing
γ	= density of the soil below footing area
c	= soil cohesion
p_0	= effective overburden pressure of soil at foundation level
N_c, N_q, N_γ	= bearing capacity factors
s_c, s_q, s_γ	= shape factors
d_c, d_q, d_γ	= depth factors
i_c, i_q, i_γ	= load inclination factors
b_c, b_q, b_γ	= base inclination factors
g_c, g_q, g_γ	= ground surface inclination factors.

Values of bearing capacity factors N_c, N_q and N_γ are expressed as a function of the internal friction angle, as illustrated in Figure 2.25. Bearing capacity calculations therefore require appropriate values of shear strength to be estimated. This can be done from the correlations introduced above in under the heading Soil properties in granular materials. A limiting factor to this approach is that q_n is extremely sensitive to friction angle. This is particularly true for large values of ϕ' where a small increase in angle will lead to large increases in N_q and N_γ. Consequently, a factor of safety applied to bearing capacity is by no means equal to a factor of safety applied to shear strength.

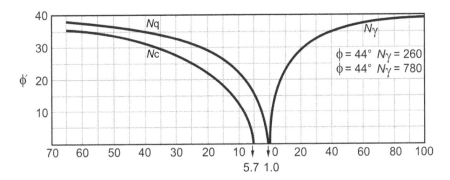

Figure 2.25 Bearing capacity factors N_c, N_q and N_γ.

Given the recognized risk of using equation (2.52) in bearing capacity formulations, engineers are guided by two distinct alternatives:

1   to place a factor of safety on the shear strength of the soil and to use this factored strength in equation (2.52) to obtain safety bearing pressures, or
2   to compare the predicted bearing capacity against allowable bearing pressures derived from experience to guarantee that predicted values do not exceed those considered suitable in local practice.

Terzaghi and Peck (1948) were the first to propose a correlation between the N-value and the allowable pressure, expressed as a function of the size of the footing and a reference value of 25 mm settlement for a deep groundwater table. After this pioneering work, various different procedures have been proposed to predict the allowable pressure of structures founded on granular soil. Many of them are incorporated into regional codes of practice.
    Settlement is usually the controlling factor in the design of shallow foundations in sand, gravels and granular fill materials. Although settlements take place almost immediately after foundation loading, and therefore reflect the elastic deformation of the soil underneath the footing, there is no practicable laboratory test procedure for measuring soil stiffness, given the difficulty of retrieving undisturbed samples in granular soils. Engineers have developed interpretative methods from in situ tests that are essentially based on reviews of the extensive database compiled from case records of the settlement of structures founded on these soils.
    Schultze and Sherif (1973) established a quick estimate of settlements ρ in sand based on the relationship between N_{SPT} values, foundation dimensions and embedment depth, expressed as:

$$\rho = \frac{s.p}{N^{0.87} * [1 + (0.4 + D/B)]} \qquad (2.53)$$

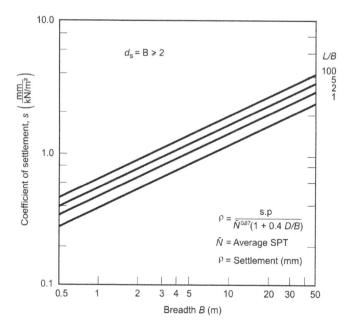

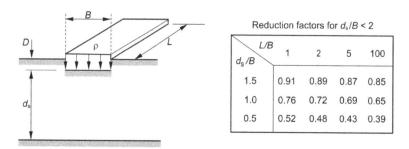

Figure 2.26 Prediction of settlements of spread footings on granular soil (after Schultze and Sherif, 1973).

where

s = coefficient of settlement (cm³/kgf)
N = average value of blow count number
q = average net applied stress (kgf/cm²)
D = foundation depth (m)
B = foundation width (m)
L = foundation length (m)
d_s = thickness of sublayer (m).

Graphical values of the coefficient of settlement are shown in Figure 2.26. The depth of influence over which the average N_{SPT} value was assumed by the authors

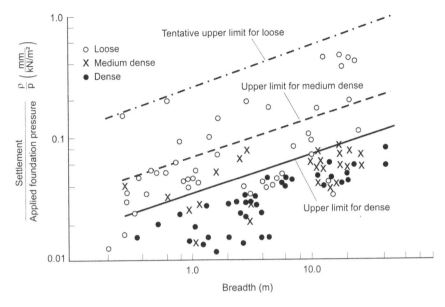

Figure 2.27 Prediction of settlements of spread footings on granular soil (after Burland et al., 1977).

is twice the foundation width. For cases where the thickness of the sublayer is greater than twice the foundation width ($d_s/B < 2$), the estimated settlement should be reduced by a factor that depends on the ratios L/B and d_s/B.

A similar approach was adopted by Burland et al. (1977) to estimate the upper limits of settlements for dense, medium and loose sand. The depth of influence adopted to calculate an average N_{SPT} value is taken as $1.5B$ below the foundation level. Settlement records are plotted in Figure 2.27, in which the ratio of settlement (mm)/applied foundation pressure (kPa) is plotted against foundation width (m). The probable settlements will be about one-half of the upper limits, expressed as:

$$\rho_{max} = q(0.32B^{0.3}) \quad \text{loose } (N < 10) \tag{2.54}$$

$$\rho_{max} = q(0.07B^{0.3}) \quad \text{medium dense } (10 < N < 30) \tag{2.55}$$

$$\rho_{max} = q(0.03B^{0.3}) \quad \text{dense } (N < 30) \tag{2.56}$$

$$\rho = \frac{1}{2}\rho_{max} \tag{2.57}$$

Burland and Burbidge (1985) produced a more detailed method that uses a database of 100 case records of settlements on sands and gravels. For normally consolidated sand, the average settlement at the end of construction is calculated as:

$$\rho = f_s f_1 f_t (q B^{0.7} I_c)$$ (2.58)

where f_s, f_1 and f_t are correction factors and I_c is a compression index that is related to the SPT N-value as shown in Figure 2.28. The average N_{SPT} is taken over the depth of influence z_1, which is also obtained from the same figure in the space (B plotted against z_1).

The authors concluded that the shape of the footing, thickness of the compressive layer and time also have significant effects on predicted settlements. These effects are expressed by the correction factors in equation (2.58). The shape factor for $L/B > 1$ is expressed as:

$$f_s = \left(\frac{1.25 L/B}{(L/B) + 0.25} \right)^2$$ (2.59)

For conditions in which the depth of influence of the applied pressure is greater than the thickness of the compressible material, H_s:

$$f_1 = \frac{H_s}{z_1} \left(2 - \frac{H_s}{z_1} \right)$$ (2.60)

The time factor for estimating settlements more than three years after construction:

$$f_t = \left(1 + R_3 + R_t \log \frac{t}{3} \right)$$ (2.61)

Conservative values of R_3 and R_t are 0.3 and 0.2 respectively for static loading and 0.7 and 0.8 respectively for fluctuating loading. The effect of the previous overconsolidation of the soil is taken into account by expressing equation (2.58) as:

$$\rho = (q - 2/3 \sigma'_{v0}) B^{0.7} I_c$$ (2.62)

where σ'_{v0} is the maximum vertical effective stress at the foundation level. Overconsolidation is known to have considerable effects on settlement and for cases in which the stress increase remains below σ'_{v0}:

$$\rho = 1/3 q B^{0.7} I_c$$ (2.63)

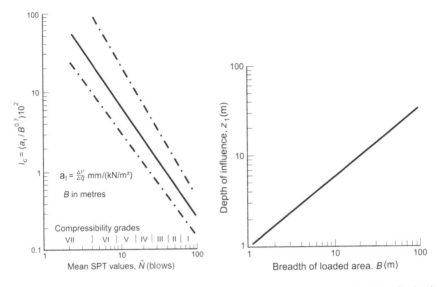

Figure 2.28 Prediction of settlements of spread footings on granular soils (after Burland and Burbidge, 1985).

It is interesting that no corrections are made to the N-values with respect to the effective overburden pressure (i.e. depth of footing and position of the water table). However, in deriving the compression index I_c two cases may require the use of corrected N-values:

1 very fine silty sand, below the water table

$$N = 15 + 0.5(N_{measured} - 15) \tag{2.64}$$

2 gravel or sandy gravel

$$N = 1.25 N_{measured} \tag{2.65}$$

A comparison of measured and predicted settlements for 12 case records published in the literature has been presented by Milititsky *et al.* (1982) and is shown in Figure 2.29. Even if the mean values fall around a 1:1 line, significant scatter is observed, with field observation being sometimes conservative and sometimes failing to achieve a margin of safety (settlements highly underestimated). Caution should always be exercised when using the statistical type of correlations – analytical and numerical approaches using appropriate stiffness as input data are always recommended for more accurate predictions of stress and strain fields below footings.

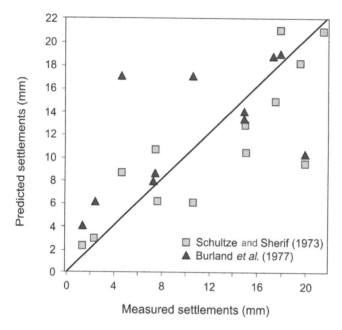

Figure 2.29 Settlement prediction (after Milititsky *et al.*, 1962).

Bearing capacity of piles

The ultimate capacity, Q, of a pile under axial load is calculated by the sum of the base capacity, Q_b, and the shaft capacity, Q_s:

$$Q = Q_b + Q_s = A_b q_b + A_s f_s \tag{2.66}$$

where

A_b = area of the pile base
A_s = area of the pile shaft
q_b = unit end-bearing resistance
f_s = average shaft resistance.

The relative magnitude of the base and shaft capacities will depend on pile geometry and soil properties, which will define both parameters q_b and f_s. A number of methods can be considered when estimating these parameters, including laboratory testing, field pile load tests and empirical correlations from in situ test data. A comprehensive review of existing correlations has been reported by Poulos (1989) and is summarized in Tables 2.12 and 2.13 for f_s and q_b respectively. Correlations between end-bearing resistance and standard penetration resistance in the vicinity of the pile tip indicate that bored and cast-in-place piles develop significantly smaller end-bearing resistance

Table 2.12 Comparison between shaft resistance and SPT value: $f_s = \alpha + \beta N$ kN/m2 (after Poulos, 1989)

Pile type	Soil type	α	β	Remarks	Reference
Driven displacement	Cohesionless	0	2.0	f_s = average value over shaft N = average SPT along shaft Halve f_s for small displacement pile	Meyerhof (1956) Shioi and Fukui (1982)
	Cohesionless and cohesive	10.0	3.3	Pile type not specified $50 \geq N \geq 3$ $f_s > 170$ kN/m²	Decourt (1982)
	Cohesive	0	10.0		Shioi and Fukui (1982)
Cast-in-place	Cohesionless	30	2.0	$f_s > 200$ kN/m²	Yamashita et al. (1987)
		0	5.0		Shioi and Fukui (1982)
	Cohesive	0	5.0	$f_s > 150$ kN/m²	Yamashita et al. (1987)
		0	10.0		Shioi and Fukui (1982)
Bored	Cohesionless	0	1.0		Findlay (1984) Shioi and Fukui (1982)
		0	3.3		Wright and Reese (1979)
	Cohesive	0	5.0		Shioi and Fukui (1982)
		10	3.3	Piles cast under pentonite $50 \geq N \geq 3$ $f_s > 170$ kN/m²	Decourt (1982)
	Chalk	−125	12.5	$30 \geq N \geq 15$ $f_s > 250$ kN/m²	Fletcher and Mizon (1984)

than driven piles. Considerable variations occur in the predicted shaft resistance f_s, particularly for bored and cast-in-place piles. These empirical type f_s correlations should be used with great caution because factors controlling pile foundation behaviour under axial loading are not directly considered and consequently large scatter is observed (e.g. Tomlinson, 1969; Poulos, 1989).

Brazilian foundation engineering practice is essentially based on statistical analysis of a large database of pile loading tests from which N-values are directly correlated to bearing capacity (Aoki and Velloso, 1975; Decourt

Table 2.13 Comparison between end-bearing resistance and SPT value: q_b = KN MN/m^2 (after Poulos, 1989)

Pile type	Soil type	K	Remarks	Reference
Driven displacement	Sand	0.45	N = average SPT in local failure zone	Martin *et al.* (1987)
	Sand	0.40		Decourt (1982)
	Silt, sandy silt	0.35		Martin *et al.* (1987)
	Glacial coarse to fine silt deposits	0.25		Thorburn and Mac Vicar (1971)
	Residual sandy silts	0.25		Decourt (1982)
	Residual clayey silts	0.20		Decourt (1982)
	Clay	0.20		Martin *et al.* (1987)
	Clay	0.12		Decourt (1982)
	All soils	0.30	for $L/D \geq 5$ if $L/D < 5$	Shioi and Fukui (1982)
		0.1 + 0.004L/D 0.06L/D	(closed-end piles) (open-ended piles)	
Cast-in-place	Cohesionless	0.15	q_b = 3.0 MN/m^2	Shioi and Fukui (1982)
			q_b => 7.5 MN/m^2	Yamashita *et al.* (1987)
	Cohesive	–	q_b = 0.009 (1 + 0.16z) z = tip depth (m)	Yamashita *et al.* (1987)
Bored	Sand	0.1		Shioi and Fukui (1982)
	Clay	0.15		Shioi and Fukui (1982)
	Chalk	0.25 3.3	N < 30 N > 40	Hobbs (1977)

and Quaresma, 1978). For routine design, Decourt and Quaresma's method was first developed for concrete driven piles and later extended to different categories of piles (Decourt *et al.*, 1996). The method reflects an average 72% of free-fall energy (Decourt *et al.*, 1996) from which the ultimate load capacity can be determined:

$$Q = \alpha K N_b A_b + P \beta \Sigma 10 \left(\frac{N_m}{3} + 1 \right) \Delta L \qquad (2.67)$$

Table 2.14 Base coefficient K (Decourt and Quaresma, 1978)

Soil type	K (kN/m²)
Clay	120
Silty clay	200
Silty sand	250
Sand	400

Table 2.15 Base coefficient α (Decourt and Quaresma, 1978; Decourt et al., 1996)

Soil/pile	Driven pile	Bored pile	Bored pile (bentonite)	Continuous hollow auger	Root piles	Injected piles (high pressure)
Clay	1.0	0.85	0.85	0.30*	0.85*	1.0*
Intermediate soils	1.0	0.60	0.60	0.30*	0.60*	1.0*
Sands	1.0	0.50	0.50	0.30*	0.50*	1.0*

Note
*Conservative values; require validation from further load testing data.

Table 2.16 Shaft coefficient β (Decourt and Quaresma, 1978; Decourt et al., 1996)

Soil/pile	Driven pile	Bored pile	Bored pile (bentonite)	Continuous hollow auger	Root piles	Injected piles (high pressure)
Clay	1.0	0.80	0.90*	1.0*	1.5*	3.0*
Intermediate soils	1.0	0.65	0.75*	1.0*	1.5*	3.0*
Sands	1.0	0.50	0.60*	1.0*	1.5*	3.0*

Note
*Conservative values; require validation from further load testing data.

where

P = pile perimeter
ΔL = soil thickness (m)
N_b = N_{72} close to pile base
N_m = average N_{72} for every soil layer ΔL
K, α and β are coefficients.

K is a coefficient listed in Table 2.14 that relates the base capacity to N_b for different soil conditions. Coefficients α and β are summarized in Tables 2.15 and 2.16 for various types of foundations.

Soil liquefaction

Liquefaction is a phenomenon in which both soil strength and soil stiffness are suddenly reduced by earthquake shaking or cyclic loading. Liquefaction occurs predominantly in saturated fine grain sands and silts, taking place when seismic shear waves pass through the saturated granular soil layer, increasing the porous water pressure and causing the pore spaces to collapse.

A simplified procedure for assessing soil liquefaction from the cyclic shear stress ratios (CSR) induced by earthquake ground motion was introduced by Seed and Idriss (1971):

$$CSR = \frac{\tau_{av}}{\sigma'_{v0}} = 0.65 \left(\frac{a_{max}}{g}\right)\left(\frac{\sigma_{v0}}{\sigma'_{v0}}\right) r_d \qquad (2.68)$$

where:

τ_{av} = average applied shear stress
σ'_{v0} = effective vertical stress
σ_{v0} = total overburden stress
a_{max} = maximum horizontal acceleration at ground surface
g = gravitational acceleration
r_d = stress reduction factor.

The coefficient 0.65 is used to convert the peak cyclic stress ratio to a cyclic stress ratio that is representative of the most significant cycles over the full duration of load. The value of CSR calculated from equation (2.68) corresponds to the cyclic stress ratio that causes liquefaction for a moment magnitude $M = 7.5$ earthquake.

Stress reduction factors, r_d, were initially presented in Seed and Idriss (1971) to express the dependency of the flexibility of a soil column for sites having sand in the upper first 15 m. Re-evaluation of the initial proposal by improved data sets and interpretation led to several expressions for r_d (e.g. Robertson and Wride, 1997; Idriss, 1999). For most practical applications, the parameter r_d could be adequately expressed as a function of depth and earthquake magnitude, as represented in Figure 2.30, in which the stress reduction factor is plotted against depth below ground surface for M ranging from 5.5 to 8.

A magnitude scaling factor, MSF, is used to adjust the induced CSR during earthquake magnitude M to an equivalent CSR for the earthquake magnitude $M = 7.5$. In this case:

$$MSF = \frac{CSR_M}{CSR_{M=7.5}} \qquad (2.69)$$

The MSF–M relationship can be approximated by the following expression (Idriss, 1999):

$$MSF = 6.9 \exp\left(\frac{-M}{4}\right) - 0.058$$

$$MSF \leq 1.8 \tag{2.70}$$

This framework established the background for the development of semi-empirical procedures designed for evaluating the liquefaction potential of saturated cohesionless soils during earthquakes on the basis of SPT blow count. Since the use of shear strength indices to assess soil liquefaction characteristics requires the effects of soil density and confining stress to be separated, the blow count number has necessarily to be normalized as $(N_1)_{60}$, as extensively discussed above under the heading Measurements and corrections.

Correlations between blow count and liquefaction dating back to the mid-1960s have subsequently been revised by many researchers in an attempt to define a boundary line to differentiate between liquefiable and non-liquefiable conditions (Seed, 1979; Seed *et al.*, 1985; Tokimatsu, 1988; Yoshimi *et al.*, 1989; Robertson and Wride, 1997). A threshold relation between the cyclic stress ratio and the N-value of the SPT test was developed by Seed (1979). Figure 2.31 illustrates recent contributions, in a plot relating the cyclic stress ratio causing initial liquefaction to the normalized N-value (after Ishihara, 1996). The stress ratio shown on the y-axis is defined as a function of the amplitude of average shear stress taken over the time history of seismic motions, as well as the amplitude of maximum shear stress required to cause liquefaction.

A summary of research efforts by Seed and his co-workers is presented in Figure 2.32, in which the liquefaction potential estimated from the CSR is plotted against $(N_1)_{60}$ for clean sands, $M = 7.5$ and $\sigma'_{v0} = 1$ atm. Differences between the various proposals are relatively modest with regard to N-values up to 25, but for the higher range of $(N_1)_{60}$ the curves diverge, being asymptotic to vertical at rather different values. This partially reflects the lack of conclusive experimental data on liquefaction in stiff materials.

Apart from penetration resistance, liquefaction potential of cohesionless deposits is significantly influenced by the grain size characteristics expressed in terms of the percentage of fines. Figure 2.33 displays the Idriss and Boulanger (2004) influence of fines content for: (a) clean sand data with fines content (FC) less than 5%; (b) silty sand data with FC equal to 35%; and (c) silt data with FC greater than 35%. The boundaries defined in this figure can be conveniently expressed using the expressions below (Idriss and Boulanger, 2004). The SPT penetration is adjusted to an equivalent clean sand value $(N_1)_{60cs}$ as:

$$(N_1)_{60cs} = (N_1)_{60} + \Delta(N_1)_{60} \tag{2.71}$$

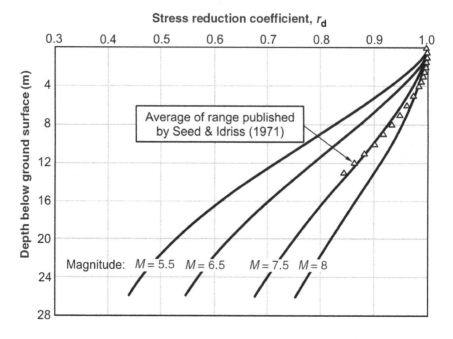

Figure 2.30 Variations of stress reduction coefficient with depth and earthquake magnitude (Idriss, 1999).

where

$$\Delta(N_1)_{60} = \exp\left(1.63 + \frac{9.7}{FC} - \left(\frac{15.7}{FC}\right)^2\right) \quad (2.72)$$

For $M = 7.5$ and $\sigma'_{v0} = 1$ atm, the value of CSR can be calculated based on $(N_1)_{60cs}$ only:

$$CSR = \exp\left\{\frac{(N_1)_{60cs}}{14.1} + \left(\frac{(N_1)_{60cs}}{126}\right)^2 - \left(\frac{(N_1)_{60cs}}{23.6}\right)^3 + \left(\frac{(N_1)_{60cs}}{25.4}\right)^4 - 2.8\right\} \quad (2.73)$$

The SPT, as a means of assessing liquefaction, fulfils the need to obtain a soil sample to determine the fines content of the soil. Poor repeatability and reliability of penetration resistance due to the lack of standardization in test procedures and unknown values of energy delivered to the SPT rod are the limiting factors that may result in misleading predictions. In this respect, CPT (Chapter 3) and DMT (Chapter 6) soundings can be viewed as practical companions to SPT in site investigation programmes.

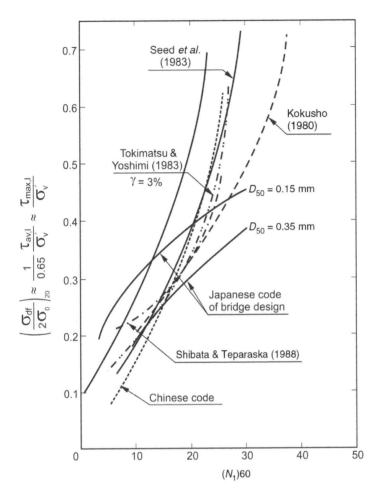

Figure 2.31 Summary chart for evaluation of the cyclic strength of sands based on the normalized SPT N-value (after Ishihara, 1996).

Concluding remarks

Some fairly practical conclusive remarks can be distilled from the preceding discussion.

- Given its simplicity, low cost and general applicability to a variety of soil formations, the SPT is a common test widely accepted throughout the world. It is a routine site investigation tool that will continue to be used in the future, especially for foundation engineering.

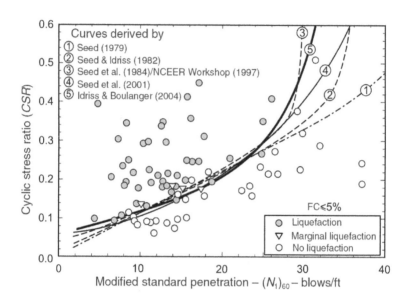

Figure 2.32 Curves relating *CSR* to $(N_1)_{60}$ published over the past 24 years for clean sands, $M = 7.5$ and $\sigma'_{v0} = 1$ atm (after Idriss and Boulanger, 2004).

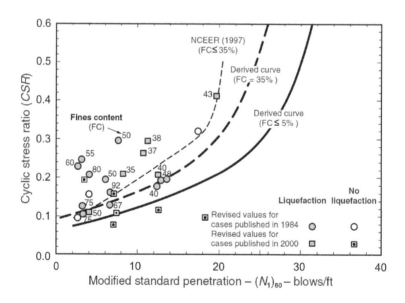

Figure 2.33 SPT case histories of cohesionless soils with FC = 35% and the NCEER Workshop (1997) curve and the recommended curves for both clean sand and for FC = 35% for $M = 7.5$ and $\sigma'_{v0} = 1$ atm (after Idriss and Boulanger, 2004).

- Since complete standardization is not likely to be achieved, given the differences in equipment and techniques adopted worldwide, it is strongly recommended that the measured blow count N-values be corrected with respect to a reference value of 60% of the potential energy of the SPT hammer as a means of quality control for SPT results. Furthermore, inaccuracies and errors in the actual measurements of the blow count number N due to the variability of procedures for tests that are not fully standardized can be minimized by observation of recommended codes of practice and better workmanship.
- Application of SPT data to estimate soil properties and to predict the behaviour of foundation elements can be very useful in engineering practice, provided that the necessary caution is exercised when manipulating the various correlations described in this chapter.
- Wave propagation analysis applied to the mechanics of dynamic penetration tests can support new and more rational methods of interpretation. Using the wave-equation approach, it is possible to evaluate the driveability of a pile with a particular piece of equipment, to relate the driving dynamic resistance to the load capacity of the pile under static loading, and to predict soil properties. This will be a trend for the near future and will be sufficient to promote energy measurements in SPT, LPT and other dynamic penetration techniques.

Notes

1 Some standards recommend a weight of 65 kg falling through a distance of 750 mm. Variations in the characteristics of equipment strengthen the need for calibrating N-values before using test results in existing empirical correlations.
2 Rock degradation generally progresses from the surface and therefore there is normally a gradation of properties with no sharp boundaries within the profile. Lateritic and saprolitic residual soil profiles are often encountered. Lateritic soils are formed under hot and humid conditions involving a high permeability profile, often resulting in a bond structure with high contents of oxides and hydroxides of iron and aluminium, and low penetration resistance – $(N)_{60} < 10$. In saprolitic profiles, the original disposition of the decomposed crystals of the parent rock is retained and penetration resistance is greater than $(N)_{60} > 15$.
3 The value of the effective internal friction angle ϕ' required for foundation design is often assumed to be the triaxial axi-symmetric value, whereas other engineering applications such as retaining walls, slopes and even strip footings clearly require the plane strain value.
4 Hardin and Black (1968) and Hardin (1978) identified σ'_{v0}, e_0, OCR, degree of saturation, grain characteristics, soil structure and temperature as important factors governing the G value (see Chapter 3 under Consolidation coefficients).
5 Energy was not specified in the original publication and the author is arbitrarily assuming a reference value of 60%.

Piezocone penetration test (CPTU)

A better understanding of the stress-strain-time behaviour of geomaterials ... will become of little benefit to our discipline if not accomplished by some useful tasks such as: (a) framing these concepts within a new generation of constitutive models, (b) implementing the constitutive models in computer codes operating in a user-friendly manner and able to solve the boundary value problem of practical interest, and (c) checking the results obtained by means of the above numerical modelling against well-documented case histories to verify coincidences and diversities between the behaviour observed in the field and that computed on the basis of laboratory (and field) measurements.

(M. Jamiolkowski, 2001)

The cone and piezocone penetration tests, commonly referred to as CPT and CPTU respectively, are now internationally recognized as established, routine and cost-effective tools for site characterization, soil profiling and assessment of the constitutive properties of geomaterials. Particularly useful in marine and lacustrine sediments in costal regions, use of the CPT has spread to peat, silt, coarse-grained soils, residual soils and a variety of hard materials such as chalk and cemented sands, as well as reclaimed land formed by hydraulic fills, dredging and mine tailings.

Although the first references to the test date back to the early 1930s in Holland (Barentsen, 1936), the technique was consolidated in the 1950s (e.g. Begemann, 1963, 1965). Its extensive use gave rise to a number of specific conferences dedicated to the theme: e.g. ASCE '81 – *Cone Penetration Testing and Experience*, CPT '95 – *International Symposium on Cone Penetration Testing*. Complete state-of-the-art reports on the interpretation of piezocone tests focusing on the several aspects of engineering practice are given by Jamiolkowski *et al.* (1985, 1988), Lunne *et al.* (1985, 1997), Meigh (1987), Robertson and Campanella (1988, 1989), Yu (2004) and Schnaid (2005). For a general review of the subject the reader is encouraged to refer to Lunne *et al.* (1997), *CPT in Geotechnical Practice*.

Equipment and procedures

The CPT principle consists of driving a tip with a 60° apex angle into the soil at a constant rate of penetration of 20 mm/s. The cone cross-sectional area is usually 10 cm², reaching up to 15 cm² for more robust and higher-load pushing equipment. While test procedures are already standardized, there are some differences between equipment, which can be classified into three categories:

- the **mechanical cone**, which is characterized by the measurement on the surface, via mechanical transfer of the rods, of the necessary thrust to push the cone into the ground;
- the **electrical cone**, where adaptation of electrically instrumented load cells allows the measurement of the cone tip resistance q_c and sleeve friction resistance f_s directly on the cone; and
- the **piezocone**, which allows the continuous monitoring of the porewater pressures u generated during driving, in addition to the electrical measurements of q_c and f_s.

The pushing equipment normally consists of hydraulic jacking built on a reaction system that can be adapted for onshore, near-shore and offshore operations, as discussed in Chapter 1. Offshore rigs with continuous pushing produced by rotating steel wheels to drive the cone rods have been used extensively since 1984 (e.g. Lunne *et al.*, 1994). Dedicated drillships available for investigations can provide useful data from the shallow near-shore environment to deepwater conditions (as previously illustrated in Figures 1.2 and 1.3 in Chapter 1), assisting the design of structures for ports, the oil and gas industry and wind turbines, among others. For onshore operations the assembly can be set up over a truck, van, rig or trailer, with capacity usually ranging from 10 to 20 t (100 to 200 kN). The power for the hydraulic jacking is supplied by a combustion motor from the truck or from an independent electric engine. A heavy duty CPT truck is shown in Figure 3.1. The interior of a typical truck is shown in Figure 3.2 where two vertical hydraulic cylinders are supporting a stiff beam that moves up and down, pushing and pulling the driving rods through an automatic hydraulic clamp designed to make operation quicker and safer. A flow rate valve enables a precise control of the rate of penetration during the test within the recommended range of 20 mm/s ± 5 mm/s. Penetration is accomplished by pushing the rods in 1 m strokes, followed by retraction of the hydraulic rams for the next stroke.

Lightweight rigs have proved themselves to be versatile pieces of equipment, used to perform tests in difficult locations, working as stand-alone units that are anchored by propeller blades into the soil to provide the required reaction force (Figure 3.3).

Figure 3.1 French CPT rig in action at Pijnacker, The Netherlands (courtesy of Fugro Ltd).

CPT equipment and test procedures fully comply with ISSMGE International Reference Test Procedure (referred to as IRTP) and well-established national codes of practice. Data acquisition systems are usually connected to CPTs by analogue to digital converters, from which simple computing programs produce the data logging of the signals, test control and storage in situ of the recorded data in a computer. A set-up showing all the electronics for data acquisition on board a ship is shown in Figure 3.4. Recently, the acoustic transmission of signals has been gaining in popularity, despite indications that it can be less reliable than cabled systems in operations where environmental frequencies can interfere with measurements. This system eliminates the need for a cable but still allows real time recording.

Figure 3.5 shows a diagram of a typical cone. Routine penetrometers have employed either one mid-face element for porewater pressure measurement (designated as u_1) or an element positioned just behind the cone tip (shoulder, u_2). The ability to measure pore pressure during penetration greatly enhances the profiling capability of the CPTU, allowing thin lenses of material to be detected. Various penetrometers, namely the miniature electric element with a cone tip area of 2 cm², a 5 cm² minicone, a standard 10 cm² piezocone and a 10 cm² piezocone with five porewater sensors, a seismic piezocone and a resistivity cone, are shown in Figure 3.6.

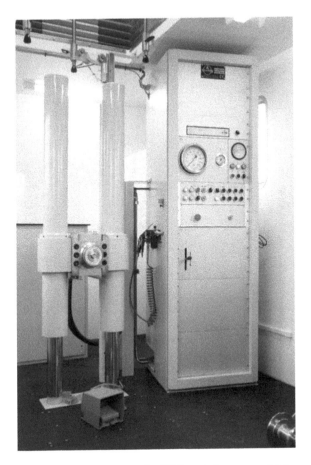

Figure 3.2 Typical interior of a CPT truck (courtesy of A.P. van den Berg).

The internal components of an open piezocone are shown in Figure 3.7, which details the pressure transducer and the two load cells for tip and shaft resistance, respectively.

New developments

CPT technology is still undergoing some important developments in the following areas:

- the combination of different sensors within a single test device, particularly the seismic cone and the cone pressuremeter (e.g. Mayne, 2001; Schnaid, 2005);

(a)

(b)

Figure 3.3 Lightweight rigs: (a) small trailer operating in Hong Kong (courtesy of A.P. van den Berg); (b) CPT testing in Leidschendam, The Netherlands (courtesy of Fugro Ltd).

Figure 3.4 Operator monitoring CPT testing on board a ship (courtesy of Fugro Ltd).

- the adaptation of additional sensors to penetration tools to enhance and expand their capabilities for geo-environmental purposes (e.g. Robertson *et al.*, 1998); and
- the changes in execution procedures and testing for advanced offshore engineering (e.g. Van Impe and Van der Broeck, 2001).

Current trends in equipment development consist of combining different techniques, coupling the cone test robustness with the additional information derived from other sensors.

Seismic cone

Geotechnical site characterization can be improved by independent seismic measurements, adding the downhole shear wave velocity (v_s) to the measured tip cone resistance (q_t), sleeve friction (f_s) and porewater pressure (u). It has now been recognized that the combination of different measurements into a single sounding provides a particularly powerful means of assessing the characteristics of natural materials.

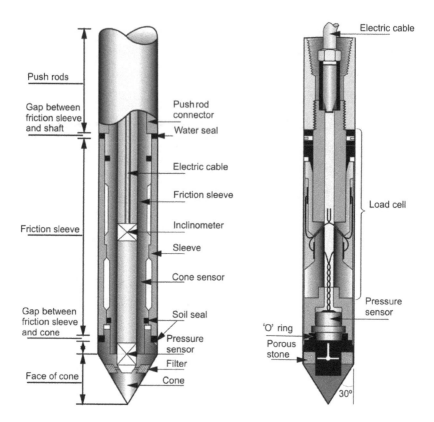

Figure 3.5 Cone diagram (adapted from Zuidberg, 1988).

The seismic cone penetration test (SCPT) is becoming a routine site investigation tool in many countries because of its facility of adding a seismic adapter equipped with an accelerometer or geophone to a standard CPT probe (e.g. Campanella *et al.*, 1986). This downhole seismic technique is the simplest, fastest and most cost-effective way of measuring shear wave velocities. The additional cost and time required for a seismic measurement in a cone test is modest and the input provided by the small strain shear modulus is essential for soil characterization and prediction of ground surface settlements under dynamic loading.

Seismic energy is generated on the surface in the form of polarized shearing waves (v_s) using a shear wave hammer. This is currently achieved by using two hammers connected to either end of a steel beam that is held to the ground by a vehicle or a heavy weight. Recording both positive and negative polarized *S*-waves enables the arrivals on the receiver shot records to be distinguished from those of *P*-waves and environmental noise, as demonstrated in Figure 3.8. In addition, various automatic wave-generating

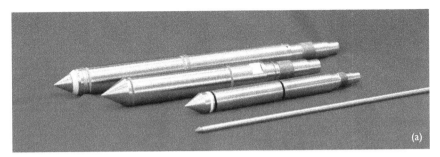

(a)

(b)

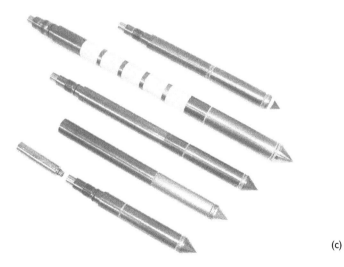

(c)

Figure 3.6 Range of CPT tools available for site investigation: (a) 2 cm^2, 5 cm^2 and 10 cm^2 cones; (b) 10 cm^2 electrical cone with five porewater pressure sensors used by the Korean technical research centre KHC at Seoul; (c) piezocone, seismic cone and resistivity cone (courtesy of A.P. van den Berg).

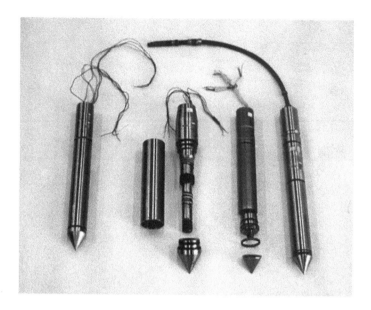

Figure 3.7 Internal components of a piezocone.

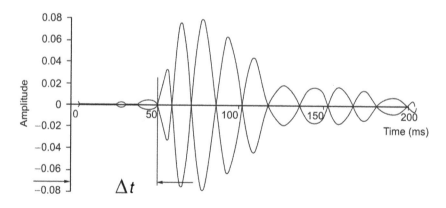

Figure 3.8 Shear wave velocities measured by a downhole test.

systems are available for downhole geophysical testing, which include pneu-matic, hydraulic and electrically driven sources. Automatic systems improve productivity, repeatability and accuracy and can be used in both onshore and offshore operations.

Since accurate arrival time estimates are critical for a proper definition of the frequency spectrum of the recorded signals, the use of two sets of sensors to measure the arrival of the same wave to directly determine Δt is always recommended. Both velocity sensors (i.e. geophones typically in the range of 1 to 300 Hz) or accelerometers (typically in the range of 1 Hz to 100 kHz) can be easily incorporated into the cone; accelerometers are available in uniaxial configuration for measuring shear wave velocity and in triaxial configuration for measuring both shear and compression wave velocities.

Data acquisition, hardware and software components are usually designed to run directly in a notebook computer connected to a signal conditioning system. The measured voltage signals are then transferred to the surface by a cable, raw data are recorded and frequency filtering of data is carried out during interpretation, after acquisition.

The theoretical basis on which seismic and other geophysical measurements are founded are beyond the scope of this book. A number of reference textbooks cover this subject area extensively, such as Richard *et al.* (1970), Sharma (1997) and Santamarina *et al.* (2001). For the purposes of this current study it is important to recall that geophysical methods rely on a significant contrast of physical properties of the materials under investigation. In this book attention is given to the measurement of shear wave velocities, from which it is possible to obtain the small strain stiffness of the soil at induced strain levels of less than about $10^{-3}\%$:

$$G_0 = \rho(v_s)^2 \tag{3.1}$$

where G_0 is the small strain shear modulus, ρ the mass density and v_s the velocity of shear waves for a linear, elastic, isotropic medium. As with the downhole tests, the SCPT enables the velocity of vertically propagating, horizontally polarized (s_{vh}) shear waves to be measured.

Cone pressuremeter

The cone pressuremeter (CPMT) is an in situ testing device that combines the 15 cm^2 cone with a pressuremeter module mounted behind the cone tip, as illustrated in Figure 3.9. The first prototype was developed in England (Withers *et al.*, 1986), followed by experiments in Canada (Campanella and Robertson, 1986), Italy (Ghionna *et al.*, 1995) and Holland (Zuidberg and Post, 1995). Since the pressuremeter test is not carried out in undisturbed ground, the effects of installation have to be accounted for and large strain analysis is required. Nevertheless, the device has certain advantages, principally in the simplicity and economy of its installation, combining some of the merits of the pressuremeter with the operational convenience of the cone: a continuous soil profile is obtained during penetration, pore pressures and friction ratio provide detailed and representative characterization of soil layers, shear strength can be assessed from cone resistance and, in addition, stiffness and strength parameters are estimated from pressuremeter data. This technique is perceived as having a great potential that has not yet been fully recognized in practice. Analysis of the test in clay is achieved by a simple geometric construction of the curve to

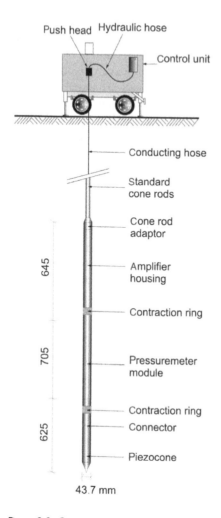

Push head   Hydraulic hose

Control unit

Conducting hose

Standard
cone rods

Cone rod
adaptor

Amplifier
housing

Contraction ring

Pressuremeter
module

Contraction ring

Connector

Piezocone

645

705

625

43.7 mm

Figure 3.9 Cone pressuremeter (after Withers *et al.*, 1986).

determine the undrained shear strength, the shear modulus and the in situ horizontal stress (Houlsby and Withers, 1988). Analysis in sand is significantly more complex and interpretation is largely based on calibration chamber tests (Nutt and Houlsby, 1992; Schnaid and Houlsby, 1992). Research over the past ten years has provided basic interpretation procedures from which engineering properties of soils are assessed, including shear strength, relative density, state parameter, friction angle and in situ stress state. A detailed description of intepretation methods is presented in Chapter 5 (Pressuremeter tests).

Installation in the ground is performed as part of the cone penetration test operation with the cone driven into the soil at a constant penetration

Figure 3.10 Envirocone measuring pH, redox potential and temperature (courtesy of A.P. van den Berg).

rate of 20 mm/s. Insertion is halted with the centre of the pressuremeter probe at the required depth to allow pressuremeter probe expansion. A continuous cone profile is obtained during penetration and is used to estimate strength parameters in both sand and clay. The pressuremeter, in turn, enables soil stiffness and strength parameters to be assessed from the measured pressure–displacement curve.

Environmental cone

Environmental cones are designed for the investigation of contaminated sites. Without having to deal with all requirements of handling and disposal involved in cutting materials in contaminated areas, the driving of a cone enables the engineer to cross-correlate the information acquired by the measurements of a piezocone – nature and sequence of layers and hydro-geologic regime – with measurements from chemical sensors such as pH, redox, conductivity, temperature and dissolved oxygen (Figure 3.10). In addition to these screening devices that log the ground profile in search of contaminants, sampling probes have been developed to sample vapour, liquids and solids.

Of particular interest for characterizing zones of electrically conductive contamination is the resistivity cone penetrometer test (RCPT), which uses a resistivity module that is located directly behind the standard piezocone.

The module, consisting of two or more electrodes separated by insulating material, operates within the principle that the potential drop across two adjacent pairs of electrodes, at a given excitation current, is proportional to the electrical resistivity of both soil and pore fluid (after Archie, 1942). Knowing that the soil electrical properties may vary in the presence of contaminating fluids, it is possible to map the extent of the contaminated areas spatially through resistivity measurements.

A review of penetration tests for geo-environmental applications is given by Lunne et al. (1997) and case studies are reported by Hornsnell (1988), Campanella and Weemees (1990), Fukue et al. (2001), Strutymky et al. (1991) and Woeller et al. (1991), among others.

Flow penetrometers

The T-bar and ball-bar penetrometers have been recently introduced in Australia (e.g. Stewart and Randolph, 1991; Randolph et al., 1998; Randolph, 2004; Einav and Randolph, 2005; Randolph et al., 2005) in order to improve the accuracy of strength profiling in soft soils through a better representation of the failure penetration mechanism. The cylindrical T-bar was first introduced following the recognition that the plane strain path for soil elements around the probe could be modelled by strain path solutions. Later, the axisymmetric ball head was suggested to reduce the potential for the load to be subjected to bending moments arising from non-symmetric resistance along the T-bar.

These flow penetrometers have been designed for the investigation of soft sediments, especially for the offshore industry. In comparison with the cone penetration test, the existing experimental database is very limited.

Specifications and standards

The inherent difficulties in comparing results obtained with different equipment led to the standardization of the test by ISSMGE (last reviewed in 1999), as well as the publication of several national and regional codes of practice (for example, Brazil: ABNT MB-3406, 1991; Netherlands: NEN 5140, 1996; Eurocode 7, Part 3, 1997; France: NFP 94-113, 1989; UK: BS 1377, 1990; USA: D 5778, 1995). Recommendations on terminology, dimensions, procedures, precise measurements and presentation of results are specified in these codes. Analyses and interpretation methods of test results presented in this chapter are applicable to tests performed strictly according to IRTP. The penetration rate should be 20 ± 5 mm/s, the friction sleeve diameter should be as large as or larger than the cone diameter (not exceeding 0.35 mm), the tip should have a 60° angle and roughness less than 1 µm. By including a slope sensor in the penetrometer, the inclination can be recorded so that verticality of the rods during penetration can be ensured to prevent damage. Generally, a 5° change in inclination over 1 m of penetration can impose detrimental push rod bending.

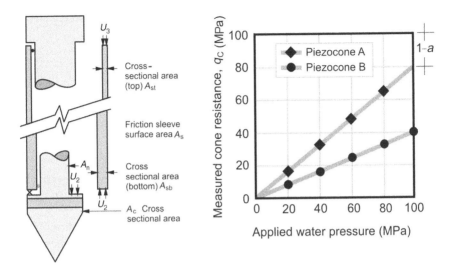

Figure 3.11 Pore pressure effects on measured cone data (after Battaglio and Maniscalco, 1983).

A standard piezocone test measures cone tip resistance, sleeve friction and porewater pressure. Porous elements for pore pressure measurements can be located anywhere in the piezocone, with the choice of a particular position not yet standardized – cone tip (u_1), shoulder (u_2) or sleeve (u_3). The approriate position in any given situation depends on the intended use of the pore pressure results (e.g. Robertson *et al.*, 1992a; Chen and Mayne, 1994; Danziger *et al.*, 1996). To enable the measured values of tip resistance to be corrected for pore pressure effects (e.g. Jamiolkowski *et al.*, 1985), the porous element should be necessarily positioned at the shoulder of the cone, immediately above the cone face (u_2). Porewater pressures acting in the joints of the penetrometer (on the shoulder area behind the cone and on the ends of the friction sleeve) affect the measured results, as illustrated in Figure 3.11. This is konwn as 'unequal area effects' and it should be taken into account when calculating the actual penetration resistance mobilized at the cone tip (Campanella *et al.*, 1982; Jamiolkowski *et al.*, 1985):

$$q_t = q_c + (1 - a)u_2 \qquad (3.2)$$

where

q_t = correct cone resistance
q_c = measured cone resistance
a = area ratio (= A_N/A_T), easily determined through calibration according to the method illustrated in Figure 3.11.

If assessment of soil properties is intended, the correction to the measured tip resistance is essential. In analogy to the correction of q_c, the corrected sleeve friction can be expressed according to the following equation:

$$f_t = f_s - \frac{u_2 A_{sb}}{A_l} + \frac{u_3 A_{st}}{A_l} \qquad (3.3)$$

where

f_t = corrected sleeve friction
f_s = measured sleeve friction
A_{sb}, A_{st} = bottom end and top end areas of friction sleeve, respectively
A_l = friction sleeve shaft area.

It is important to observe that the cone tip resistance should be corrected in all tests, following IRTP recommendations, requiring the pore-water pressure u_2 to be monitored during driving. This correction can be especially important in soft clay deposits where it can make a significant difference to the results, depending on the geometry of the penetrometer.

There are no significant differences in test procedures between a CPT and a CPTU, except for those required for the saturation of the piezocone. The saturation fluid can be de-aired water, silicon oil or glycerin oil (e.g. Robertson and Campanella, 1989; Danziger, 1990; Larsson, 1992). Mineral oil and grease have also been used successfully. The use of de-aired water may cause difficulties in maintaining saturation of the porous element during penetration above groundwater level. Filter elements are saturated by applying vacuum inside a calibration chamber, where immediate responses to stress increments indicate full saturation. This process is usually performed in a laboratory prior to the actual test, keeping the porous element immersed until driving.

An alternative approach to a 'standard' filter is to use a 'slot' filter. In this arrangement, the pore pressure is measured by an open system with a 0.3 mm slot directly behind the cone tip (Larsson, 1995). The slot communicates with the pressure chamber, which is saturated by de-aired water or other liquid, whereas the slot and channels are saturated with gelatine, silicon grease or a similar product, reducing the time required for preparation of the probe. One important benefit is that the slot system can maintain its saturation when passing through unsaturated zones in the soil upper layers. Limitations are that the pore pressure response may not be as accurate as standard type filters.

In conclusion, the main features of the CPT and CPTU are the continuous sounding capability and the good repeatability of the measured q_c, f_s and u, which enables a detailed description of subsoil stratigraphy. Despite being a fully automatic process, skill and experience in daily practice by qualified personnel is the miminum criterion to be satisfied as a matter of good practice.

Test results

In this section, some typical results will be presented with the aim of:

- demonstrating the requirements for final presentation, and
- familiarizing the reader with the sort of information that is obtained from test results.

In the CPTU, the continuous measurements of q_c, f_s and u (and calculated values of R_f and B_q) are plotted along with depth. $R_f (= f_s / q_c \, 100\%)$ is the friction ratio, a parameter that is considered useful for soil classification. The piezocone relies on the highly accurate pore pressure measurements taken during penetration, expressed by the pore pressure parameter ratio B_q:

$$B_q = \frac{(u_2 - u_0)}{(q_t - \sigma_{v0})} \qquad (3.4)$$

where u_0 represents the equilibrium pore pressure and σ_{v0} the total overburden stress. The continuously recorded measurements of cone tip resistance, combined with both sleeve friction and pore pressure, enable the characterization of soil stratigraphy and the precise identification of soil stratum, including the identification of thin drained layers.

A typical example of a piezocone profile is presented in Figure 3.12, in which a 15 m thick soft clay layer is identified by low values of q_t combined with very high pore pressures ($u \sim q_t$; $B_q \sim 1$). A thin layer of sand at approximately 5.5 m depth is identified by the localized increase of q_t and a significant reduction in pore pressure towards the hydrostatic value ($\Delta u = 0$).

Soil classification

Several attempts have been made to combine the piezocone measurements in order to produce classification charts designed to describe soil type for engineering applications (Douglas and Olsen, 1981; Senneset and Janbu, 1985; Robertson et al., 1986; Robertson, 1990; Jefferies and Davies, 1991). Published CPTU soil classification charts are typically constructed following a few basic concepts:

- tip resistances are relatively higher in sands and decrease as the fines content increases;
- inversely, sleeve resistances are relatively lower in sands and increase as fines content increases; and
- pore pressures are lower in sands due to high permeability and increase as fines content increases.

Most recognized charts used in routine engineering practice are shown in Figure 3.13 for comparison. Following early recommendations from Wroth

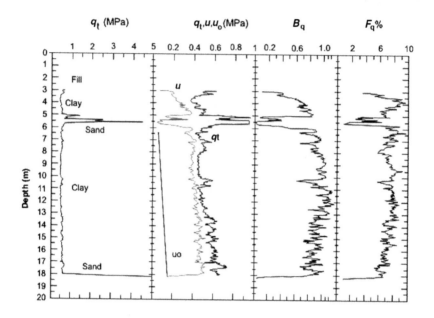

Figure 3.12 Typical piezocone profile in a soft clay deposit.

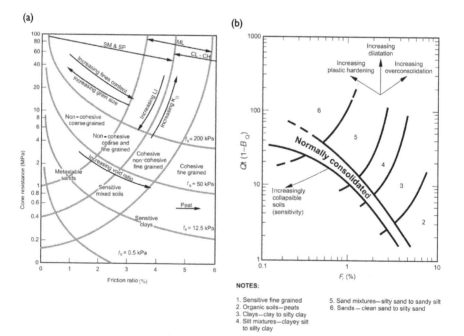

NOTES:

1. Sensitive fine grained
2. Organic soils—peats
3. Clays—clay to silty clay
4. Silt mixtures—clayey silt to silty clay

5. Sand mixtures—silty sand to sandy silt
6. Sands — clean sand to silty sand

Figure 3.13 (Continued on next page).

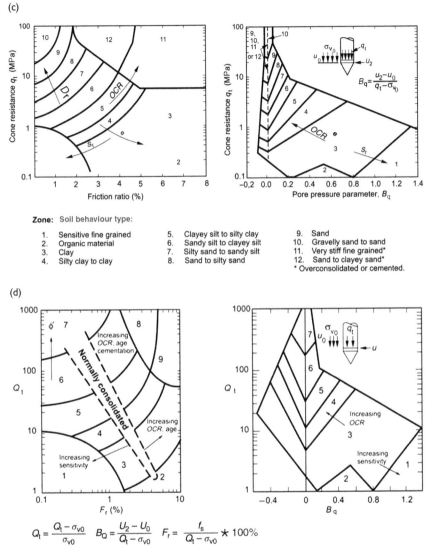

(c)

Zone: Soil behaviour type:

1. Sensitive fine grained
2. Organic material
3. Clay
4. Silty clay to clay
5. Clayey silt to silty clay
6. Sandy silt to clayey silt
7. Silty sand to sandy silt
8. Sand to silty sand
9. Sand
10. Gravelly sand to sand
11. Very stiff fine grained*
12. Sand to clayey sand*
* Overconsolidated or cemented.

(d)

$$Q_t = \frac{q_t - \sigma_{vo}}{\sigma_{vo}} \quad B_Q = \frac{U_2 - U_0}{q_t - \sigma_{vo}} \quad F_r = \frac{f_s}{q_t - \sigma_{vo}} * 100\%$$

Figure 3.13 Soil type classification charts: (a) Douglas and Olsen, 1981; (b) Jefferies and Davies, 1991; (c) Robertson et al., 1986; (d) Robertson, 1990.

(1988), experience suggests that it is prudent to express classification charts as a function of normalized parameters in order to correct for overburden stress effects.

The original chart by Douglas and Olsen (1981) uses q_c versus normalized friction ratio F_r, whereas more recent work, based predominantly on data obtained from piezocone tests, makes extensive use of the accurate measurements of pore pressure. CPT classification charts do not provide accurate predictions of soil type based on grain size distribution, but are able to provide a guide for soil behaviour type.[1]

Interpretation methods

Results from CPTs and CPTUs have three main applications in engineering practice:

- to assess soil stratigraphy and identify soil type;
- to predict engineering parameters for geotechnical design;
- to provide direct empirical assessment of foundation performance and soil liquefaction.

For predicting soil parameters in both clay and sand, interpretation of piezocone test results has evolved from early empirical approaches to rather sophisticated and rigorous methods of analysis. However, in any theoretical solution for penetration tests, considerable idealization is necessary in order to describe the stress–strain path around a penetration probe. Relatively accurate solutions are obtained in very few cases, for example the strain path method applied to soft clay (Baligh, 1985). The strain path method integrates the velocity along stream lines to determine deformation fields in deep penetration problems. These deformation fields are kinematically constrained and combined with independently constitutive parameters of the soil surrounding the probe to enable penetration stresses to be estimated. An example of octahedral and shear induced pore pressures generated during penetration is illustrated in Figure 3.14. It shows that in normally consolidated clay the octahedral stresses control both the penetration resistance and the excess pore pressures. The pore pressure distribution along the cone in Boston clay illustrates the fact that pore pressures decrease dramatically along the shaft and gradually away from the tip (Figure 3.15, after Baligh and Levadoux, 1986; Campanella and Robertson, 1988).

The stress path method and other analytical and numerical solutions are useful in assisting the understanding of deep penetration problems and in estimating strength and consolidation parameters. An enormous variety of literature is now available because of the strong theoretical basis on which the penetration mechanism is modelled. Some potential uses of CPTs and CPTUs and key references are shown in Tables 3.1 and 3.2.

Soil properties in cohesive materials

This section focuses on the interpretation of cone penetration tests in clay. A critical review is presented and specific recommendations are given to derive undrained shear strength, soil stiffness, stress history and consolidation coefficients.

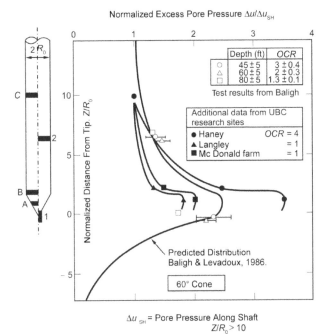

Figure 3.14 Distribution of measured and predicted normalized excess pore pressures during penetration (Baligh and Levadoux, 1986; Campanella and Robertson, 1986).

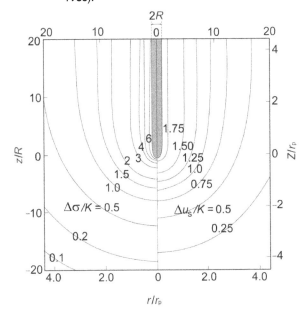

Figure 3.15 Octahedral and shear induced pore pressures (Baligh, 1986).

Table 3.1 Potential use of CPTs and CPTUs (modified from Battaglio et al., 1986)

	CPTU	CPT
Soil profile	High	High
Soil structure	Low	Moderate
Stress history	Low	Moderate
Mechanical properties	Moderate to high	High (clay)
		Moderate (other materials)
Consolidation parameters	–	High
Water table	–	High
Liquefaction potential	Moderate	High
Economy in investigation costs	High	High

Table 3.2 List of soil parameters interpreted from piezocone data in clays (modified from Chen and Mayne, 1994)

Soil parameters	Key references
Soil classification	Douglas and Olsen (1981); Senneset and Janbu (1985); Robertson et al. (1986); Robertson (1990)
In situ stress state (K_0)	Kulhawy and Mayne (1990); Kulhawy et al. (1985); Brown and Mayne (1993); Masood and Mitchell (1993)
Effective friction angle (ϕ')	Senneset and Janbu (1985); Kulhawy and Mayne (1990); Sandven (1990)
Constrained modulus (M)	Duncan and Buchignani (1975); Kulhawy and Mayne (1990)
Shear modulus (G_{max})	Rix and Stokes (1992); Mayne and Rix (1993); Tanaka et al. (1994); Simonini and Cola (2000); Powell and Butcher (2004); Watabe et al. (2004); Schnaid (2005)
Stress history (σ'_p, OCR)	Schmertmann (1978); Senneset et al. (1982); Jamiolkowski et al. (1985); Konrad (1987); Larsson and Mulabdic (1991); Mayne (1991, 1992); Chen and Mayne (1994)
Sensitivity (S_t)	Robertson and Campanella (1988)
Undrained strength (s_u)	Vesic (1975); Aas et al. (1986); Konrad and Law (1987); Teh and Houlsby (1991); Yu et al. (2000); Su and Liao (2002)
Hydraulic conductivity (k)	Robertson et al. (1992a)
Coefficient of consolidation (C_h)	Torstensson (1977a); Baligh (1985); Baligh and Levadoux (1986); Teh and Houlsby (1991); Robertson et al. (1992a)
Unit weight (γ)	Larsson and Mulabdic (1991)
Effective cohesive intercept (c')	Senneset et al. (1988)

Undrained shearing strength

The undrained shear strength is a key parameter in the design of natural and man-made structures in soft clay. The undrained shear strength of clays, in common with all soils, depends on the mode of failure, rate of shearing, soil anisotropy and stress history. For a clay that conforms with critical state soil mechanics, relationships between undrained shear strength, overconsolidation ratio and critical state parameters can be obtained (Wroth, 1984):

$$\frac{s_u}{\sigma'_{v0}} = \frac{M}{2} \left(\frac{OCR}{r}\right)^{\Lambda} \tag{3.5}$$

where r is the spacing ratio (= 2 in modified Cam clay) and M and Λ are critical state parameters. Under some simplifying assumptions and supported by experimental data in clay, Ladd et al. (1977) used a superscript m instead of Λ and suggested an average value of $m = 0.8$ to fit results for soils with plasticity index varying from 0.21 to 0.75. A normally consolidated Cam clay type of soil should then yield a ratio s_u / σ'_p from 0.25 to 0.30 depending on shearing mode, where σ'_p is the preconsolidation pressure.

The CPTU cannot directly measure the undrained shear strength and therefore CPTU assessment of s_u may have to rely on a combination of theory and empirical correlations. Since penetration tests in clay are generally undrained, and therefore excess pore pressures are generated, the cone tip resistance q_c can be related to s_u as follows:

$$q_c = N_c s_u + \sigma_0 \tag{3.6}$$

where N_c is a theoretical cone factor and σ_0 is the in situ total stress (either vertical or mean total stress). The theoretical solutions available for determining N_c can be grouped as (e.g. Yu and Mitchell, 1998; Yu, 2004):

- bearing capacity theory (BCT),
- cavity expansion theory (CET),
- strain path method (SPM),
- finite element method (FEM).

The cone factor may be determined using simple bearing capacity formulations, whereby a failure mechanism is postulated and a failure pressure is calculated assuming a rigid-plastic stress–strain relationship. Several different solutions have been proposed for distinct shapes of the plastic zone (Terzaghi, 1943; Meyerhof, 1951; Caquot and Kerisel, 1956). Available solutions based on cavity expansion or strain path methods have proven adequate to model the response of normally consolidated soils. The strain path method developed by Baligh (1985) is considered a breakthrough in our understanding of the penetration mechanism in clay. This original concept has been adopted by many authors in an attempt to overcome the equilibrium problem of the strain path method. In addition, Yu (2004) pointed out that, while each theory may be used alone for cone penetration analysis, better predictions of cone penetration mechanisms may be achieved if some of the methods are used in combination (Teh and Houlsby, 1991; Yu and Whittle, 1999; Abu-Farsakh et al., 2003). A combination of strain path analysis and finite element calculations was used by Teh and Houlsby (1991) to model cone penetration in a Von Mises soil. The solution includes the influence of the rigidity index (= G/s_u), cone and shaft roughness factor (α, α_f) and in situ stress anisotropy, $\Delta = (\sigma_{v0} - \sigma_{h0})/2s_u$. Yu and Whittle (1999) proposed a cone factor estimated from both strain path analysis and cavity expansion methods. In this approach, the strain path solution developed by Baligh (1986) was used to estimate the size of the plastic zone produced by penetration. Once the plastic zone is established, spherical cavity expansion is used to determine the stress distribution and therefore cone resistance.

Solutions generally assume isotropy of strength and stiffness and radially symmetric initial stresses. A solution that can account for effects of anisotropy that result from both structure and rotation of principal stresses around an advancing cone was proposed by Su and Liao (2002). Cone factors are expressed as a function of the strength anisotropy ratio $A_r = (s_u)_{tc}/(s_u)_{te}$, with the undrained shear strength $(s_u)_{tc}$ and $(s_u)_{te}$ measured from CK_0U compression and CK_0U extension triaxial tests respectively. The effect of anisotropy is not always significant; maximum differences in cone factors of up to 20% have been obtained for normally to slightly overconsolidated clay with moderate to high strength anisotropy.

Table 3.3 summarizes the most significant theoretical solutions for N_c. The calculation requires the rigidity index I_r to be estimated, I_r being the ratio of shear stiffness to shear strength, G/s_u. In natural clay deposits, it ranges from 50 to 500, decreasing with the increase in OCR and, for the same OCR, increasing with the plasticity index reduction.

Although theoretical solutions have been contributing to the understanding of the fundamental mechanics of cone penetration, empirical correlations are still widely used in practice to estimate s_u from cone resistance. The most widely used correlation is:

$$q_t = N_{kt}s_u + \sigma_{v0} \tag{3.7}$$

where N_{kt} is an empirical cone factor and σ_{v0} is the total in situ vertical stress. Values of N_{kt} range from 10 to 20 and are influenced by soil plasticity, overconsolidation ratio, sample disturbance, strain rate and scale effects, as well as the reference test from which s_u has been established (e.g. Aas et al., 1986; Mesri, 1989, 2001; Lunne et al., 1997). Laboratory triaxial, s_u(CIU) or s_u(CK_0U), and field vane, s_u(FVT), should preferably be adopted as reference tests to measure

Table 3.3 Theoretical cone factor solutions

N_c	Penetration model	Reference
7.0 to 9.94	BCT	Caquot and Kerisel (1956); de Beer (1977)
$3.90 + 1.33\ln I_r$	CET	Vesic (1975)
$12 + \ln I_r$	CET	Baligh (1975)
$\left(1.67 + \dfrac{I_r}{1500}\right)(1 + \ln I_r) + 2.4\lambda_c - 0.2\lambda_s - 1.8\Delta$	SPM+ FEM	Teh and Houlsby (1991)
$0.33 + 2\ln I_r + 2.37\lambda - 1.83\Delta$	FEM	Yu *et al.* (2000)
$\dfrac{1+A_r}{\sqrt{1+2A_r}}\ln I_r + \dfrac{1-A_r}{3} + R\left(1 + \dfrac{1+A_r}{\sqrt{1+2A_r}} + 0.52A_r^{1/8}(1+A_r)\right) - 2\Delta$	CET+ FEM	Su and Liao (2002)
$2.45 + 1.8\ln I_r - 2.1\Delta$	CET+ FEM	Abu-Farsakh *et al.* (2003)

Note

$I_r = G/s_u$; $\Delta = \dfrac{\sigma'_{v0}(1-K_0)}{2s_u}$; λ = cone roughness; A_r = strength anisotropy ratio.

s_u, and therefore to calibrate N_{kt}. Stress history affects strength, stiffness, strain softening and strain to failure, which in turn changes N_{kt}.

Comparisons between theoretical and empirically predicted N_{kt} factors are shown in Figure 3.16, for I_r ranging from 50 to 500 and strength anisotropy ratio of 0.5 to 0.9. In practice, the choice of average I_r values is arbitrary and N_{kt} values from 12 to 15 are often adopted in engineering design problems, and are generally higher than theoretical predicted values for rigidity indexes between 50 and 200. For deposits where little experience is available, a conservative approach of estimating s_u using measured cone tip resistance q_t and cone factor values close to the upper limit (~18 to 20) is a sensible recommendation.

A typical example of predicting an N_{kt} factor representative of a soft clay deposit by way of the relation between cone and vane tests is presented in Figure 3.17. These data have been obtained from a comprehensive site investigation programme in the Porto Alegre soft clay deposit. Considerable scatter is observed in the measured values, which can be attributed to the list of factors previously mentioned: test procedure, penetration rate, soil disturbance, soil variability, shear anisotropy, rigidity index and plasticity index. Although this is a relatively homogeneous deposit in which previous experience is vast, N_{kt} values

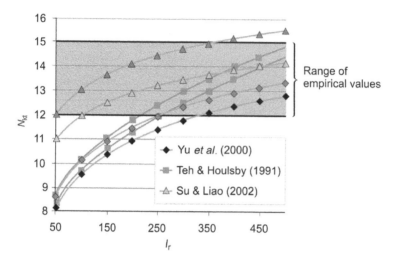

Figure 3.16 Theoretical and empirically predicted N_{kt} factors.

range between 8 and 16 (with an average value of 12). It is interesting to note that Teh and Houlsby's (1991) method yielded values of N_c of the same order of magnitude, which suggests a mean value of 12 as representative of the cone factor for the Porto Alegre clay deposit.

Given the background information that is available from the theoretical predictions, it is no longer necessary to attribute variations of N_{kt} to scatter. For large projects, site-specific correlations are required, and the crossreference between existing tests should account for the soil rigidity index, the reference test adopted in the calibration and even strain rate effects, as well as soil anisotropy. All existing information compiled together should provide a reliable assessment of N_{kt}. If, after a detailed examination of the evidence, N_{kt} is still outside the predicted range, this may be an indication that other important soil characteristics are not being considered; for example, stress history, partial drainage during penetration (i.e. consolidation effects) and soil type, among others.

Empirical approaches available for the interpretation of s_u from CPTU can also rely on correlations using 'effective' cone resistance and excess pore pressure. Senneset et al. (1982) suggested deriving s_u from the difference between the tip cone resistance and porewater pressure, measured at the cone shoulder (u_2):

$$s_u = \frac{(q_t - u_2)}{N_{ke}} \tag{3.8}$$

Experimental studies indicate that values of N_{ke} are scattered around a mean value of 9 and appear to correlate with the pore pressure parameter

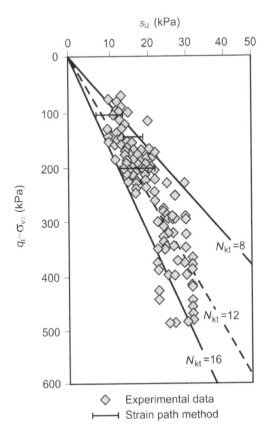

Figure 3.17 Predicted N_{kt} factors representative of the Porto Alegre soft clay deposit.

B_q (Lunne *et al.*, 1985). Alternatively, s_u can be expressed directly as a function of excess porewater pressure as (e.g. Battaglio *et al.*, 1981):

$$s_u = \frac{(u_2 - u_0)}{N_{\Delta u}}$$
(3.9)

where $N_{\Delta u}$ is shown from cavity expansion to fall within the range of 2 to 20.

Several attempts have been made to correlate the cone factor N_{kt}, N_{ke} and $N_{\Delta u}$ to the pore pressure parameter B_q in an attempt to narrow the band of predicted values. Karlsrud *et al.* (1996) explored this possibility by producing the correlations presented in Figure 3.18, in which measured values of s_u are obtained from CIU laboratory tests on high-quality Sherbrooke samples. Based on the author's own experience, these correlations produce more scattered data. Furthermore, the theoretical experience built around $N_c(N_{kt})$ is

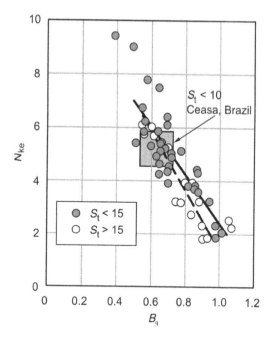

Figure 3.18 N_{ke}–B_q relationship (modified from Karlsrud et al., 1996).

much more extensive and supports the use of equation (3.7) to derive s_u, with the recommendations previously described.

Recent developments using full flow probes, including T-bar, spherical ball and plate penetrometers (e.g. Randolph, 2004), have been used with the aim of overcoming the uncertainties in determining the undrained shear strength in soft clay. Uncertainties emerge from corrections of unequal pore pressure area effects and the complicated mechanism around the cone that reflects on N_{kt}. In a flow bar penetrometer, lower and upper bound solutions can be found, as the problem of considering an intrusive volume introduced into the soil during cone penetration is avoided. The bearing capacity factor can be conveniently expressed as the ratio of an average penetration resistance q_u to the shear strength s_u, so that $N = q_u/s_u$. A theoretical bearing capacity solution of 10.5 for the T-bar N factor was first presented by Randolph and Houlsby (1984). Recently, Randolph (2004) introduced new upper and lower bounds for this penetration mechanism, which are summarized in Figure 3.19 in a plot that relates the N factor to the interface friction ratio α at the T-bar–soil interface. Long and Phoon (2004) summarized recent published data using the T-bar and demonstrated that, on average, the experimental values are close to the theoretical bearing capacity number of 10.5 from Randolph and Houlsby (1984) and within the range of solutions presented by Randolph (2004).

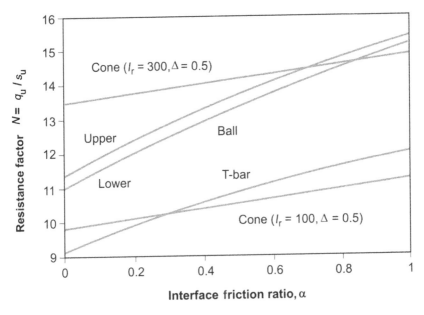

Figure 3.19 Theoretical bearing capacity factors for flow penetrometers (Randolph, 2004).

The discussion above focuses on penetration problems in normally consolidated and slightly overconsolidated clays where modified Cam clay reproduces the characteristic features of clay behaviour with reasonable accuracy. Heavily overconsolidated clay cannot be successfully modelled and therefore theoretical values of N_{kt} yield unrealistic undrained shear strength values. Engineers should rely solely on empirical correlations which indicate typical N_{kt} values in the range of 15 to 30 (e.g. Powell and Quarterman, 1988; Lunne *et al.*, 1997).

Stress history

Stress history is expressed by the overconsolidation ratio OCR, defined as the ratio of the maximum past effective mean stress and the currently applied stress. In practice, the current stress is taken as the present effective overburden stress which characterizes a mechanically overconsolidated soil. For clay, OCR is a key property required to characterize its mechanical behaviour. Unfortunately, cone penetration (being a measurement of strength) is not very sensitive to stress history and correlations between these two quantities should be viewed as no more than an indication of the soil stress history.

Acknowledging the approximate nature of correlations between q_c and OCR, a recommended practice to guide engineering judgement is to estimate OCR from the undrained shear strength ratio (s_u/σ'_{v0}), as suggested by

Schmertmann (1978) and Lunne *et al.* (1997). The method consists simply of estimating s_u from CPT data and σ'_{v0} from the soil profile to compute $s_{u/}\sigma'_{v0}$. The ratio is then compared to the corresponding normally consolidated undrained shear strength, adopting Shansep's method of approach for comparison (Ladd *et al.*, 1977). Since this is an important concept, it deserves a more detailed examination. An experienced engineer would identify a normally consolidated clay deposit (NC) by the ratio between s_u/σ'_{v0} of 0.25 to 0.30 (e.g. Bjerrum, 1973). Values greater than about 0.3 would be an indication of overconsolidation, as established in the concepts of critical state theory (Schofield and Wroth, 1968; Ladd *et al.*, 1977):

$$\frac{[s_u/\sigma'_{v0}]}{[s_u/\sigma'_{v0}]_{nc}} = OCR^\Lambda$$

(3.10)

where Λ is the critical state parameter (see the section on Undrained shearing strength above) obtained in laboratory tests. Equation (3.10) can be simplified and rewritten as (Jamiolkowski *et al.*, 1985):

$$\frac{s_u}{\sigma'_{v0}} = 0.23OCR^{0.8}$$

(3.11)

or (Mesri, 1975):

$$s_u = 0.22\sigma'_p$$

(3.12)

where σ'_p is the preconsolidation pressure. Figure 3.20 shows a chart that may be readily used to estimate OCR from the undrained shear strength ratio (after Andresen *et al.*, 1979).

Prediction of OCR directly from piezocone data can be made from several different solutions, both theoretically and empirically based (Senneset *et al.*, 1982; Wroth, 1984; Konrad and Law, 1987; Tavenas and Leroueil, 1987; Mayne, 1991). Mayne (1991) suggested an approach based on cavity expansion theory and critical state theory expressed as:

$$OCR = 2\left[\frac{1}{1.95M}\left(\frac{q_c - u_1}{\sigma'_{v0}}\right)\right]^1 \quad \text{mid-face pore pressure element } (u_1) \quad (3.13)$$

and

$$OCR = 2\left[\frac{1}{1.95M+1}\left(\frac{q_c - u_2}{\sigma'_{v0}}\right)\right]^{1/\Lambda} \quad \text{shoulder pore pressure element } (u_2) \quad (3.14)$$

At low values of OCR, the model is not particularly sensitive to critical state parameters M or Λ and, therefore, typical representative values of

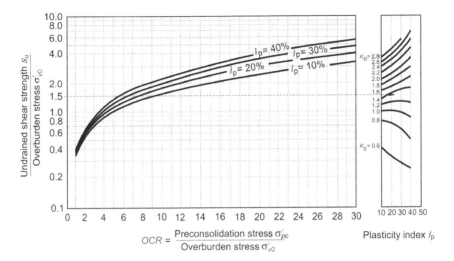

$$OCR = \frac{\text{Preconsolidation stress } \sigma'_{pc}}{\text{Overburden stress } \sigma'_{v0}}$$

Figure 3.20 Chart to estimate OCR from the undrained shear strength ratio (after Andresen et al., 1979).

$M = 1.2$ (corresponding to $\phi' = 30°$) and $\wedge = 0.75$ can be used for routine practice (Mayne, 2001). In addition, empirical estimates of the preconsolidation pressure can be obtained as (Chen and Mayne, 1996):

$$\sigma'_p = 0.305(q_t - \sigma_{v0}) \tag{3.15}$$

or

$$\sigma'_p = 0.65(q_t - \sigma_{v0})(I_p)^{-0.23} \tag{3.16}$$

where I_p is the clay plasticity index. Other correlations in the form of $\sigma'_p = k(u_2 - u_0)$ or $\sigma'_p = k(q_1 - u_2)$ can also be adopted, bearing in mind that they are all site-specific correlations that should be validated locally.

Examples of application are presented in Figure 3.21, in which values of OCR predicted from CPTU are compared with values measured in laboratory and other field tests (after Mayne, 2001). Predictions of OCR using data from both mid-face and shoulder piezometric elements are presented for four sites: lightly to normally consolidated soft clay at the Bothkennar research site in Scotland (Powell et al., 1988; Jardine et al., 1995); medium-stiff moderately overconsolidated deposit of lean sensitive clay at Haga, Norway (Lunne et al., 1986); hard microfissured and cemented clay at Taranto, Italy (Jamiolkowski et al., 1985; Battaglio et al., 1986); and heavily overconsolidated and fissured London clay at Brent Cross (Lunne et al., 1986). As observed, the CPTU-yielded values of OCR are similar to other techniques and appear to be consistent with the geological characterisitics of each site.

In another approach, the existing similarity between the pore pressure parameter B_q and the parameter A of Skempton (1954) seems to suggest that changes in B_q may reflect possible variations in OCR (Wroth, 1984; Houlsby, 1988; Chen and Mayne, 1996). Results of tests carried out in Brazil were compiled to evaluate the applicability of this concept. Results are shown in Figure 3.22, in which a tendency is observed of reducing B_q with increasing OCR. However, the degree of scatter in the experimental data does not encourage a direct use of this correlation in estimating OCR. Work in Norway (Karlsrud et al., 1996) suggests that scatter is mainly produced by sample disturbance. Data from high-quality specimens retrieved from Sherbrooke block samples yielded the correlations presented in Figure 3.23, showing significant levels of consistent trends of changes in cone factor as a function of B_q.

No matter how precise the predicted OCR values are, caution must be exercised when estimating OCR from piezocone test data due to the following considerations:

- the empirical nature of correlations,
- q_t is known to be fairly insensitive to changes in OCR, and
- cementation and aging effects cannot be accounted for because OCR reflects mechanically overconsolidated soils only.

Engineers should always bear in mind that these approaches are not rigorous and therefore there is always a need for validation using site-specific correlations.

Stress state

The vertical geostatic stress of the soil is simply calculated as $\sigma_v = \gamma z$, considering that the state of stress is produced by the soil's own weight and the surface of the soil is horizontal. In the equation above, γ is the volume weight of the soil and z is the current depth. The ratio between the horizontal and vertical normal in situ stresses is defined by the so-called coefficient of earth pressure at rest, K_0, expressed in terms of effective stresses as:

$$K_0 = \frac{\sigma'_{h0}}{\sigma'_{v0}} \tag{3.17}$$

Simple constitutive equations based on Hooke's law for isotropic material can be used to determine K_0 from Poisson's ratio. However, the coefficient K_0 exhibits considerable variability in nature, much larger than depicted from elastic theory, due to complex deposition processes and stress histories that leave their traces on the soil structure. In practice, the magnitude of K_0 should be estimated initially from empirical approaches using, for example, Jaky's equation (1944) for normally consolidated deposits:

$$K_0 = 1 - sin\phi' \tag{3.18}$$

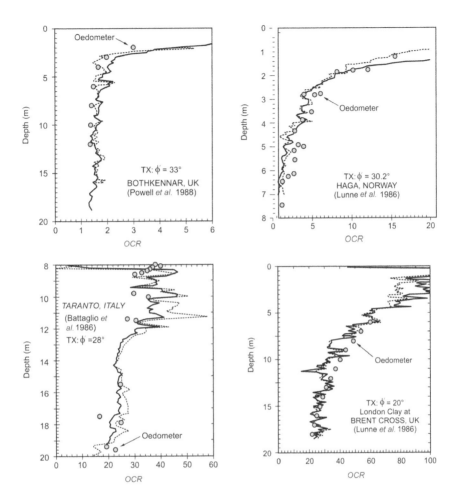

Figure 3.21 Predictions of OCR (after Mayne, 1991).

where ϕ' is the soil effective peak angle of internal friction. For overconsolidated deposits, the magnitude of the coefficient K_0 depends also on the overconsolidation ratio OCR and can be expressed in a more generic form (Mayne and Kulhawy, 1982):

$$K_0 = (1 - sin\phi')OCR^{sin\phi'} \qquad (3.19)$$

This leads to uncertainties in estimating K_0 because a practising engineer must investigate the variation of OCR and ϕ' in order to evaluate the stress state of the soil. In this case, values of ϕ' and OCR can be measured in

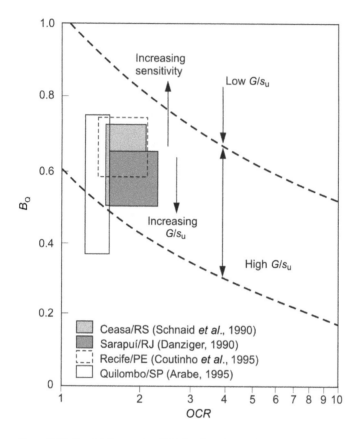

Figure 3.22 B_q with increasing *OCR* for Brazilian soft clays.

laboratory tests, predicted by in situ tests or estimated from correlations with Atterberg limits (see Chapter 7).

Currently, there is no reliable method of determining K_0 from piezocone tests because the state of stress is impacted by the stress and strain history of the soil and, in contrast, the measured tip cone resistance is relatively unaffected by *OCR*. Therefore, empirical correlations such as those proposed by Kulhawy and Mayne (1990) produce no more than a rough estimate of K_0:

$$K_0 = 0.1 \frac{q_t - \sigma_{v0}}{\sigma'_{v0}} \tag{3.20}$$

Figure 3.24 illustrates the applicability of this approach to the normally consolidated Porto Alegre soft clay deposit, where the CPTU predictions from equation (3.20) are compared to other well-known approaches: pressuremeter test results (see Chapter 5) and triaxial test data (equation 3.18).

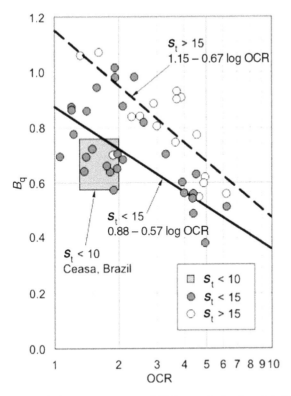

Figure 3.23 B_q with increasing *OCR* for a soft clay deposit in Norway (modified from Karlsrud *et al.*, 1996).

Although this particular example yielded measured and predicted values of K_0 that are comparable among themselves and compatible with the geological characteristics of the deposit, this approach should be viewed only as a guide and engineering judgement is required when selecting representative values of the coefficient of earth pressure at rest.

Stiffness

Soil stiffness depends on interactions of state (bonding, fabric, degree of cementation, stress level), strain level (and the corresponding effects of destructuration), stress history and stress path, time-dependent effects (aging and creep) and type of loading (monotonic or dynamic). Despite these complex interactions, the characteristic response of clay with respect to small strain stiffness, small strain anisotropy and stiffness non-linearity can be directly assessed from in situ crosshole and downhole tests. As for the cone test, cone tip resistance is insensitive to most of these factors and correlations between tip resistance and soil stiffness are unreliable.

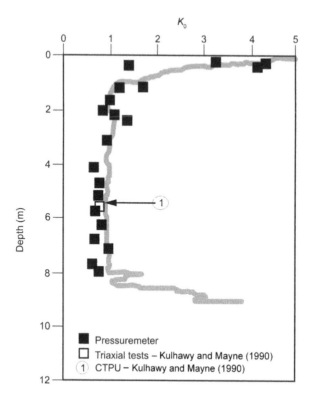

Figure 3.24 Prediction of K_0 in a normally consolidated soft clay deposit.

Assessment of soil stiffness is not a matter of simply selecting appropriate correlations; it is, in fact, a demanding area that requires a thorough knowledge of material behaviour. Hardin and Black (1968) and Hardin (1978) identified various factors governing the G value and proposed a general expression:

$$G = f(\sigma'_{v0}, e_0, OCR, S_r, C, K, T) \tag{3.21}$$

where:

σ'_{v0} = effective vertical stress
e_0 = initial void ratio
OCR = overconsolidation ratio
S_r = degree of saturation
C = grain characteristics
K = soil structure
T = temperature.

In reconstituted clay, laboratory resonant column tests carried out to characterize G_0 have shown that stiffness is a function of mean stress and that the effect of OCR is embedded into the effect of void ratio. An empirical correlation to describe G_0 can then be written as:

$$G_0 = SF(e)(\sigma'_{v0}\sigma'_h)^n p_a^{(1-2n)} \tag{3.22}$$

where S and n are experimental constants and $F(e)$ is the void ratio function. $F(e)$ can take the form of $(1/e)$, $(1 + e)$ or $(2.17 - e)^2/(1 + e)$, leading to the following representative correlations (Hardin, 1978):

$$G_0 = 625 \frac{1}{(0.3 + 0.7e_0^2)} (p'_0)^{0.5} OCR^k (\text{kPa}) \quad k = I_p^{0.72/50} < 0.5 \tag{3.23}$$

and (Jamiolkowski *et al.*, 1995a):

$$G_0 = 480(e)^{-1.43}(\sigma'_{v0})^{0.22}(\sigma'_h)^{0.22}(p_a)^{1-2(0.22)} \tag{3.24}$$

where p_0 = mean effective stress, p_a = atmospheric pressure and I_p = plasticity index. Coefficients in the above equations represent average recommended values. The need to express G_0 as a function of the mean stress is recognized; the effects of vertical and horizontal stresses being investigated as separate variables by Jamiolkowski *et al.* (1995b). Since σ_h is not usually accurately known, Shibuya *et al.* (1997a) considered it more practicable to express G_0 as a function of σ'_{v0}, with due recognition that its validity is restricted to normally consolidated clays.

$$G_0 = 24000(1 + e_0)^{-2.4}(\sigma'_{v0})^{0.5} (\text{kPa}) \tag{3.25}$$

For preliminary projects, various authors proposed to estimate G_0 directly from tip cone resistance q_c or q_t (e.g. Rix and Stokes, 1992; Mayne and Rix, 1993; Tanaka *et al.*, 1994; Simonini and Cola, 2000; Powell and Butcher, 2004; Watabe *et al.*, 2004). Mayne and Rix (1993) suggested a correlation that explicitly considers the dependency of G_0 on void ratio:

$$G_0 = 406q_c^{0.695}e_0^{-1.130} \tag{3.26}$$

Watabe *et al.* (2004) developed the following relationship between the small stiffness measured from the seismic cone and net cone resistance:

$$G_0 = 50(q_t - \sigma_{v0}) \tag{3.27}$$

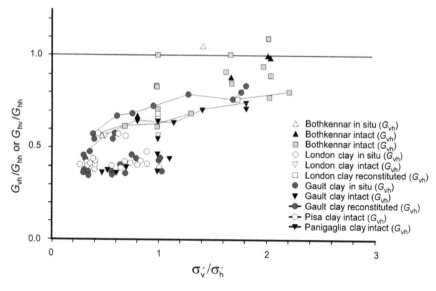

Figure 3.25 Anisotropy of small strain stiffness in natural and reconstituted soft and stiff clays (Leroueil and Hight, 2003).

Care must always be taken when using equations (3.26) and (3.27). Strictly speaking, a small strain value cannot be derived from an ultimate strength measurement and, therefore, this correlation should be seen just as an indication of stiffness that does not replace the need for direct measurements of shear wave velocities.

Information about the anisotropy of small strain stiffness (inherent and stress induced) from seismic measurements is becoming more readily available and is a generally recommended technique. The crosshole test (CHT) and downhole test (DHT) enable the velocity of horizontally propagated, vertically polarized (s_{hv}), vertically propagated, horizontally polarized (s_{vh}) and horizontally propagated, horizontally polarized (s_{hh}) shear waves to be measured. Leroueil and Hight (2003) presented a compilation of data on the anisotropy of small strain stiffness in natural and reconstituted soft and stiff clays, shown in Figure 3.25, expressed in terms of the ratio G_{hv}/G_{hh} or G_{vh}/G_{hh} versus consolidation stress ratio σ'_v / σ'_h. In Bothkennar clay, the fabric has rendered a stiffness ratio ranging from 0.8 to 1.0 in an unconfined state. Jamiolkowski *et al.* (1995a) found G_{vh}/G_{hh} ratios under isotropic stress conditions of 0.7 and about 0.65 for the natural Pisa and Paniaglia clays respectively. In situ data for the stiff and very stiff London and Gault clays are very similar, although under isotropic stress conditions the Gault clay shows a lower ratio for G_{vh}/G_{hh}.

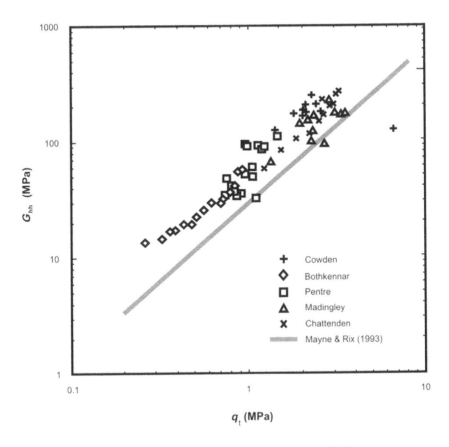

Figure 3.26 Correlation between G_{hh} and q_t (Powell and Butcher, 2004).

Due to anisotropy, G_0 is not independent with regard to the direction of propagation and polarization of shear waves, and correlations with penetration resistance should preferably be expressed as a function of G_{vh}, G_{hv} or G_{hh}. Contrary to many advocated approaches established from down-hole tests, Powell and Butcher (2004) have recently suggested that there is no single correlation between q_t and G_{vh} for all clays. Furthermore, the authors have found a strong correlation between G_{hh} and q_t, which is partially explained by the strong dependency of q_t on the horizontal stress. Figure 3.26 shows a set of results in a log × log plot that, given the scatter, should also be viewed as an indication of G_0 only.

Another approach that is often adopted in engineering practice is to correlate q_t with an operational Young's modulus which is representative of a given strain or stress level. Practitioners adopt these operational moduli for direct estimation

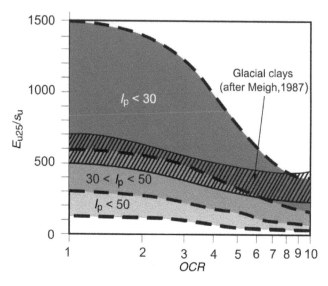

Figure 3.27 Stiffness ratio E_{u25}/s_u as a function of OCR (Duncan and Buchignani, 1976).

of settlements according to local practice and experience. The principle for this approach is the recognition that a material which is stiffer in deformation may be stronger in strength, yielding the empirical relationships of the form:

$$E_u = ns_u$$

(3.28)

The approach proposed by Duncan and Buchignani (1975), shown in Figure 3.27, provides the necessary link between undrained shear strength (estimated from CPT) and soil stiffness through soil stress history and soil plastic index.

By analogy, it is possible to estimate the constrained modulus from empirical approaches such as (Kulhawy and Mayne, 1990):

$$M = 8.25(q_t - \sigma_{v0})$$

(3.29)

Consolidation coefficients

Coefficients of consolidation can be assessed in situ from observations of settlements under embankments or directly from in situ test results, preferably from piezocone dissipation tests. A dissipation test consists of monitoring the decay of pore pressure with time after stopping penetration at a given depth.

Analytical and numerical procedures have been developed to provide an estimate of the coefficient of consolidation C_h from piezocone dissipation tests

Table 3.4 Time factor T* for consolidaton analysis (Houlsby and Teh, 1988)

Degree of consolidation (%)	Filter position			
	Cone face (u_1)	Cone shoulder (u_2)	Five radii above cone shoulder	Ten radii above cone shoulder
20	0.014	0.038	0.294	0.378
30	0.032	0.078	0.503	0.662
40	0.063	0.142	0.756	0.995
50	0.118	0.245	1.110	1.460
60	0.226	0.439	1.650	2.140
70	0.463	0.804	2.430	3.240
80	1.040	1.600	4.100	5.240

in which the decay of excess pore pressure with time is monitored. Methods rely either on one-dimensional cavity expansion (Torstensson, 1977a; Randolph and Wroth, 1979) or two-dimensional strain path method (Baligh and Levadoux, 1986; Levadoux and Baligh, 1986; Teh and Houlsby, 1991; Burns and Mayne, 1998). The analysis of dissipation tests is mostly predicted on the basis of uncoupled consolidation theory and requires two distinct stages. The first considers the penetration of a cone into an elastic-perfectly plastic isotropic, homogeneous material, viewed as a steady flow of a soil past a static cone, leading to the prediction of total stresses and porewater pressures in the soil around the probe. The second stage of the analysis takes these pore pressures as the initial values for a Terzagui uncoupled consolidation process, and calculates the subsequent dissipation around a stationary probe. Although the precise values of the pore pressures as a function of time depend on the specific initial pore pressure distribution, Teh and Houlsby (1989) showed that this form could be generalized by the definition of a suitable dimensionless time factor for the consolidation process T^*, expressed as:

$$T^* = \frac{c_h t}{R^2 \sqrt{I_r}}$$

(3.30)

where

R = probe radius
t = dissipation time (normally adopted as t_{50})
I_r = rigidity index (= G/s_u)
G = shear modulus.

Thus, for a given T^* and t values, calculated C_h values are directly proportional to the square root of the I_r value. In Table 3.4 the time factor T^* values for consolidation analysis are listed according to the degree of consolidation $(1 - u)$.

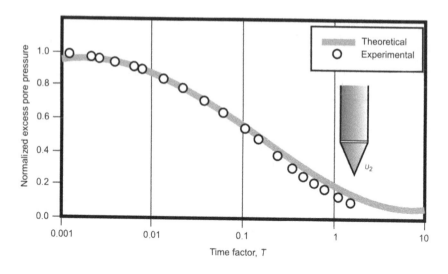

Figure 3.28 Typical dissipation test in soft clay.

Three independent sets of data from both field and laboratory testing programmes reported by Schnaid *et al.* (1997), in which the pore pressures were measured at four different locations on the piezocone, suggest that equation (3.30) should be applied to pore pressures measured at the shoulder immediately above the cone face. A comparison between an experimental result and the analytical solution obtained for a typical dissipation test is presented in Figure 3.28, in which it is possible to observe that the theory reproduces measured behaviour. The theoretical curves provide a less precise match with the experimental dissipation records measured at other locations (u_1, u_3 and u_4).

Based on equation (3.30), the horizontal coefficient of consolidation C_h is determined from a simple procedure that uses the dissipation data behind the cone, u_2, following recommendations from Robertson *et al.* (1992a):

1   plot the dissipation curve as dimensionless excess pore pressure dissipation, $U = (u_t - u_0)(u_i - u_0)$, against time in a log or $\sqrt{t}$ scale, where u_t, u_i and u_0 are the current, initial and hydrostatic (final) pore pressures, respectively;

2   define u_0 from available data on groundwater level;

3   calculate the distance between the initial pore pressure u_i and the hydrostatic pore pressure u_0, and from these measurements define the time for 50% dissipation, $u_{50\%} = (u_i - u_0)/2$;

4   use t_{50} and T^* values in Table 3.4 to calculate C_h from equation (3.30).

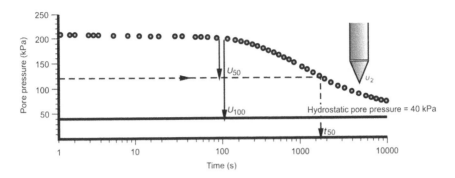

Figure 3.29 Interpretation of CPTU dissipation tests.

The procedure illustrated in Figures 3.28 and 3.29 is only applicable to undrained penetration of normal to lightly overconsolidated clays ($OCR < 3$) in which the excess pore pressure distribution predicted by the strain path method falls close to the pore pressure measurements during piezocone penetration.[2] Its applicability should follow a number of recommendations which are:

- the precise determination of u_i is essential for the correct determination of C_h;
- allow for 50% dissipation of excess pore pressure to provide sufficient data for interpetation, since at this degree of dissipation the calculated C_h from piezocone data should correspond to the in situ consolidation state of the soil (e.g. Torstensson, 1977a; Baligh and Levadoux, 1986; Robertson *et al.*, 1992a);
- a constant I_r is used in the solution, although in fact the value of the shear modulus will depend on the shear strain amplitude, which is shown by strain path calculations to vary in a complex manner around a 60° penetrometer. The success of the analysis in predicting C_h depends on the use of appropriate I_r values, which in turn depends on the shear modulus G. For natural clay deposits, I_r ranges from less than 50 to more than 600; it is known to decrease with increasing OCR and, for the same OCR, it increases with decreasing I_p. It is recommended that the Houlsby and Teh theoretical solution be used with an I_r value calculated from triaxial tests, in which the shear stiffness G is taken at a level of 50% of the yield stress (Schnaid *et al.*, 1997). Triaxial CIU, CK_0U and UU tests produced the following representative values: I_r of 70 for Oxford calibration chamber tests (Nyirenda, 1989), 44 for the Sarapui site in Brazil (Danziger *et al.*, 1996) and 135 for the Porto Alegre site in Brazil (Schnaid *et al.*, 1997).

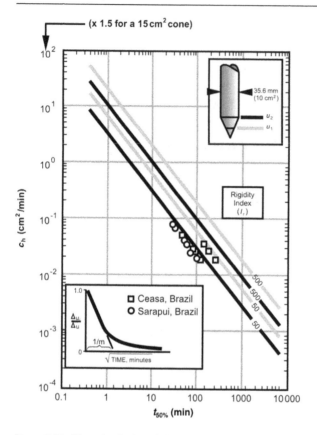

Figure 3.30 Chart for finding C_h from t_{50} (after Robertson et al., 1992a).

Robertson *et al.* (1992a) produced a simplified diagram (shown in Figure 3.30) using the Houlsby and Teh solution (Teh and Houlsby, 1991) that can be adapted to obtain the coefficient of consolidation from the actual time taken as 50% consolidation t_{50}. Predictions from dissipation tests measured both in the field and in calibration chamber tests are also shown in the figure to illustrate the application of the proposed methodology.

The value of C_h obtained from this procedure is representative of problems involving horizontal flow in the OC range (Baligh, 1986; Baligh and Levadoux, 1986). Assessment of the coefficient of consolidation C_h in the NC range can be attained from the empirical rule proposed by Jamiolkowski *et al.* (1985):

$$C_h(NA) = \frac{RR}{CR} C_h(Piezocone) \tag{3.31}$$

with the recompression to virgin compression ratio RR/CR ranging from 0.13 to 0.15.

Table 3.5 Permeability ratios for soft clay (Jamiolkowski et al., 1985)

Nature of clay	k_h/k_v
No macrofabric, or slightly developed macrofabric, essentially homogeneous deposits	1 to 1.5
From fairly well to well-developed macrofabric, e.g. sedimentary clays with discontinuous lenses and layers of permeable material	2 to 4.0
Varved clays and other deposits containing embedded and more or less continuous permeable layers	3 to 15

The coefficient of consolidation for vertical flow C_v can be obtained using the horizontal to vertical permeability ratio:

$$C_v = \frac{k_v}{k_h} C_h \tag{3.32}$$

Permeability k_h/k_v ratios characteristic of the in situ anisotropy of clays are given in Table 3.5.

Granular soils

Without the possibility of retrieving undisturbed samples, engineering design in granular soils should rely on the interpretation of in situ tests. There are a number of important basic concepts that an engineer should have in mind when attempting to interpret results in sands:

- drained response prevails under monotonic load, where the relatively high hydraulic conductivity of sands and gravels means that a 'quasi static' cone penetration test is generally drained;
- the behaviour of granular soils prior to the achievement of the critical state is largely controlled by the state parameter. A concept introduced by Wroth and Basset (1965) and developed by Been and Jefferies (1985), the state parameter Ψ is defined as the difference between current void ratio e and critical state void ratio e_c, at the same mean stress (Figure 3.31). It can be conveniently expressed as:

$$\Psi = e + \lambda \ln\left(\frac{p'}{p_1'}\right) - \Gamma \tag{3.33}$$

- where λ and Γ are the critical state parameters and p_1' is the reference mean effective stress taken as 1 kPa. In a manner similar to sedimented clays, which are accurately characterized by the overconsolidation ratio, sands can be characterized by the state parameter following the recognition that it combines the effects of both relative density and stress level;

- most natural soils and soft rocks are microstructured so that, at a given void ratio, they can sustain stresses higher than could the same material without microstructure (e.g. Burland, 1990; Leroueil and Vaughan, 1990; Leroueil and Hight, 2003). Although this is evident for hard soils there is also evidence of microstructure in sands (Mitchell and Solymar, 1984; Schmertmann, 1991; Fernandez and Santamarina, 2001). Given the difficulty of sampling granular soils, evidence of bonding has to be obtained from field tests. This not only allows soil behaviour to be characterized but also permits the influence of microstructure on the small strain shear modulus to be quantified (e.g. Eslaamizaad and Robertson, 1997; Schnaid, 1997; Schnaid et al., 2004);
- the shear strength envelope is non-linear, with the curvature increasing with increasing relative density and grain crushability.

Cementation, aging and crushability are three phenomena recognized as important in the behaviour of granular materials. However, interpretation of cone penetration tests deals almost exclusively with non-cemented cohesionless soils, because current experience has been built on calibration against a database from large laboratory chamber tests on fresh sands.

Finally, it is recognized that there are no reliable methods based on penetration tests for determining the stress state and stress history in granular soils.

Characterization of aging and cementation

In principle, a material that is stiffer in deformation may be stronger in strength, yielding the following empirical relationship that appears to be valid for various geomaterials ranging from soft soils to hard rocks (e.g. Tatsuoka and Shibuya, 1991; Shibuya et al., 2004):

$$E_{max}/q_{max} = 1000 \pm 500 \qquad (3.34)$$

This type of correlation aims at evaluating the spatial variability of stiffness from measured strength obtained in conventional laboratory testing. Since the characteristic structure of natural sands cannot be reproduced by reconstituted samples in the laboratory, it is necessary to develop a methodology capable of identifying the existence of distinctive behaviour emerging from aging or cementation from field tests. Following recent recommendations that soil characterization and mechanical properties should preferably be based on the combination of measurements from independent tests (Schnaid et al., 2004; Schnaid, 2005), an approach for characterizing aging and cementation has been developed on the basis of the ratio of elastic stiffness to ultimate strength (G_0/q_c, G_0/N_{60}). This is achieved by combining cavity expansion theory and critical state concepts with the variables that control the small strain stiffness of sand (Schnaid and Yu, 2007). From this approach it is possible to calculate

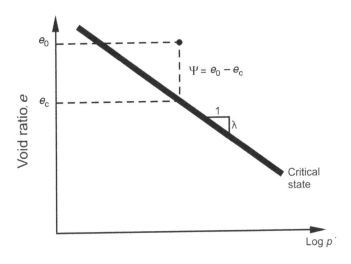

Figure 3.31 State parameter Ψ.

theoretical values of q_c and from these values to obtain the G_0/q_c and q_{c1} relationship from a set of given parameters: e, M, λ, Γ and p'. The approach makes
use of a normalized dimensionless parameter q_{c1}, defined as:

$$q_{c1} = \left(\frac{q_c}{p_a}\right)\sqrt{\frac{p_a}{\sigma'_{v0}}}$$

(3.35)

where p_a is the atmospheric pressure.
    The theoretical correlation between G_0/q_c and q_{c1} is shown in Figure 3.32 in a
relationship representative of unaged uncemented soils. Results were calculated
in a stress interval between 50 kPa and 500 kPa, which should cover the range
of applications encountered in geotechnical engineering practice. The computed
values are shown to be insensitive to both the initial stress state and soil compressibility, which fully justifies the use of this approach in soil characterization.
For comparison with natural sands, the upper and lower boundaries empirically
established by Schnaid et al. (2004) are presented together with the theoretical
derived database. The boundaries that represent the variation of G_0 with q_c for
fresh unaged, uncemented sand deposits are expressed as (Schnaid et al., 2004):

$$G_0 = \alpha \sqrt[3]{q_a \sigma'_{v0} p_a}$$

(3.36)

where α is a parameter that ranges from 110 to 280. Note that these
boundaries have roughly the same slope as the one produced by the analytical solution.

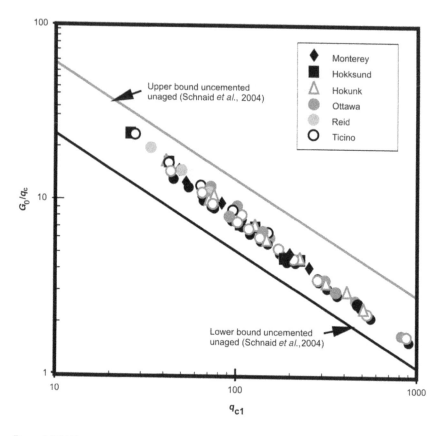

Figure 3.32 Theoretical correlation between G_0/q_c and q_{c1}.

Two independent sets of data from natural deposits have been used to validate this theoretical approach (Figure 3.33): compilation of a database for natural sands by Eslaamizaad and Robertson (1997) and a variety of sand types from Western Australia (Schnaid *et al.*, 2004). Fresh uncemented sand characterizes a well-defined region in the G_0/q_c versus q_{c1} space, whereas values of q_c and G_0 profiles that fall outside and above this region suggest possible effects of stress history, degree of cementation and aging, as demonstrated by Eslaamizaad and Robertson (1997).

Significant scatter observed in natural sand deposits probably results from soil anisotropy, degree of cementation and aging. Due to anisotropy, G_0 is not independent in the direction of propagation and polarization of shear waves and correlations with penetration resistance should preferably be expressed as a function of G_{vh}, G_{hv} or G_{hh}. Scatter can also be partially attributed to the influence of the horizontal stress on both initial stiffness and tip cone resistance (e.g. Schnaid and Houlsby, 1992). Equation 3.36

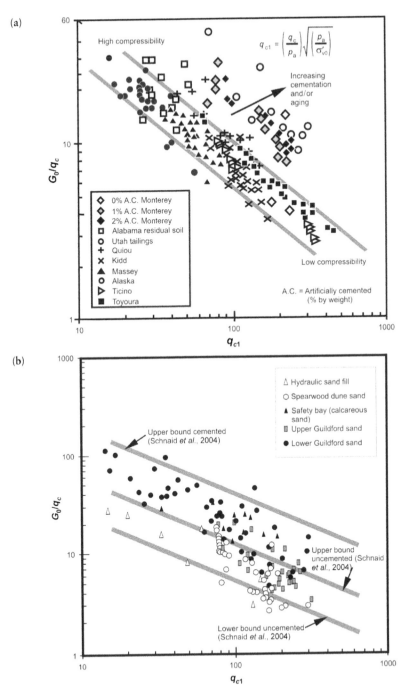

(a)

$$q_{c1} = \left(\frac{q_c}{p_a}\right)\sqrt{\left(\frac{p_a}{\sigma'_{v0}}\right)}$$

High compressibility

Increasing
cementation
and/or
aging

Low compressibility

◇ 0% A.C. Monterey
◇ 1% A.C. Monterey
◆ 2% A.C. Monterey
□ Alabama residual soil
○ Utah tailings
+ Quiou
× Kidd
▲ Massey
○ Alaska
▷ Ticino
■ Toyoura

A.C. = Artificially cemented
(% by weight)

G_0/q_c

q_{c1}

(b)

△ Hydraulic sand fill
○ Spearwood dune sand
▲ Safety bay (calcareous sand)
▣ Upper Guildford sand
● Lower Guildford sand

Upper bound cemented
(Schnaid et al., 2004)

Upper bound
uncemented (Schnaid
et al., 2004)

Lower bound uncemented
(Schnaid et al., 2004)

G_0/q_c

q_{c1}

Figure 3.33 Database for natural sands: (a) Eslaamizaad and Robertson (1997); (b) Schnaid et al. (2004).

should ideally be referred to horizontal stress or mean in situ stress rather than to vertical stress. The preference for σ'_{v0} is justified by the impossibility of determining with reasonable accuracy the value of the horizontal stress in most natural deposits, because they have undergone complex stress history, cementation and aging effects that are difficult to reconstruct.

In conclusion, the G_0/q_c ratio is useful because it provides a measure of the ratio of elastic stiffness to ultimate strength, which may be expected to increase with sand age and cementation, primarily because the effect of these on G_0 is stronger than on q_c. The ratio is fairly insensitive to changes in mean stress, relative density and sand compressibility.

Shear strength

Bishop (1971) defined cohesionless soils as geomaterials in which intrinsic interparticle forces or bonds make a negligible contribution to their mechanical behaviour. In cohesionless soils, the strength parameter of major interest is the internal friction angle ϕ'.

Two different approaches can be adopted for interpretation of the CPT in sand:

- analysis based on bearing capacity and cavity expansion theories which, given the complexities of modelling penetration in sand, can only be regarded as approximate (Vesic, 1972; Durgunoglu and Mitchell, 1975; Salgado et al., 1997); and
- methods based on results from large laboratory calibration chamber tests (e.g. Jamiolkowski et al., 1985; Bellotti et al., 1996).

Analytical solutions have been useful as a means of understanding the penetration mechanism, as well as identifying the influence of different constitutive parameters on the penetration process (Durgunoglu and Mitchell, 1975; Vesic, 1977; Salgado et al., 1997). Durgunoglu and Mitchell (1975) presented a well-known bearing capacity solution for deep cone penetration. A simple expression is obtained from a plane strain failure mechanism coupled with an empirical shape factor to account for the axisymmetric geometry of the cone, for cases in which the soil–cone interface friction angle is half of the soil friction angle:

$$N_q = \frac{q_c}{\sigma'_{v0}} = 0.194 \exp(7.63 \tan \phi') \tag{3.37}$$

where N_q is the cone factor in sand.

Relationships between cone resistance and friction angle from large calibration chamber tests are summarized by examples presented in Figures 3.34 and 3.35 (after Robertson and Campanella, 1983). This database, not

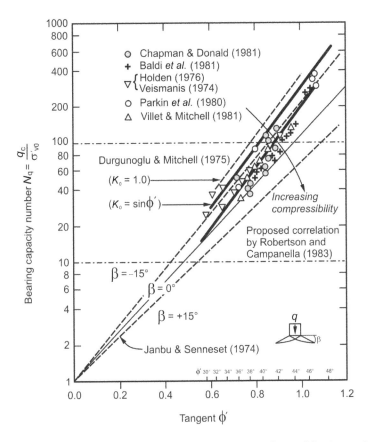

Figure 3.34 Relationship between bearing capacity number and friction angle (Robertson and Campanella, 1983).

corrected for boundary size effects, can be conveniently expressed as (Mayne, 2006):

$$\phi' = \arctan\left[0.1 + 0.38\log(q_{t/\sigma'_{v0}})\right] \tag{3.38}$$

A more diverse approach in engineering practice is to estimate relative density D_r from cone tip resistance adopting empirical correlations (e.g. Jamiolkowski *et al.*, 1985, 2003; Houlsby, 1998). Jamiolkowski *et al.* (2003) suggested a correlation between cone resistance, relative density and mean effective stress in the form:

$$D_r = \frac{1}{C_2}\ln\left(\frac{q_c}{C_0 p^{C_1}}\right) \tag{3.39}$$

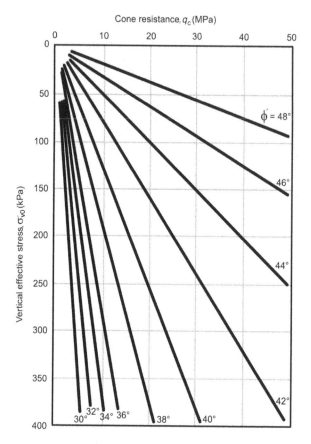

Figure 3.35 Relationship between cone resistance, vertical effective stress and friction angle (Robertson and Campanella, 1983).

According to the authors, vertical stress can be substituted for mean stress p in equation (3.39), provided that the soil is normally consolidated having a K_0 constant with depth. In this case, parameters C_0, C_1 and C_2 can be taken as 17.68, 0.50 and 3.10 respectively.

Approaches to predicting relative density that account for the influence of compressibility are presented in Figures 3.36 and 3.37. Jamiolkowski *et al.* (1985) recommended the following formula to estimate D_r (Figure 3.36):

$$D_r = -98 + 66\log_{10}\frac{q_c}{(\sigma'_{v0})^{0.5}} \tag{3.40}$$

where q_c and σ'_{v0} are expressed in t/m^2. This approach provides a relative density estimate with precision of ±20% (uncertainty range inherent to the

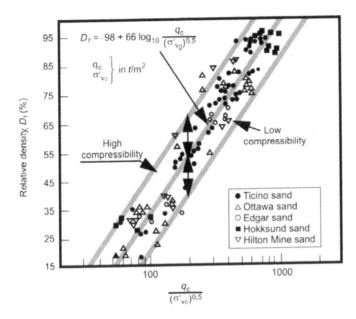

Figure 3.36 Relationship between normalized cone resistance and relative density (Jamiolkowski *et al.*, 1985).

method) and, since it is established in calibration chambers, it should be corrected for chamber size and boundary conditions (e.g. Schnaid and Houlsby, 1992). In general, the correlations are acceptable for NC soils, while for OC deposits the value of σ'_{v0} must be replaced by the horizontal effective stress σ'_{h0} in equation (3.40).

Despite the fact that there is a wealth of experience in using these semi-empirical correlations in practice – and experience provides a reassuring way of approaching engineering problems – these methods are prone to produce far too large a scatter in predicted density. Calibration chamber data in sand have clearly shown that, for a given density, cone resistance depends primarily on the in situ horizontal stress and therefore σ'_{h0} must be accounted for in a rational interpretation of field tests (Houlsby and Hitchman, 1988; Schnaid and Houlsby, 1992). The dependency of relative density (D_r) on q_c and σ'_{h0} is illustrated in Figure 3.38 and can be expressed in percentage terms as (Houlsby, 1988):

$$\log_{10}\left(\frac{q_c - \sigma_{h0}}{\sigma'_{h0}}\right) = 1.51 + 1.23 D_r \qquad (3.41)$$

It is clear from Figure 3.38 that even well-controlled calibration chamber tests lead to considerable scatter for different soils and soil conditions. This is likely

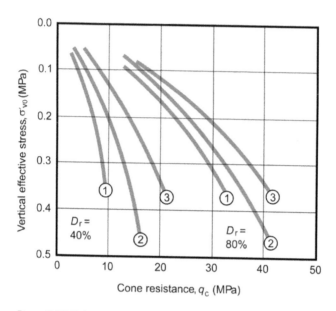

Figure 3.37 Relationship between normalized cone resistance and relative density (Robertson and Campanella, 1983).

to be a result of the influence of soil stiffness and has prompted the need to develop correlation with the state parameter (see the following section).

Values of D_r, estimated from the equations above, can be combined with operational stress levels to produce an estimate of peak friction angles. The method has previously been demonstrated in Chapter 2 under the heading Soil properties in granular materials (e.g. Bolton, 1986):

$$\phi'_p - \phi'_{cv} = m[D_r (Q - 1\mathrm{n}\, p') - R] \qquad (2.32)\mathrm{idem}$$

where p' is the mean effective stress (in kPa), R is an empirical factor found equal to 1 as a first approximation and Q is a logarithm function of grain compressive strength, known to range from about 10 for silica sand to 7 for calcareous sand (e.g. Jamiolkowski, 2001; Randolph, 2004).

State parameter

The seismic cone (see the section under the heading Seismic cone at the beginning of this chapter) provides a straightforward means of assessing both soil shear strength and soil stiffness. Since G_0 and q_c are controlled by void ratio, mean stress, compressibility and structure and are therefore different functions of the same variables, it is possible to anticipate that, as a ratio, these two measurements can be useful in predicting soil

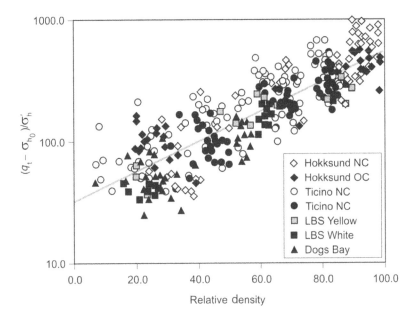

Figure 3.38 Normalized q_c–D_r correlation from calibration chamber testing data (Houlsby, 1988).

properties. In this book, it is systematically demonstrated that a set of critical state parameters combined with initial state conditions can be used to calculate both G_0 and q_c. Acknowledging that critical state parameters and initial soil state are at the root of the so-called 'state parameter', Ψ, an obvious approach is to correlate the G_0/q_c ratio and Ψ (after Schnaid and Yu, 2007).

The stiffness of sand has been long recognized as being controlled by the confining stress and void ratio (e.g. Hardin, 1978; Lo Presti *et al.*, 2001), which has prompted the establishment of many useful correlations for predicting G_0 adopting slightly different $F(e)$ functions (e.g. Lo Presti *et al.*, 1997; Shibuya *et al.*, 2004). Lo Presti *et al.* (1997) suggest the following expression:

$$\frac{G_0}{p_a} = Ce^{-x}\left(\frac{p'}{p_a}\right)^n \tag{3.42}$$

where p' is the mean effective stress and C, n and x are average values for the material parameters. For solutions of cone penetration in sand, the significant volume change occurring in shear has to be captured. Given its complexity, the existing methods of interpretation of the penetration mechanism in sand can only be regarded as approximations and solutions have

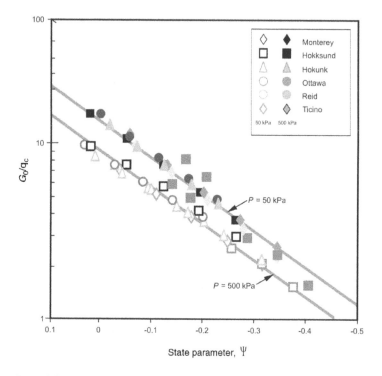

Figure 3.39 Theoretical ratio of elastic stiffness to cone resistance and *in situ* state parameter.

to be calibrated against experimental data from calibration chamber tests. In one solution, the effective cone resistance q_c was estimated directly from the spherical cavity limit pressure σ_s (Ladanyi and Johnston, 1974):

$$q_c = \sigma_s(1 + \sqrt{3}\tan\phi'_{ps})$$

(3.43)

where ϕ'_{ps} is the plane strain peak friction angle of the soil, as deep cone penetration may be assumed to occur under plane strain conditions in the penetration direction.

A theoretical relationship between G_0/q_c with Ψ can be obtained from equations (3.33), (3.42) and (3.43). Results are plotted in Figure 3.39 for the six reference sands and for a stress interval between 50 and 500 kPa. For a given mean stress, the ratio of G_0/q_c decreases with decreasing Ψ (i.e. G_0/q_c decreases with increasing relative density). Although the correlation is not very sensitive to variations in sand strength and stiffness, for a given

Ψ the G_0/q_c ratio reduces with increasing mean stress. The theoretical database generated by this approach can be represented by the following expression:

$$\Psi = \alpha \ln\left(\frac{p'}{p_a}\right)^{\beta} + \chi \ln\left(\frac{G_0}{q_c}\right) \tag{3.44}$$

where $\alpha = -0.520$; $\beta = -0.07$ and $\chi = 0.180$ are average coefficients obtained from calibration chamber data. In practice, it is recommended that these coefficients be obtained from site-specific correlations. Predictions of the state parameter from the measured G_0/q_c ratio would therefore require an independent assessment of the in situ horizontal stress to calculate the mean stresses in equation 3.45, which, for a normally consolidated sand, can rely on Jaky's equation.

Validation of the proposed method from a number of case studies was carried out by Schnaid and Yu (2007) on the basis of centrifuge and field tests. Penetration and seismic testing data from the Canlex site (e.g. Wride *et al.*, 2000) are very useful in evaluating the proposed correlation, since there is no other area in which the freezing technique has been adopted so extensively to retrieve undisturbed samples in sand from which the state parameter was assessed. Results are presented in Figure 3.40, in which data from two locations (Mildred Lake and J-pit sites) are summarized. Substantial scatter is observed in this plot, which reflects the actual scattered data reported by the authors. The two sets of measured data fall in rather distinct regions in the G_0/q_c versus Ψ space, with the J-pit data falling consistently above the data reported at Mildred Lake as a result of the different mean in situ stresses at the two locations. The J-pit data follow a line which has a slope similar to that predicted from equation (3.44) and the state parameter can be predicted with reasonable accuracy. Data from Mildred Lake is much more scattered and does not show a clear trend of reducing G_0/q_c with reducing Ψ.

From critical state soil mechanics, it is possible to demonstrate that constitutive parameters in sand normalize well to the state parameter. It is important to recall that the state parameter has been successfully correlated with triaxial friction angles ϕ'_{tc} using an empirical relationship of an exponential type (Been and Jefferies, 1985):

$$\phi'_{tc} - \phi'_{cv} = A[\exp(-\Psi) - 1] \tag{3.45}$$

where A is a curve fitting parameters ranging from 0.6 to 0.95 depending on the type of sand. This approach is recommended for practising engineers when attempting to narrow the uncertainty attributed to predictions of shear strength from cone resistance only.

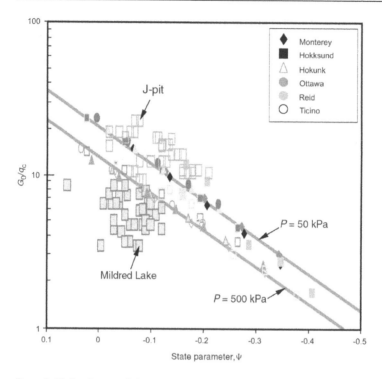

Figure 3.40 Prediction of the state parameter at Canlex site.

Soil stiffness

Assessment of reliable stress–strain relationships in soils requires an accurate evaluation of both the small strain stiffness and the reduction in stiffness with shear strain amplitude. In an earlier section, the variation of G_0 with q_c observed for natural sands was summarized in Figure 3.32. Equations (3.46) and (3.47) can match the range of recorded G_0 values and, despite the fact that this equation was originally proposed to distinguish cemented and unce-mented soils, it is likely that practitioners may be tempted to employ it to estimate G_0. For preliminary design, and in the absence of direct measure-ments of shear wave velocities, the proposed lower bounds are recommended for an evaluation of the small strain stiffness from q_c (Schnaid *et al.*, 2004):

$$G_0 = 280 \sqrt[3]{q_c \sigma'_{v0} p_a}: \text{ lower bound, cemented}$$

$$(3.46)$$

$$G_0 = 110 \sqrt[3]{q_c \sigma'_{v0} p_a}: \text{ lower bound, uncemented}$$

$$(3.47)$$

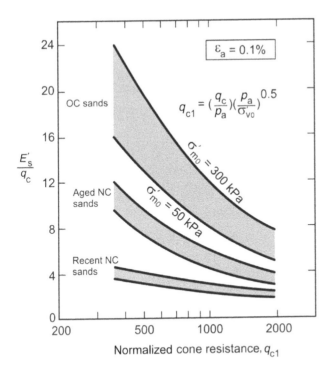

Figure 3.41 Young's modulus for silica sand (Bellotti et al., 1989).

These equations predict values of G_0 that are not far from previously published relationships developed for sands (Bellotti et al., 1989; Rix and Stokes, 1991; Jamiolkowski et al., 1995a). However, the effect of natural cementation and aging is quantified here and is shown to produce a marked increase in the G_0/q_c ratio. Given the considerable scatter observed for different soils, these correlations are only approximate indicators of G_0 and do not replace the need for in situ shear wave velocity measurements.

Alternatively, engineers can adopt a simple approach for a first estimation of an operational modulus (e.g. Schmertmann, 1970; Simons and Menzies, 1977; Robertson and Campanella, 1983; Meigh, 1987; Bellotti et al., 1989). For uncemented sands, secant Young's modulus values, E_s, for an average axial strain of 0.1% can be readily correlated to stress history and aging, as shown in Figure 3.41. Both OCR and aging are shown to have significant effects on the measured modulus that, taken into consideration, can provide a representative stiffness for most foundation problems.

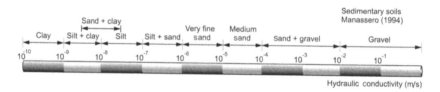

Figure 3.42 Soil permeability (Manassero, 1994).

For the one-dimensional case, there are correlations between tip cone resistance and the drained constrained modulus M (Lunne and Christoffersen, 1983):

NC sand

$$M_0 = 4q_c \qquad\qquad \text{for } q_c < 10 \text{ MPa}$$
$$M_0 = 2q_c + 20 \text{ (MPa)} \qquad \text{for } 10 \text{ MPa} < q_c < 50 \text{ MPa}$$
$$M_0 = 120 \text{ MPa} \qquad\quad \text{for } q_c > 50 \text{ MPa}$$

OC sand, with OCR > 2

$$M_0 = 5q_c \qquad\qquad \text{for } q_c < 50 \text{ MPa}$$
$$M_0 = 250 \text{ MPa} \qquad\quad \text{for } q_c > 50 \text{ MPa} \qquad\qquad (3.48)$$

where M_0 is the reference constrained deformation modulus at the in situ effective vertical stress.

Properties in silty soils

Many soils exhibit a complex macro and micro structure and may have a scattered grain size distribution and variations in mineralogy and clay content. These features have a dominant effect on soil permeability, hence also on in situ behaviour at given loading rates, producing geomaterials in the so-called intermediate permeability range of 10^{-5} to 10^{-8} m/s, as illustrated in Figure 3.42. For intermediate soils, the simplest idealized approach of broadly distinguishing between drained (gravels and sand) and undrained (clay) conditions for the interpretation of in situ tests cannot be applicable since test response can be affected by partial consolidation, and consequently existing analytical, numerical or empirical correlations can lead to unrealistic assessment of geotechnical properties (e.g. Schnaid *et al.*, 2004). Unfortunately, there are no standardized recommendations to guide engineers on how to perform in situ tests or interpret test results in these materials.

Assessment of the possible effects of partial drainage is essential and, currently, it seems far from being satisfactorily resolved. There are two important questions to be addressed before deriving the undrained strength for soils, which are:

- How is partial drainage avoided during penetration? If it happens, how is it recognized and what are the possible consequences on the derived s_u values?
- How are in situ test results normalized in a sensible way for engineering applications?

The evaluation of drained effects in penetration tests is mainly empirically based and follows recommendations derived from field observations (e.g. Bemben and Myers, 1974; Danziger and Lunne, 1997). McNeilan and Bugno (1985) suggested that, for hydraulic conductivities greater than 10^{-4} m/s, cone penetration is fully drained, whereas for hydraulic conductivities less than 10^{-8} m/s an undrained penetration will take place. Alternatively, partially drained penetration can be evaluated directly from the pore pressure parameter B_q; if B_q is less than 0.4 a direct use of N_k factors for the assessment of undrained strength is questionable (Senneset et al., 1988). Hight et al. (1994) found that a relationship between B_q, $(q_t - \sigma_{v0})\,\sigma'_{v0}$ and clay content could be a useful approach for interpreting results under fully undrained conditions. Their analysis appears to suggest that penetration is fully undrained for values of B_q greater than 0.5.

An example is given on how to identify the effect of partial drainage during penetration on a normally consolidated silty deposit (Figure 3.43). Results are expressed in terms of the normalized parameter $(q_t - \sigma_{v0})\,\sigma'_{v0}$, B_q and the ratio of undrained strength. A normally consolidated Cam clay type of soil

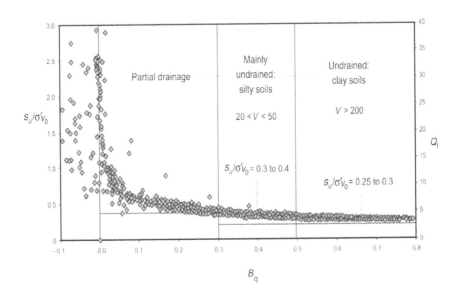

Figure 3.43 Effects of partial drainage.

should yield a ratio s_u/σ'_{v0} of between 0.25 to 0.30, depending on shearing mode. Deviation from this pattern is related to (a) overconsolidation, (b) partial drainage or (c) a characteristic behaviour of silty soils that does not fully comply with Cam clay models. Although separating out these effects is not straightforward, it is evident that the s_u/σ'_{v0} ratio reduces significantly with increasing B_q value, and at a B_q value of approximately 0.5 the *undrained strength ratio* reaches a plateau at a constant value of about 0.25; this value is coincidentally of the same order as that given by equation (3.5). For values of B_q ranging between 0.3 and 0.5, s_u/σ'_{v0} ratios vary within the range of 0.3 to 0.4, with a tendency to increase slightly with reducing B_q. The calculated values of s_u/σ'_{v0} are consistent with the measured range of ϕ' values, suggesting that these investigated layers are predominantly silty soils and that partial drainage is not dominant. A B_q value of 0.3 seems to give a lower boundary, below which the undrained strength ratio exhibits considerable scatter and a marked increase to unrealistic values in a normally consolidated deposit. It is then reasonable to consider that partial drainage prevails and that the derived values of undrained shear strength are overestimated.

Recent studies developed to identify drainage during penetration place emphasis on the normalization of penetration results that are represented by an analytical 'drainage characterization curve' (Randolph, 2004; Schnaid et al., 2004; Schnaid, 2005). In this alternative method, normalized penetration results are represented in a space of a non-dimensional parameter V (Randolph and Hope, 2004) versus degree of drainage U:

$$V = \frac{vd}{C_v} \tag{3.49}$$

and

$$U = \frac{(qc - q\text{cund})}{(q_{\text{cdr}} - q_{\text{cund}})} \tag{3.50}$$

where

  d = probe diameter
  v = penetration rate
  C_v = coefficient of consolidation
  q_{cund} = cone resistance under undrained conditions
  q_{cdr} = cone resistance under drained conditions.

The 'drainage characterization curve' can be adopted to identify the transition from drained to partially drained to undrained penetration, which has been achieved from solutions of the limit pressures of spherical and cylindrical cavities, experimental results on centrifuge tests and in situ tests in both clay and silty soils (Schnaid et al., 2005; Schneider et al., 2007).

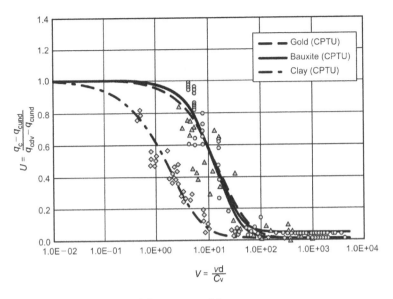

$$V = \frac{vd}{C_v}$$

Figure 3.44 Identification of drainage conditions.

Figure 3.44 shows penetration tests in clay (Randolph and Hope, 2004), gold tailings and alumina tailings. Since no substantial excess pore pressures have been recorded during penetration for normalized velocities ranging from 10^{-2} to 10^{-1}, it is reasonable to suggest that measured tip resistances q_c are representative of a fully drained penetration. A normalized $V \approx 10^{+2}$ corresponds to a condition that matches a fully undrained penetration. The transition from drained to partially drained to undrained is notably different for different geomaterials since the rate of dissipation has been shown to be fairly sensitive to soil stiffness and is therefore a function of the rigidity index and overconsolidation ratio.

Based on these studies, it appears reasonable to suggest that a drainage characterization curve should be established before carrying out an extensive site investigation in silt deposits. A general recommendation is to avoid penetration ratios that yield dimensionless velocities within the range of 10^{-1} to 10^{+2}. In this range, partial drainage is expected to occur and properties assessed from field test interpretation can be overestimated, in particular the undrained shear strength.

Properties in bonded geomaterials

The characterization of bonded geomaterials will be carried out with close reference to the experience recently accumulated in the interpretation of in situ tests in residual soils. A ground investigation in the so-called 'residual soils' often reveals weathered profiles exhibiting high heterogeneity on both vertical and horizontal directions, complex structural arrangements, expectancy of pronounced metasta-bility due to decomposition and lixiviation processes, presence of rock block, boulders, etc. (e.g. Vargas, 1974; Novais Ferreira, 1985). The process of in situ weathering of parent rocks (which creates residual soils) gives rise to a profile containing material ranging from intact rocks to completely weathered soils. Rock degradation generally progresses from the surface and therefore there is normally a gradation of properties with no sharp boundaries within the profile.

Site investigation campaigns in residual soils are generally implemented from a mesh of boreholes associated with either SPT or CPT, to depths defined by the capacity of the penetration tool, and followed by continuous rotational coring below the soil–rock interface for a global geological char-acterization of weathering patterns. To enhance consistency, it is recom-mended that geophysical surveys are considered simultaneously. In this highly variable environment, a combination of different in situ test tech-niques assists in characterizing stress-strain and strength properties on residual soil profiles.

It follows from the foregoing on granular materials that a bonded/cemented structure produces G_0/q_c and G_0/N_{60} ratios that are systematically higher than those measured in cohesionless soils. These ratios therefore provide a useful means of assisting site characterization. Typical results from residual soils are presented in Figure 3.45, in which the G_0/q_c ratios are plotted against normalized parameter q_{c1} for CPT data (after Schnaid, 1999; Schnaid et al., 2004). The bond structure generates normalized stiff-ness values that are considerably higher than those for uncemented soils and, as a result, the datapoints for residual soils fall outside and above the band proposed for sands by Eslaamizaad and Robertson (1997) and theo-retically determined by Schnaid and Yu (2007).

Soil stiffness

As extensively shown throughout this chapter, considering the natural variation observed in bonded soils, it is preferable to express correlations between strength and stiffness in terms of lower and upper boundaries designed to match the range of recorded G_0 values. The variation of G_0 with q_c can be expressed as (Schnaid et al., 2004):

$$
\left.
\begin{aligned}
G_0 &= 800 \sqrt[3]{q_c \sigma'_{v0} p_a} \quad && \text{upper bound: cemented} \\
G_0 &= 280 \sqrt[3]{q_c \sigma'_{v0} p_a} \quad && \begin{aligned}&\text{lower bound: cemented}\\&\text{upper bound: uncemented}\end{aligned} \\
G_0 &= 110 \sqrt[3]{q_c \sigma'_{v0} p_a} \quad && \text{lower bound: uncemented}
\end{aligned}
\right\} \quad (3.51)
$$

Given the considerable scatter observed for different soils, correlations such as those given in equation (3.51) do not replace the need for in situ shear wave velocity measurements.

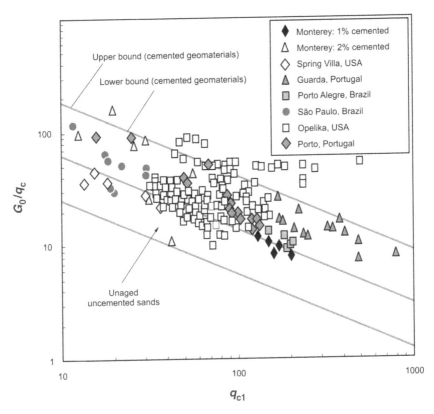

Figure 3.45 G_0/q_c ratios plotted against normalized parameter q_{c1} for CPT data.

Shear strength

Since the natural structure of bonded soils has a dominant effect on their mechanical response and the bonded strength is recognized as a net sign of this structure, identification of the two components of strength (c' and ϕ') is crucial in geotechnical design problems. The use of triaxial tests on high-quality samples to quantify the cohesive-frictional parameters of bonded soil is always advisable, but given the acknowledged difficulties of maintaining the relic structure during sampling, which may mask the effects of bonding on stiffness and strength, in situ tests remain a viable option in engineering practice.

Penetration tests such as SPT, CPT and DMT, which are extensively used in transported soils, are also commonly adopted in the investigation of structured cemented deposits, with empirical correlations for assessing soil properties being locally adapted to meet standards that reflect regional engineering practice. For shear strength measurements, the preceding discussion under the heading Properties in silty soils applies. A limited ground investigation based on CPT will not produce the necessary database for any rational assessment of soil properties, because two strength parameters (c' and ϕ') cannot be derived from a single measurement (q_c). In cohesive frictional soils, engineers tend to ignore the c' component of strength and correlate the in situ test parameters with the internal friction angle ϕ'. Average c' values may be assessed later from previous experience and back-analysis of field performance.[3]

Pile bearing capacity

As already discussed in Chapter 2, the ultimate capacity, Q, of a pile under axial load is calculated by the sum of the base capacity, Q_b, and the shaft capacity, Q_s:

$$Q = Q_b + Q_s = A_b q_b + A_s f_p \qquad (2.66)\text{idem}$$

where

> A_b = area of the pile base
> A_s = area of the pile shaft
> q_b = unit end-bearing pressure
> f_p = average shaft resistance.

One of the earliest applications of the cone was to predict the bearing capacity of piles, following the recognition that the penetration mechanism of a model cone pushed into the ground is similar to a prototype pile. Conceptually, the average tip cone resistance q_c measured in the vicinity of the pile tip and the summation of the cone shaft resistance f_s measured along the length of the pile are used to calculate the values of q_b and f_p respectively. Since sleeve resistances are acquired over larger surface areas behind the tip and are considered to be less reliable and of lower resolution than tip resistance, in practice measured q_c alone is used for the calculations of both pile end-bearing and pile shaft resistance.

Three CPT methods are recommended to predict the bearing capacity of piles and are outlined below: the LCPC method (Bustamante and Gianeselli, 1982), Eurocode 7-3 method and DIN 4014 method.

LCPC method

In this method, the unit end-bearing pressure and average shaft resistance are calculated directly from the tip cone resistance q_c using the following equations (Bustamante and Gianeselli, 1982):

$$q_b = k_c q_{c,avg} \qquad (3.52)$$

$$f_p = \frac{q_{c,z}}{\alpha} \qquad (3.53)$$

Table 3.6 Load capacity factors k_c and α (Bustamante and Gianeselli, 1982)

Nature of soil	q_c (MPa)	k_c	α	$f_{p,max}$ (MPa)
Soft clay	$q_c < 1$	0.40	30	15
Moderately compact clay	$1 < q_c < 5$	0.35	40	80
Compact to stiff clay, compact silt	$q_c > 5$	0.45	60	80
Silt and loose sand	$q_c < 5$	0.40	60	35
Moderately compact sand and gravel	$5 < q_c < 12$	0.40	100	120
Compact to very compact sand and gravel	$q_c > 12$	0.30	150	150

where:

$q_{c,avg}$ is the average value of $q_{c,z}$ between the depth of $1.5D$ above and $1.5D$ below the pile tip

$q_{c,z}$ is q_c at depth z

k_c, α are factors depending on pile and soil type and measured cone resistance (see Table 3.6)

D is the diameter of the pile.

The value of $q_{c,avg}$ should be calculated in three stages:

1 calculate q_{ac} as the mean value of q_c between $1.5D$ and $-1.5D$;
2 eliminate q_c values higher than $1.3q_{ac}$ and lower than $0.7q_{ac}$;
3 calculate $q_{c,avg}$ within the range defined in step 2 and input this value into equation (3.52).

The process is illustrated in Figure 3.46.

Eurocode 7-3 method

In this method the unit end-bearing pressure and average shaft resistance are calculated directly from the following equations:

$$q_b = \alpha_p 0.5 \left(\frac{q_{cI,mean} + q_{cII,mean}}{2} \right) + q_{cIII,mean} \tag{3.54}$$

$$f_p = \alpha_s q_{c,z} \tag{3.55}$$

where:

$q_{cI,mean}$ is the mean of the q_{cI} values over the depth running from the pile base level to a level (critical depth) which is at least 0.7 times and

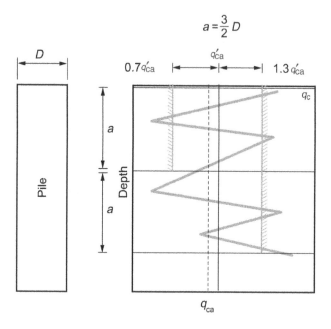

$$a = \frac{3}{2} D$$

Figure 3.46 Calculations of equivalent average cone resistance (Bustamante and Gianeselli, 1982).

at most 4 times deeper than the pile base diameter (critical depth is where the calculated value of q_b becomes a minimum)

$$q_{cI,mean} = \frac{1}{d_{crit}} \int_0^{d_{crit}} q_{cI} dz$$

(3.56)

$$0.7D \leq d_{crit} \leq 4D$$

$q_{cII,mean}$ is the mean of the lowest q_{cII} values over the depth going upwards from the critical depth to the pile base

$$q_{cII,mean} = \frac{1}{d_{crit}} \int_{d_{crit}}^0 q_{cII} dz$$

(3.57)

$q_{cIII,\,mean}$ is the mean of the q_{cIII} values over the depth interval running from the pile base level to a level 8 times higher than the pile base diameter. This procedure starts with the lowest q_{cII} value used for the computation of $q_{cII,mean}$

$$q_{cIII,mean} = \frac{1}{8D} \int_0^{-8D} q_{cIII} dz$$

(3.58)

Table 3.7 Friction coefficient α_s (Eurocode 7–3)

Soil type	Relative depth z/D	α_s
Fine to coarse sand		0.006
Very coarse sand		0.0045
Gravel		0.003
Clay/silt ($q_c \geq 1$ MPa)	$5 < z/D < 20$	0.025
Clay/silt ($q_c \leq 1$ MPa)	$z/D \geq 20$	0.055
Clay/silt ($q_c > 1$ MPa)	not applicable	0.035
Peat	not applicable	0

Table 3.8 Calculation of unit end-bearing pressure in non-cohesive soils (DIN 4014)

Average tip cone resistance q_c (MPa)	Unit end-bearing pressure (MPa)
10	2.0
15	3.0
20	3.5
25	4.0

Table 3.9 Calculation of unit shaft resistance in non-cohesive soils (DIN 4014)

Average tip cone resistance q_c (MPa)	Unit shaft resistance (MPa)
0	0.0
5	0.04
10	0.08
15	0.12

$q_{c,z} = q_c$ at depth z

α_p = pile class factor

α_s = factor depending on the pile class and soil conditions (see Table 3.7).

DIN 4014 method

The German Standard makes a distinction between cohesive and non-cohesive soils. For piles in non-cohesive soils, the end-bearing pressure and shaft resistance are evaluated directly from the measured cone resistance values summarized in Tables 3.8 and 3.9. For end-bearing pressures, the mean q_c value is determined over a zone of three times the pile diameter under the tip.

Table 3.10 Calculation of unit end-bearing pressure in cohesive soils (DIN 4014)

Average undrained shear strength s_u (MPa)	Unit end-bearing pressure (MPa)
0.1	0.80
0.2	1.50

Table 3.11 Calculation of unit shaft resistance in cohesive soils (DIN 4014)

Average undrained shear strength s_u (MPa)	Unit shaft resistance (MPa)
0.025	0.025
0.1	0.04
0.2	0.06

In the case of cohesive soils, unit pressures are expressed in terms of the undrained shear strength s_u using Tables 3.10 and 3.11. Values of s_u can be determined from methods described throughout this chapter, using an average q_c value over a zone of three times the pile diameter under the tip, in analogy with non-cohesive materials.

Liquefaction

In discussing the requirements for evaluating ground hazards from soil liquefaction and in order to produce sound risk assessments of seismic events, it was indicated in Chapter 2 that some of the difficulties with semi-empirical SPT procedures can be avoided by the use of CPT probes. This section attempts to summarize the current and evolving developments in CPT-based liquefaction analysis methods and to demonstrate their applicability to geotechnical design problems.

In Chapter 2, soil liquefaction was associated with large earthquakes, inducing a loss of strength in saturated cohesionless soils caused by the reduction of effective stress from increased pore pressures, u. In this section, the term liquefaction is related to different mechanisms of induced ground failures under both monotonic and cyclic loading of saturated soils, described as flow liquefaction and dynamic liquefaction (e.g. Robertson, 2004):

- *Flow liquefaction* is the result of any mechanism that takes place when loading conditions are sufficient to increase the pore pressure to its critical level, reducing the effective stress to values close to the limit condition. Various mechanisms can trigger the collapse of the inter-particle structure in flow liquefaction, either under monotonic or cyclic loading. For example, rate of construction faster than the pore pressure dissipation, loading due to vibrations or explosions and increase in groundwater levels. This is typical of geomaterials placed in a loose state with the in situ void ratio higher than the critical void ratio at the same mean stress.

- *Dynamic liquefaction* is the mechanism triggered by seismic earthquake events in which undrained shear stress reversal occurs under cyclic loading, leading to zero shear stress.

The stress state leading to flow liquefaction can be determined from consolidated undrained triaxial compression tests sheared with full saturation. Specimens showing a contractive response to loading, accompanied by a pronounced strain softening, to shearing stresses much lower than the peak value, are clearly susceptible to liquefaction. It is not uncommon to see reported data on very loose sands or tailings in which the material has completely lost shear resistance as the effective stress drops to zero.

Critical state soil mechanics and the concept of state parameter are useful frames of reference to identify this mechanism (see under State parameter above). The critical state line (CSL) defines the relationship between the ultimate void ratio and the mean effective stress p' (where $p' = (\sigma'_1 + 2\sigma'_3)/3$) for which deformation takes place under constant stress and constant volume. In this space, the state parameter Ψ is defined as the difference between current void ratio e and critical state void ratio e_c, at the same mean stress (see Figure 3.31). In situ soil at a void ratio above the CSL will undergo strain softening (contractive response) when subjected to undrained loading, whereas a void ratio below the critical state line results in strain-hardening, dilative behaviour. The acknowledged difficulties involved in defining the critical state condition of a soil due to uncertainties in void ratio determination, both in situ and in laboratory test samples, has prompted the development of methods to assess the state parameter directly from in situ test measurements. The practising engineer is recommended to adopt correlations based on the combination of measurements from independent tests, such as the ratio of the elastic stiffness to cone resistance G_0/q_c (using the seismic cone, following concepts described in the section above entitled State parameter) and the ratio of cone resistance to pressuremeter limit pressure q_c/p_L (using the cone pressuremeter, as described in Chapter 5).[4] In these approaches, positive values of Ψ assessed directly from the G_0/q_c or q_c/p_L ratios indicate the potential for soil liquefaction.

Evaluation of potential liquefaction during earthquakes can be obtained by semi-empirical procedures designed to differentiate between liquefiable and non-liquefiable soil conditions. Initial proposals have been based on correlations between the SPT and the CPT to convert the available SPT-based charts (Chapter 2 under the heading Soil liquefaction) to equivalent CPT charts (Douglas and Olsen, 1981; Seed and Idriss, 1981). In recent years, the expanded database has prompted the development of various other correlations (Robertson and Campanella, 1985; Shibata and Teparaksa, 1988; Ishihara, 1993; Stark and Olson, 1995; Suzuki *et al.*, 1995, 1997; Olsen, 1997; Robertson and Wride, 1997; Zhang and Tumay,

1999; Moss, 2003; Cetin *et al.*, 2004). Examples of graphically presented solutions are given in Figures 3.47 and 3.48.

The CPT-based approach is established by relating the measured cone resistance to the cyclic shear stress ratio (CSR) introduced by Seed and Idriss (1971):

$$CSR = \frac{\tau_{av}}{\sigma'_{v0}} = 0.65 \left(\frac{a_{max}}{g}\right)\left(\frac{\sigma_{v0}}{\sigma'_{v0}}\right) r_d$$

(2.68)idem

where τ_{av} is the average applied shear stress, σ'_{v0} the effective overburden stress, σ'_{v0} the total overburden stress, a_{max} the maximum horizontal acceleration at ground surface, g the gravitational acceleration and r_d the stress reduction factor (defined in Chapter 2 under the heading Soil liquefaction).

Identification of liquefaction potential from penetration resistance requires the effects of soil density and effective confining stress to be separated. Consequently, the cone resistance q_c is normalized by mean stress and atmospheric pressure and is conveniently expressed as a dimensionless quantity q_{c1} from equation (3.35) or similar.

Robertson and Wride (1997) proposed a CPT-based liquefaction analysis method that is based on expressions correlating the cyclic stress ratio that causes liquefaction, CRR, with normalized penetration resistance:[5]

$$CRR_{7.5} = 93\left(\frac{(q_{c1N})_{cs}}{1000}\right)^3 + 0.08$$

$$if\ 50 \leq (q_{c1N})_{cs} \leq 160$$

(3.59a)

$$CRR_{7.5} = 0.83\left(\frac{(q_{c1N})_{cs}}{1000}\right) + 0.05$$

$$if\ (q_{c1N})_{cs} < 50$$

(3.59b)

where

$$q_{c1N} = \left(\frac{q_c - \sigma_{v0}}{p_a}\right)\left(\frac{p_a}{\sigma'_{v0}}\right)^n$$

(3.60)

$(q_{c1N})_{cs}$ is the stress-normalized cone tip resistance, corrected for apparent fines content based on a similar concept to that presented for the SPT where:

$$(q_{c1N})_{cs} = K_c(q_{c1N})$$

(3.61)

The correction factor K_c is calculated through the soil behaviour type index I_c, which is a function of q_c and R_f only:

$$I_c = \left[(3.47 - \log Q_t)^2 + (\log F_r + 1.22)^2\right]^{0.5}$$

(3.62)

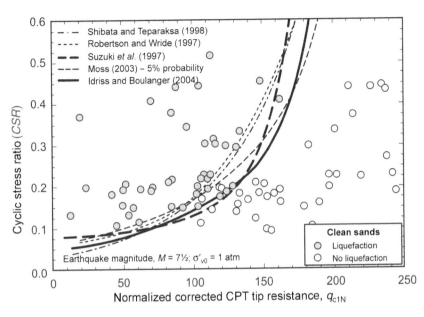

Figure 3.47 CPT-based case histories and relationships for identification of liquefaction in clean sand (after Idriss and Boulanger, 2004).

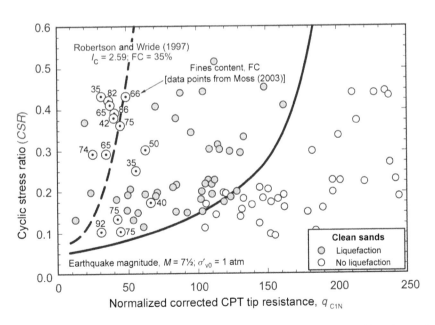

Figure 3.48 CPT-based case histories and relationships for identification of liquefaction in sand–silty soils (after Idriss and Boulanger, 2004).

Q_t and F_r being the normalized cone resistance and friction ratio respectively (defined in Figure 3.13) and n a stress exponent. An interactive procedure is required when calculating K_c. Assume an initial stress exponent $n = 1.0$ and calculate Q_t and F_r and then I_c. Estimate n as:

$$if..I_c < 1.64..........................n = 0.5$$
$$if..I_c > 3.30.........................n = 1.0 \qquad (3.63)$$
$$if..1.64 < I_c < 3.30.............. n = (I_c - 1.64)0.3 + 0.5$$

Iterate until the change in the stress exponent is less than 0.01. The relationship between K_c and I_c can then be established as:

$$K_c = 1.0.....if\ I_c \le 1.64$$
$$K_c = -0.403I_c^4 + 5.58I_c^3 - 21.63I_c^2 + 33.75I_c - 17.88..........if\ I_c > 1.64 \quad (3.64)$$

A revised approach to CPT-based charts was presented by Idriss and Boulanger (2004) for clean sand with a fines content FC ≤ 5%. The proposed correlation is expressed as:

$$CRR = \exp\left\{ \left(\frac{q_{c1N}}{540}\right) + \left(\frac{q_{c1N}}{67}\right)^2 - \left(\frac{q_{c1N}}{80}\right)^3 + \left(\frac{q_{c1N}}{114}\right)^4 - 3 \right\} \qquad (3.65)$$

In the Idriss and Boulanger (2004) method, the normalized (q_{c1N}) is calculated as:

$$q_{c1} = \left(\frac{q_c}{p_a}\right)\left(\frac{p_a}{\sigma'_{v0}}\right)^\beta \qquad (3.66)$$

$$\beta = 1.338 - 0.249(q_{c1N})^{0.264} \qquad (3.67)$$

with $(p_a/\sigma'_{v0})^\beta$ limited to a maximum value of 1.7. Note that these relations have reference to magnitude $M = 7.5$ earthquakes and an effective vertical stress $\sigma'_{v0} = 1$ atm. A magnitude scaling factor, MSF, used to adjust the induced CRR during earthquake magnitude M to an equivalent CRR for the earthquake magnitude $M = 7.5$, was introduced in Chapter 2.

These CRR–q_{c1N} relationships for clean sand are compared in Figure 3.47 (after Idriss and Boulanger, 2004) showing a marginal difference in predicted liquefaction potential, with the curve proposed by equation (3.65) being more conservative than that of equation (3.59) in clean sand. The influence of fines content on CPT-based charts is represented in Figure 3.48, in which the Robertson and Wride (1997) method for FC = 35% is compared to the Idriss and Boulanger (2004) approach for FC ≤ 5%. Also shown in the figure are the data-points reported by Moss (2003) for cohesionless soils with FC ≥ 35%, suggesting that calculations based on I_c are

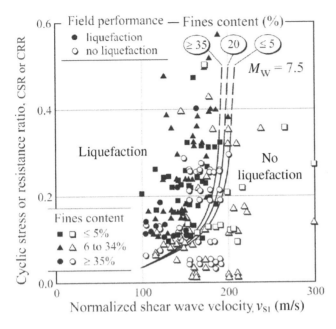

Figure 3.49 Case studies and recommended curves for evaluating CRR from shear wave velocity v_s for clean, uncemented soils (after Andrus *et al.*, 2004).

slightly less than conservative. A recommendation is made for direct soil sampling as the primary means of determining grain characteristics for the purpose of liquefaction evaluation.

In addition to semi-empirical procedures established for evaluating soil liquefaction from the CPT, it is necessary to draw attention to shear wave velocity, v_s, based correlations. Given the fact that seismically induced shear strains are sensitive to the liquefaction response of soils (Dobry *et al.*, 1982), a theoretical basis of cyclic resistance ratio curves was developed using shear wave velocity data (e.g. Andrus and Stokoe, 2000). The cyclic resistance ratio for overburden stress normalized shear wave velocity, v_{s1}, can be represented as (Andrus *et al.*, 1999; Andrus and Stokoe, 2000):

$$CRR_{7.5} = a\left(\frac{v_{s1}}{100}\right)^2 + b\left(\frac{1}{v_{s1}^* - v_{s1}} - \frac{1}{v_{s1}^*}\right) \qquad (3.68)$$

where $v_{s1} = v_s/(\sigma'_{v0})^n$ is the stress normalized shear wave velocity using a stress exponent $n = 0.25$, σ'_{v0} the vertical effective stress in atmospheres, v_{s1}^* is the limiting upper value of v_{s1} for liquefaction occurrence, and a and b are curve fitting parameters equal to 0.022 and 2.8 respectively.

The limiting value of shear wave velocity in sandy soils has been estimated to be:

$$v_{s1}^* = 215 \text{m/s} \dots\dots\dots\dots\dots\dots\dots\dots\dots \text{FC}(\%) \leq 5$$
$$v_{s1}^* = 215 - 0.5(\text{FC} - 5)\text{m/s} \dots\dots\dots 5 < \text{FC}(\%) < 35 \qquad (3.69)$$
$$v_{s1}^* = 200 \text{m/s} \dots\dots\dots\dots\dots\dots\dots\dots\dots \text{FC}(\%) \geq 35$$

Curves for evaluating CRR from v_s for clean uncemented soils and $M_w = 7.5$ magnitude earthquakes are presented in Figure 3.49. Magnitude scaling factors should be applied for different magnitudes whenever required.

Notes

1 Statistical fuzzy and neural network approaches towards CPT soil classification have recently been advocated (e.g. Zhang and Tumay, 1999), addressing the probability of mis-identification of soil types in situ by standard methods. These approaches combine strength, permeability and compressibility with soil characterization in attempting to improve our assessment of soil types by incorporating logical connections between the site and the project-specific geotechnical needs.

2 In overconsolidated and fissured clays, a dissipation test may first show an increase in pore pressure that, after reaching a peak value, decreases with time. Determination of u_i in tests exhibiting this initial dilatory response is not obvious and thus the approach described here for predicting C_h is not applicable. Burns and Mayne (1998) introduced a mathematical solution applied to both monotonic and dilatory pore pressure decay with time. It follows the concept that the measured excess pore pressure produced by the advancing cone results from a combination of changes in octahedral normal stresses due to cavity expansion and shear induced stresses, as postulated by the strain path method prediction by Baligh (1986) (see Figure 3.15).

3 The pressuremeter test (Chapter 5) is the only in situ test technique that allows for a proper characterization of strength parameters in cohesive-frictional materials. The SPT offers an empirical first assessment of ϕ' (Chapter 2).

4 In addition, methodologies have been developed to evaluate liquefaction directly from the normalized cone penetration resistance (Olson and Stark, 1998).

5 The correlation of the cyclic stress ratio required to cause liquefaction is designated as CRR to distinguish it from the cyclic stress ratio CSR induced by earthquake ground motions.

Chapter 4

Vane test

> Any successful relationship that can be used with confidence outside the immediate context in which it was established should ideally be (a) based on a physical appreciation of why the properties can be expected to be related, (b) set against a background theory, however idealised this may be, and (c) expressed in terms of dimensionless variables so that advantage can be taken of the scaling laws of continuum mechanics.
>
> (C.P. Wroth, 1984)

General considerations

Vane tests are used primarily to determine the undrained shear strength s_u of saturated clay deposits with strength generally up to 200 kPa. They can also be used in fine-grained soils such as silts, organic peats, tailings and other geomaterials where a prediction of the undrained shear resistance is required. The method is not applicable to sand, gravel and other highly permeable soils, which may allow for partial drainage at standard shearing rates. Previous knowledge of the drainage characteristics of soil layers in which the test is to be carried out is therefore necessary in order to guarantee sufficiently low permeability for the test response to be described as undrained.

The test originated in 1919 in Sweden, developed by John Olsson (in Flodin and Broms, 1981). Several improvements were made in the 1940s, giving the test the basic characteristics adopted today (Cadling and Odenstad, 1948; Carlsson, 1948; Skempton, 1948). The field vane test consists of four rectangular blades fixed at 90° angles to each other that are pushed into the ground to the desired depth, followed by the measurement of the torque required to produce rotation of the blades and hence the shearing of the soil. The torque can be analytically converted in the unit shearing resistance of a cylindrical failure surface adopting limit equilibrium analysis. Since an exact closed-form solution for s_u is easily obtained, given the symmetries of the failure mechanism, the vane shear strength is considered to be a reference against which other penetration tests are calibrated. Complementary test results may provide assessment for the soil stress history expressed by the overconsolidation ratio, OCR.

Several important publications reviewed the development and interpretation of the test and acknowledged factors affecting results (e.g. Richardson and Whitman, 1963; Bjerrum, 1973; Larrson, 1980; Walker, 1983; Wroth, 1984; Aas *et al.*, 1986; Biscontin and Pestana, 2001; Cerato and Lutenegger, 2004). In 1987, ASTM held a speciality conference with the objectives of documenting the use of vane tests, standardization and developing guidelines for interpretation. In 1988, a state-of-the-art review was presented by Chandler.

Equipment and procedures

Characteristics of the vane apparatus and basic testing procedures are presented in this chapter. It details recommendations described by National Codes of Practice: American ASTM D 2573-01, British BS 1377-90 and Brazilian NBR 3122-89, among others. These recommendations are critically reviewed on the basis of fundamental research developed to examine their validity and applicability.

The field vane test consists of placing and rotating a cruciform four-bladed vane in the undisturbed soil. Two basic penetration probe testing methods are used for land operations:

- testing in boreholes: the vane is inserted into the ground from the bottom of a borehole. In order to reduce the effect of ground disturbance on the measured strength inherently produced during boring operations the vane should be advanced below the bottom of the excavation to a depth of at least five times the outside diameter of the borehole;
- driving inside a protective cover: in this case driving must stop above the test depth with the vane inside the cover. The vane blades are then withdrawn from the protective cover for penetration into undisturbed ground for a depth of at least five times the outside vane housing diameter.

Details of these two configurations are shown in Figures 4.1 and 4.2. Figure 4.1 shows the vane blade mounted in a frame with a drive engine designed to perform vane tests in predrilled holes. The blade can be also integrated within a standard CPT unit that drives the cone to selected depths, without the need for predrilling (Figure 4.2). Specific design for the protective casing and bearings is necessary to reduce friction (Figure 4.3), but calibration of friction losses is mandatory and should be carried out by inserting the torque rods at every test depth, without the vane affixed, to measure the torque required for rotation. In both cases the vane is connected to the surface by steel torque rods and a torque measurement device is positioned at the top end.

Figure 4.1 Vane blades and driving frame for testing in boreholes (courtesy of A.P. van den Berg).

This device can accomplish a torque resolution of ±1 kPa in the computed shear strength.

Although a variety of vane sizes and shapes are still in use, it is now standard practice to use a vane with a height-to-diameter ratio (H/D) of 2. The height of the blades normally ranges from 70 to 200 mm. Blades 130 mm long by 65 mm are often suitable for soft clays with shear strength up to 50 kPa. For stiffer soils, blades of 70 to 100 mm in length are large

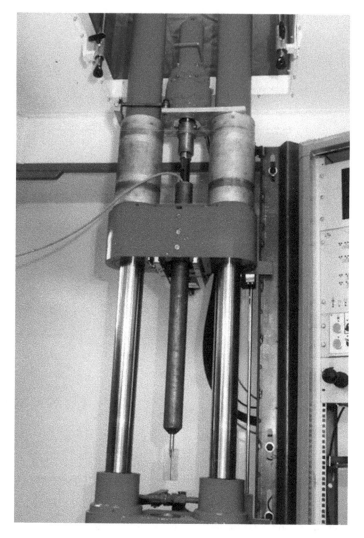

Figure 4.2 Vane blade connected to a standard CPT unit (courtesy of A.P. van den Berg).

enough to provide good torque resolution. Details of a standard penetrating vane blade are illustrated in Figure 4.4.

The blades are made of different alloys of steel and are often submitted to hardening processes in order to reduce the blade thickness. The blade should be relatively thin to minimize the disturbance produced by pushing the vane into the soil, which may seriously reduce the in situ undrained shear strength. For this reason, average blade thickness should be 2 mm and

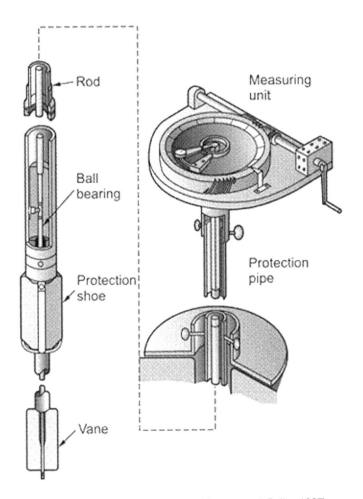

Figure 4.3 Details of the field vane test (Ortigão and Collet, 1987).

should not exceed 3 mm (ASTM D 2573-01). The influence of the blade thickness on the measured strength has been investigated by La Rochelle *et al.* (1973) and Cerato and Lutenegger (2004). This influence is shown to be significant, as illustrated in Figure 4.5, in which the measured values of s_u are plotted against the perimeter ratio α:

$$\alpha = \frac{4e}{\pi D} \tag{4.1}$$

where e is the blade thickness and D is the vane diameter. For every depth, the strength is shown to reduce with increasing blade thickness. A reference

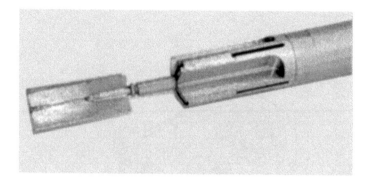

Figure 4.4 Standard vane blade.

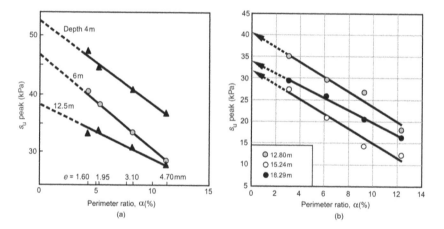

Figure 4.5 Blade thickness effect on s_u: (a) La Rochelle *et al.* (1973) and (b) Cerato and Lutenegger (2004).

strength representative of the in situ undisturbed, undrained shear strength may be obtained by extrapolating the experimental measurements back towards the y-axis for $\alpha = 0$ (i.e. for a vane of zero blade thickness). In this case a standard thickness of 2 mm ($\alpha = 3.1\%$) would underestimate the in situ undrained shear strength by about 20%.

The time from the end of vane penetration to the beginning of rotation should not exceed five minutes. After insertion, a continuous dissipation of pore pressure, induced during penetration, occurs in the soil surrounding the blades. This consolidation process produces an increase of effective stress and, consequently, an increase in shear strength that is proportional to the

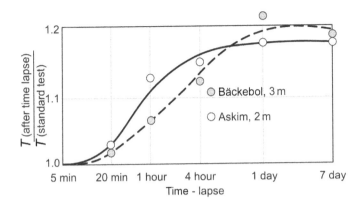

Figure 4.6 Consolidation effects (Torstensson, 1977a).

time at rest between installation and testing. For a conservative estimate of shear strength, the increase in strength due to consolidation cannot exceed the soil disturbance produced by the action of pushing the vane into the soil. This effect has been experimentally examined by Torstensson (1977a) and results are illustrated in Figure 4.6. In this figure, the time-lapse between installation and testing is plotted against the increase in shear strength for two different clay sites. Each datapoint represents the average result of three tests. The change in measured strength can be as high as 20% for a seven-day time-lapse and less than 5% for a time-lapse of half an hour. Based on this evidence, a period of no longer than five minutes was taken as the standard to attain a conservative estimate of shear strength from vane tests.

Once in position and after the time-lapse, the vane is rotated at a rate of 0.1°/s (±0.05 to 0.2°/s) and the relationship between torque and angular rotation is recorded. The maximum torque is used to calculate the peak undrained shear strength of the soil. After completion of the test the vane is extensively rotated, so that the soil along the failure surface becomes completely remoulded, for the residual shear strength to be measured and soil sensitivity to be calculated.

Interpretation

The vane test measures the shear strength by using rotating blades which create both horizontal and vertical shear surfaces. The geometrical symmetry of the cylindrical shear surface is well-defined and, in this case, a closed-form solution can be extracted from simple limit equilibrium analysis (see the Class I solution described in Chapter 1).

The undrained shear strength is usually derived on the basis of the assumptions that no drainage occurs during shear, shear stress distribution is uniform on failure surfaces and soil around the vane is isotropic and homogeneous. Since these assumptions are not necessarily correct, careful investigation has been carried out to examine these effects on s_u (e.g. Wroth, 1984). This section describes how these assumptions can support the theoretical interpretation of vane tests.

Donald *et al.* (1977) determined the stress distribution along the shear surface from a three-dimensional finite element analysis for elastic materials. Later, Menzies and Merrifield (1980) carried out experiments by instrumenting a vane in sand and in overconsolidated clay. Both numerical and experimental data gave fairly similar distributions of stresses, as illustrated in Figure 4.7, and provided evidence for the validation of analytical modelling of vane tests. The assumption of a rectangular shear stress distribution on the vertical cylindrical failure surface was confirmed as reasonable, despite the fact that some stress concentration was observed at the edges of the blades. In contrast, there is a major difference between the observed stress distribution on the horizontal failure surfaces and the assumed condition of uniformity. The shear stress increases from the centre of the vane towards the edges, and thus the actual stress variation differs from the idealized uniform stress distribution.

Based on these observations, Wroth (1984) proposed to express the shear stress at the horizontal top and bottom surfaces by a simple polynomial equation:

$$\frac{\tau_H}{\tau_{mH}} = \left(\frac{r}{D/2}\right)^n \tag{4.2}$$

where

r = radial distance from the central line of the blades
τ_H = value of shear in the horizontal failure surface at a distance r from the central line
τ_{mH} = maximum value of shear in the horizontal failure surface that occurs at the perimeter of the vane, at a distance $R = D/2$ from the central line.

The conventional assumption of a uniform distribution of shear stress is a special case of equation (4.2) for $n = 0$. For the London clay data recorded by Menzies and Merrifield (1980), Wroth (1984) suggested an approximate value of $n = 5$.

Once the boundary stresses are defined and the cylindrical failure geometry is assumed, it is possible to calculate the undrained shear strength of the soil from the measured torque required to rotate the vane. The torque

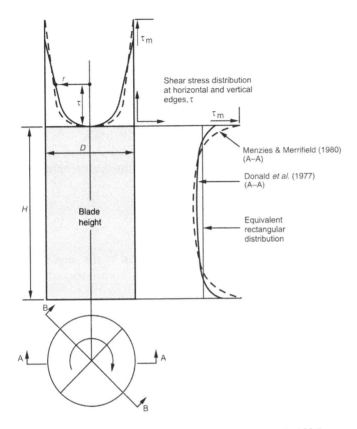

Figure 4.7 Shear stress distribution on a vane (after Wroth, 1984).

(or moment) mobilized on both top and bottom horizontal shear surfaces for a vane height H and radius $R = D/2$ is expressed as:

$$T_H = 2 \int_0^R 2\pi r^2 \tau_H dr \qquad (4.3)$$

Isolating τ_H in equation (4.2) and substituting in (4.3):

$$T_H = \left(\frac{4\pi \tau_{mH}}{n}\right) \int_0^R r^{(n+2)} dr \qquad (4.4)$$

Integrating equation (4.4) with the limits 0 to R and substituting R by $D/2$:

$$T_H = \frac{\pi D^3 \tau_{mH}}{2(n+3)} \tag{4.5}$$

Equation (4.5) represents the contribution of the torque at the horizontal surfaces and considers the polynomial expression proposed by Wroth (1984) to describe the shear stress distribution along the vane.

Since the conventional assumption of a uniform distribution of shear stress is valid along the vertical surface, the vertical torque component is simply expressed as:

$$T_V = \pi D H \tau_{mV} \frac{D}{2} = \frac{\pi D^2 H \tau_{mV}}{2} \tag{4.6}$$

The external torque applied to the vane – and resisted by the surrounding soil – is given by the sum of shearing mobilized on both horizontal and vertical surfaces:

$$T = T_H + T_V \tag{4.7}$$

The analysis considers the soil shear strength mobilized at failure when the maximum value of shear is applied, i.e. $\tau_{mH} = s_{uh}$ and $\tau_{mV} = s_{uv}$, being s_{uv} = undrained shear strength on the vertical surface, and s_{uh} = undrained shear strength on horizontal surfaces. Soil anisotropy is expressed from these two strength values:

$$b = s_{uv}/s_{uh} \tag{4.8}$$

and reflects the anisotropy of fabric developed during deposition and one-dimensional consolidation of natural soft clay deposits.

A general formula for deriving the shear strength from a vane test is obtained by incorporating equations (4.5), (4.6) and (4.8) within (4.7):

$$s_{uH} = \frac{n+3}{D + Hb(n+3)} \frac{2T}{\pi D^2} \tag{4.9}$$

Equation (4.9) considers a non-uniform shear strength distribution along the top and bottom shear surfaces, soil anisotropy and any geometrical H/D ratio.

The standard method of interpretation of the vane test assumes uniform shear strength on failure surfaces ($n = 0$) and isotropic shear strength of the soil ($b = 1$) in a rectangular vane of $H/D = 2$, leading to the following text-book expression which is derived from equation (4.9):

$$s_u = \frac{6T_m}{7\pi D^3} \qquad (4.10)$$

where s_u = undrained shear strength from the vane and T_m = maximum value of measured torque corrected for apparatus and rod friction.

From the academic point of view, it is interesting to observe that equation (4.9) can be rewritten to match any other equation previously proposed in the literature. By substituting $(n + 3)$ for a and taking the assumption of isotropy of shear strength into account, Jackson (1969) suggested:

$$s_u = \frac{2T_m}{\pi D^2 (H + D/a)} \qquad (4.11)$$

where

$a = 3.0$ (uniform stress distribution)
$a = 3.5$ (parabolic stress distribution)
$a = 4.0$ (triangular stress distribution).

Aas (1967) assumed uniform shear stress distribution ($n = 0$) combined with soil anisotropy ($b \neq 1$), leading to:

$$\frac{2}{\pi H D^2} T_m = s_{uV} + \frac{D}{3H} s_{uH} \qquad (4.12)$$

Table 4.1 summarizes possible equations that can be adopted in the interpretation of vane tests, considering the influence of vane geometry, distribution of stresses and soil structure.

The background theory outlined above represents an underlying model that supports a rational interpretation of vane tests. In fact, this model is far too simplistic to account for all the important effects controlling the rotation of the blades, due to uncertainties in representing the boundary conditions of the test, the hypothesis adopted to represent the shear strength and the impossibility of considering time-dependent effects implicitly embedded within test procedures. Given the impossibility of adequately assessing and representing most of these effects, what prevails in practice is the simplest approach of calculating the undrained shear strength from equation (4.10), assuming isotropy of the soil and uniform shear stress distribution around the vane. This calculated value of vane shear strength is often subjected to empirical corrections that are applied to vane data to account for the influence of strain rate, soil disturbance and anisotropy on shear strength, as discussed in the following section.

Table 4.1 Interpretation of vane tests (Lund et al., 1996)

Dimensions H/D	Isotropy/ anisotropy	Stress distribution – horizontal planes	Equation
H = D	Isotropic (b = 1)	Uniform (n = 0)	$s_u = 1.50 \dfrac{T_{max}}{\pi D^3}$
		Parabolic (n = 1/2)	$s_u = 1.56 \dfrac{T_{max}}{\pi D^3}$
		Triangular (n = 1)	$s_u = 1.60 \dfrac{T_{max}}{\pi D^3}$
	Anisotropic (b ≠ 1)	Uniform (n = 0)	$s_{uH} = \dfrac{6}{(3b+1)} \dfrac{T_{max}}{\pi D^3}$
		Parabolic (n = 1/2)	$s_{uH} = \dfrac{14}{(7b+2)} \dfrac{T_{max}}{\pi D^3}$
		Triangular (n = 1)	$s_{uH} = \dfrac{8}{(4b+1)} \dfrac{T_{max}}{\pi D^3}$
H = 2D	Isotropic (b = 1)	Uniform (n = 0)	$s_u = 0.86 \dfrac{T_{max}}{\pi D^3}$ *
		Parabolic (n = 1/2)	$s_u = 0.88 \dfrac{T_{max}}{\pi D^3}$
		Triangular (n = 1)	$s_u = 0.89 \dfrac{T_{max}}{\pi D^3}$
	Anisotropic (b ≠ 1)	Uniform (n = 0)	$s_{uH} = \dfrac{6}{(6b+1)} \dfrac{T_{max}}{\pi D^3}$
		Parabolic (n = 1/2)	$s_{uH} = \dfrac{7}{(7b+1)} \dfrac{T_{max}}{\pi D^3}$
		Triangular (n = 1)	$s_{uH} = \dfrac{8}{(8b+1)} \dfrac{T_{max}}{\pi D^3}$

Note
*Interpretation according to standards (see ASTM D 2573-01; BS 1377-90).

Shear strength

Standardized calculation of the undrained shear strength for a rectangular vane of H/D = 2 gives:

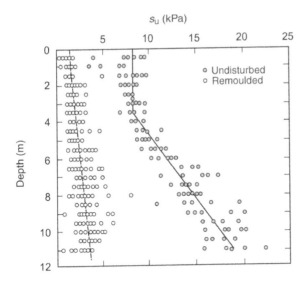

Figure 4.8 Undrained shear strength from vane tests (Ortigão and Collet, 1987).

$$s_u = \frac{6T_m}{7\pi D^3}$$ (4.10)idem

Equation (4.10) represents the peak undrained shear strength, s_u, calculated from the maximum torque measured during the first loading cycle of the vane test, just after insertion. The remoulded shear strength s_{ur} can also be calculated from equation (4.10), with the torque being measured after rotating the vane rapidly through a minimum of five to ten revolutions. A typical example of peak and residual undrained shear strength measurements is shown in Figure 4.8 for the Sarapui Rio de Janeiro clay (Ortigão and Collet, 1987). A clear pattern of strength variation with depth is observed in this figure, with the peak strength estimated from the vane showing a linear increase with depth, which is a characteristic feature of normally consolidated clay deposits. This layer is overlain by an OC crust near the surface. Residual strength is 4 to 6 times lower than peak strength, the difference increasing with depth.

Once s_u and s_{ur} are determined, the sensitivity of the soil (s_t) can be estimated as:

$$s_t = s_u/s_{ur}$$ (4.13)

Skempton and Northey (1952) classified the sensitivity of soils as shown in Table 4.2. In contrast to very sensitive Norwegian clays, Brazilian clays show low to moderate sensitivity (Table 4.3).

Table 4.2 Clay sensitivity (Skempton and Northey, 1952)

Sensitivity	S_t
Insensitive clays	1.0
Low sensitivity	1–2
Medium sensitivity	2–4
Sensitive	4–8
Extra sensitivity	>8
Quick clays	>16

Table 4.3 Sensitivity of Brazilian soft clay deposits (updated from Ortigão, 1995)

Site	Average sensitivity	Range	References
Santa Cruz, RJ (coast)	3.4		Aragão, 1975
Santa Cruz, RJ (offshore)	3.0	1–5	Aragão, 1975
Rio de Janeiro, RJ	4.4	2–8	Ortigão and Collet, 1987
Sepetiba, RJ	4.0		Machado, 1988
Cubatão, SP		4–8	Teixeira, 1988
Florianópolis, SC	3.0	1–7	Maccarini *et al.*, 1988
Aracaju, SE	5.0	2–8	Ortigão, 1988
Porto Alegre, RS	4.5	2–8	Soares, 1997

Since the field vane test gives a reliable, repeatable strength measurement at relatively low cost, it is worth evaluating effects that might impact the use of this measured value in ultimate limit state geotechnical calculations. Rate effects, soil anisotropy and partial drainage are among the most important factors influencing the undrained shear strength measured by the vane test.

Rate effects

With the strain rate associated with the vane test being several orders of magnitude higher than typical rates attained in geotechnical design, an evaluation of rate effects is necessary to ensure that the vane undrained strength is representative of the static shear strength (e.g. Bjerrum, 1972; Perlow and Richards, 1977; Torstensson, 1977a; Walker, 1983; Chandler, 1988; Biscontin and Pestana, 2001; Randolph, 2004). Typically, the relationship between rate of shear and shear strength is expressed by logarithmic or power laws:

$$s_u = s_{u,ref} \left[1 + \alpha \left(\frac{\dot{w}}{\dot{w}_{ref}} \right) \right] \tag{4.14}$$

or

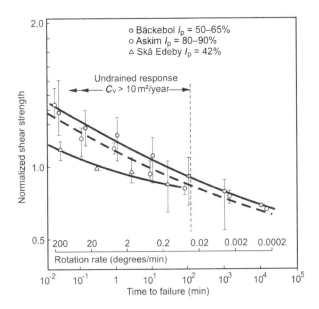

Figure 4.9 Effects of the rate of rotation on shear strength (Torstensson, 1977a).

$$s_u = s_{u,ref} \left(\frac{\dot{w}}{\dot{w}_{ref}} \right)^{\beta} \tag{4.15}$$

where $s_{u,ref}$ is a reference strength measured at an angular strain rate $\dot{w}_{ref}$, and α and β are soil-dependent material parameters. Measured values of parameters α and β are given in Table 4.4, along with a summary of previous investigations into rate effects using vane tests.

Table 4.4 reports early work by Torstensson (1977a), detailed in Figure 4.9 in a plot that shows the normalized undrained shear strength as a function of both angular rotation rate and time to failure. Values of s_u are shown to decrease with decreasing rotation rate. For the highest rate (~0.01°/min) the measured shear strength was in the order of 30 to 40% higher than that determined at a standard shear rate of 6°/min.

Biscontin and Pestana (2001) reported a detailed study of the effect of the vane rotation rate on the deduced shear strength for tests carried out with a bentonite–kaolinite soil mixture manufactured in the laboratory. Their results are shown in Figure 4.10, expressed in terms of the peripheral velocity v_p, which is simply the velocity at the edge of the blade (= $\dot{w}D/2$). The rate of increase in vane strength is represented by equations (4.14) and (4.15) (with peripheral velocity replacing strain rate) with α and β values corresponding to 0.10 and 0.055 respectively. In a similar plot, the author summarized existing data documenting changes in undrained shear strength of

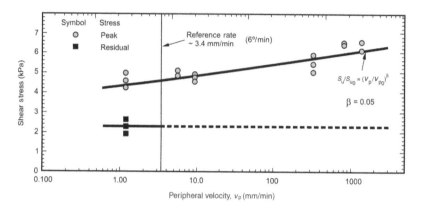

Figure 4.10 Undrained peak and residual strength measured from vane tests (Biscontin and Pestana, 2001).

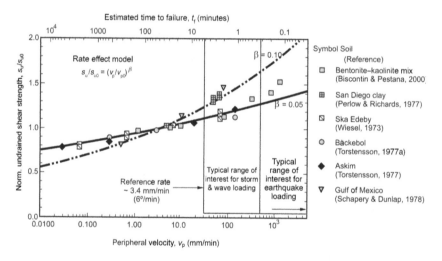

Figure 4.11 Normalized undrained shear strength as a function of peripheral velocity (Biscontin and Pestana, 2001).

soft soils as a function of peripheral velocity (Figure 4.11). A power law relationship was found to give the best fit within the limit established for $\beta = 0.05$ and 0.10 over a wide range of shearing rates.[1]

These studies reveal the influence of the shear rate on the measured undrained shear strength, with the increase in s_u values per load cycle being in the order of 10%. From a very practical perspective, this influence becomes significant only at very high shearing rates representative of earthquake loading.

Table 4.4 Studies of rate effects on undrained shear strength (after Biscontin and Pestana, 2001)

Clay	I_p (%)	Shear vane test details			Time effect		Reference
		H/D	D (mm)	Rate (°/min)	α (%)	β	
Grangemouth	≈22	1.5	50–75	6–300	5–6	0.025	Skempton (1948)
Bromma	≈31	2.5	80	6–600	20	0.086	Cadling and Odenstad (1948)
Aserum, Drammen, Lierstranda, Manglerud	8–9	¼–4	65–130	6–60	1–2	0.006	Aas (1965)
Silts and clay (disturbed and remoulded)	–	2	12.7	6–672 / 6–720	≈16 / ±5	0.05 / ±0.01	Migliori and Lee (1971); Halwachs (1972)
Ska Edeby	50–100	¼–1	65–130	0.06–600	3–6	0.02	Wiesel (1973)
Bäckebol	50–65	2	65	≈0.006–300	12	0.05	Torstensson (1977a)
Askim	80–90					±0.01	
Gulf of Mexico clay	–	2	12.7	21–79	13–27	0.05 / 0.10	Smith and Richards (1975)
Exuma Sound, Bahamas							
San Diego silt I	64	1.2	12.7 / 76,101	72–79	36 / 21	0.13 / 0.08	Perlow and Richards (1977)
San Diego silt II							
Gulf of Mexico clay	78	1.2	5.1, 12.7	21–79	60	0.20	Schapery and Dunlap (1978)
Gulf of Mexico clay	–	2		4.8–708	33	0.107	
Pierre shale	103	1.1, 5.2	12.7	4.8–107	12	0.05	Sharifounnasab and Ullrich (1985)
Saint-Louis de Beaucours	13–19 / 6–18	2	65	6–120	2–3	0.01	Roy and LeBlanc (1988)
Saint-Alban clay							
Bentonite–kaolinite mix	73	2	55	2–3000	1 / 15	0.004 / 0.055	Biscontin and Pestana (2001)

Partial consolidation

Radial consolidation cavity expansion solutions show that for soft clay coefficients of consolidation greater than, let's say, 0.3 mm²/s (> 10 m²/ yr), a time of at least 200 s for $d = 65$ mm is necessary before significant dissipation starts to be observed. Thus, excess pore pressure changes around the vane will be small for low permeability clay during the 5 min time-lapse between installation and testing, but may be significant for silty soils.

A further and more severe partial drainage effect can take place during shear under standard rotation rates in intermediate permeability soils. As already discussed in Chapter 3 under the heading Properties in silty soils, the degree of drainage in intermediate permeability soils can be represented by an analytical 'drainage characterisation curve' (Randolph, 2004; Schnaid et al., 2004; Schnaid, 2005). This method, initially applied to piezocone data, can easily be adapted to vane tests in the V versus U space, where V is a non-dimensional shear rate:

$$V = \frac{vd}{C_v}$$

(3.49)idem

and U is the degree of drainage:

$$U = \frac{(T_c - T_{und})}{(T_{dr} - T_{und})}$$

(4.16)

where

d = probe diameter
v = rate of rotation
C_v = coefficient of consolidation
T_{und} = undrained torque
T_{dr} = drained torque
T_c = torque at any rotation rate.

Vane tests carried out by Blight (1968) at different rotation rates are presented in Figure 4.12 in order to identify the pattern of drainage during shear in intermediate permeability soils. Despite the scatter produced by experimental data, a distinctive curve is observed for each material with the transition point from drained to partially drained response, starting at a normalized velocity $V \approx 10^{-1}$, and the transition from partially drained to undrained at a normalized velocity of around 10^{+2}. Once the effects of partial drainage become significant, unrealistically high s_u values are obtained.

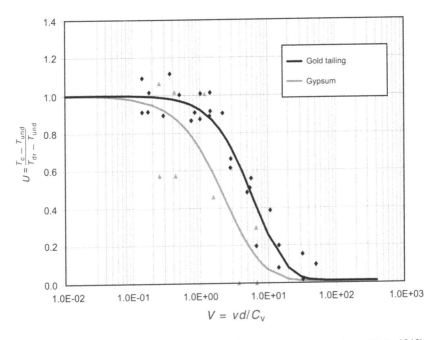

Figure 4.12 Effect of partial consolidation around a rotating vane (data from Blight, 1968).

Anisotropy

The importance of anisotropy in engineering problems has been considered by Ladd *et al.* (1977), Jardine *et al.* (1999), Tatsuoka *et al.* (1997), Hight (1998) and Hight and Leroueil (2003), among others. Soft clay deposits exhibit anisotropy of undrained shear strength resulting from both the structural features arising from the depositional processes (inherent anisotropy) and the strain experienced at any given time after deposition (induced anisotropy).

Although conventional analysis assumes that the shear strength is isotropic, a study of the in situ large strain anisotropy of clays can be undertaken using vane tests, as detailed above under the heading Equipment and procedures. The undrained shear strength can be determined from the measured torque T by the following general expression previously deduced:

$$s_{uH} = \frac{n+3}{D+Hb(n+3)} \frac{2T}{\pi D^2}$$

(4.9)idem

where H/D is the aspect ratio and n is the power law describing the shear strength distribution on the top and bottom planes. The vane torque on the

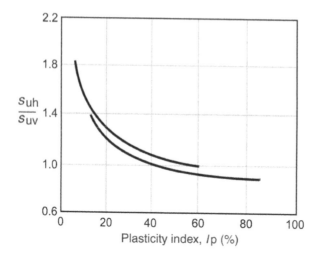

Figure 4.13 Anisotropy ratio versus plasticity index for soft clay (Bjerrum, 1973).

vertical and horizontal planes can be separated, provided that a series of vane test data with different H/D is available, allowing assessment of soil anisotropy. In this case, the angular rotation rate should be adjusted to avoid the peripheral velocity exerting an influence on measurements of undrained shear strength (e.g. Biscontin and Pestana, 2001).

Normally consolidated low plasticity clay deposits exhibit values of s_{uh}/s_{uv} varying from 1.5 to 2.0, the s_{uh}/s_{uv} ratio reducing with increasing soil plasticity (Richardson and Whitman, 1963; Aas, 1965). These findings have been used by Bjerrum (1973) to demonstrate the relationship between the s_{uh}/s_{uv} ratio and the plasticity index for soft clay (Figure 4.13).

Corrections

As extensively discussed, the undrained shear strength measured by the vane is affected by soil disturbance, strain rate, structure anisotropy and partial consolidation. Although these effects might partially compensate for each other, the measured strength is different from the value obtained in the field, as already recognized by Bjerrum (1972), and reduction coefficients based on the plasticity index are often adopted by practising engineers to correct the vane strength profile.

The original correlation proposed by Bjerrum (1972) for estimating the correction factor μ_r, which accounts for strain rate effects, is shown in Figure 4.14 and is expressed as:

$$s_u(field) = \mu_r s_u(vane) \tag{4.17}$$

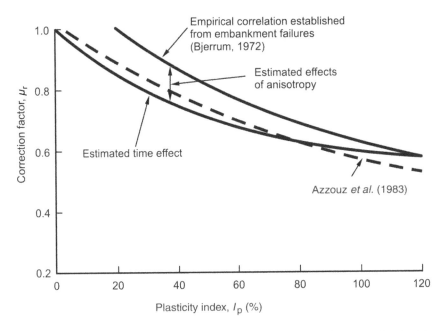

Figure 4.14 Empirical strength correction factor based on correlations with plasticity index (Chandler, 1988).

Also presented in the figure are correlations established after the re-examination of the original cases reported by Bjerrum, including that from Azzouz *et al.* (1983) in a study that included end effects calculated by a numerical three-dimensional finite element analysis. In addition, Aas *et al.* (1986) presented results where the effect of overconsolidation was considered and Chandler (1988) suggested a correction which accounts for the actual time-to-failure period. As an example, it can be seen that for a plasticity index of 80, which is representative of the Porto Alegre clay, a correction of about 0.6 is recommended to be applied to the measured vane strength.

Since these correction factors have been empirically determined, they can only be applied with relative confidence to stability analysis of embankments and excavations in soft clays. There is little justification for adopting this correction in the design calculations of other geotechnical structures that do not share the characteristic database used by Bjerrum (1972).

Stress history

As previously discussed in this book (see, for example, Chapter 3), an accurate determination of the preconsolidation pressure is the single most important piece of information in geotechnical design calculations in clay, impacting the prediction of long-term consolidation settlements as well as short-term stability problems. The preconsolidation pressure σ_p' is often normalized by the in situ effective overburden pressure σ_v' to denote the stress history of the soil expressed as an overconsolidation ratio $OCR = \sigma_p'/\sigma_v'$.

The oedometer test is the standard method of determining preconsolidation pressure in the laboratory. In addition, σ_p' or OCR can be assessed directly from field tests such as the piezocone and vane following research efforts developed during the 1980s (Wroth, 1984; Konrad and Law, 1987; Mayne, 1987; Crooks et al., 1988; Mayne and Bachus, 1988; Mayne and Mitchell, 1988; Sandven et al., 1988; Sully et al., 1988). This research was supported by the framework of critical state soil mechanics (Schofield and Wroth, 1968) and the SHANSEP method (Ladd et al., 1977), from which it is possible to demonstrate that the normalized undrained shear strength increases with the overconsolidation ratio according to the general expression:

$$\frac{[s_u/\sigma_{v0}']}{[s_u/\sigma_{v0}']_{nc}} = OCR^\Lambda \tag{3.10)idem}$$

where Λ is a constant obtained in laboratory tests. Ideally, if the specific values of $(s_u\,\sigma_{v0}')_{nc}$ and Λ have been determined for a given clay deposit, then shear strength measured by the vane can be used as reference to establish a profile of OCR against depth for a given site.

These concepts have been investigated by Mayne and Mitchell (1988) in a research programme comprising data from 96 sites where laboratory oedometer test data, index properties and field vane strength were compiled for normally consolidated ($OCR = 1$) to heavily overconsolidated clays (OCR up to 40). A direct comparison between the laboratory-measured OCR and the normalized field vane strength $s_u\,\sigma_{v0}'$ is presented in Figure 4.15, the average trend assuming a log–log relationship expressed as (Mayne and Mitchell, 1988):

$$OCR = 3.55\left(s_u/\sigma_{v0}'\right)^{0.66} \tag{4.18}$$

which yields mean values of $(s_u\,\sigma_{v0}')_{nc} = 0.146$ and $\Lambda = 1.52$. If, for simplicity, a value of $\Lambda = 1$ is assumed, the assessment of OCR from normalized undrained strength to overburden ratio may be generalized as (Mayne and Mitchell, 1988):

$$OCR = \alpha_{FV}\left(\frac{s_{uFV}}{\sigma_{v0}'}\right) \tag{4.19}$$

where parameter α_{FV} should be obtained from site-specific correlations but can be adopted as a first estimate as:

$$\alpha_{FV} = 22(I_p)^{-0.48} \tag{4.20}$$

with I_p being the plasticity index (Figure 4.16).

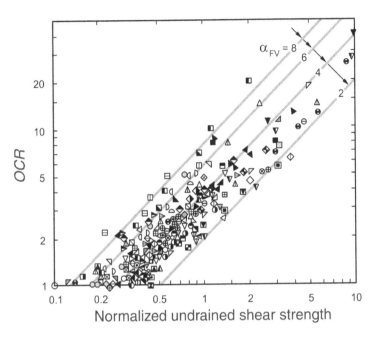

Figure 4.15 OCR versus normalized undrained shear strength (Mayne and Mitchell, 1988).

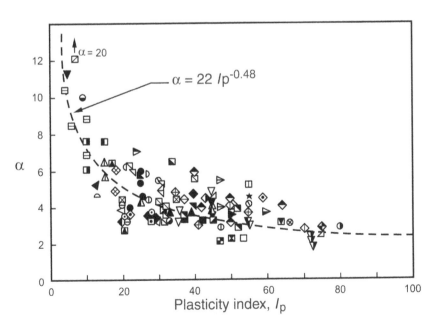

Figure 4.16 Relationship between α and I_p (Mayne and Mitchell, 1988).

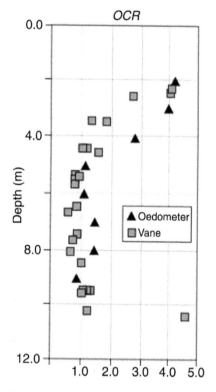

Figure 4.17 Comparison of OCR values derived from laboratory and field vane tests at the Porto Alegre clay deposit, Brazil.

A comparison between OCR values measured in laboratory oedometer tests and estimated from field vane tests in the Porto Alegre soft clay deposit is shown in Figure 4.17. This comparison shows that field and laboratory tests give comparable results, indicating a NC deposit below 4.0 m depth.

Note

1 Randolph (2004) calculated the effect of vane rotation on torque and concluded that it is the rotation rate rather than the peripheral velocity that controls the rate dependency of the torque.

Chapter 5

Pressuremeter tests

A realization in the latter part of the Twentieth Century that improved under-standing and quantification of many problems in soil and rock mechanisms could be obtained by application of cavity expansion theory marks a mile-stone in the development of our field.

(James K. Mitchell, 2001)

General considerations

Pressuremeters are cylindrical devices designed to apply uniform pressure to the wall of a borehole by means of a flexible membrane. Both pressure and deformation at the cavity wall are recorded and interpretation is provided by cavity expansion theories under the assumption that the probe is expanded in a linear, isotropic and elastic-perfectly plastic soil.

This unique and innovative technology, introduced by the French engi-neer Louis Ménard in 1955, established its reputation as a site investigation tool on the basis of what became known as the Ménard Empirical Design Rules. Conceived to assist foundation design, this set of empirical rules became available to aid engineers in relating the pressuremeter results to the behaviour of soil structures. Later, with the increasing ability to extract plausible, rational solutions from pressuremeter testing, empirical design approaches have been gradually replaced by the use of theoretical analysis. The probe contains sufficient geometric symmetries to reduce the problem to a relatively simple form, so that the measured pressure–displacement curve can be used to back-calculate the mechanical properties of the tested soil, adopting cavity expansion solutions combined with constitutive mod-els capable of representing soil behaviour (e.g. Gibson and Anderson, 1961; Palmer, 1972; Hughes *et al.*, 1977; Carter *et al.*, 1979; Wroth, 1984). Parameters derived from the test are the shear modulus, undrained shear strength in clay, internal friction angle and dilation angle in sand, in situ horizontal stress, porewater pressure and permeability.

In the past few decades, no other in situ test device has been subjected to such extensive academic work as the pressuremeter, to an extent that the current theoretical and practical research provides an assessment of all aspects of the mechanical behaviour of geomaterials. A thorough overview of the state of the art of pressuremeter testing has been tackled in several textbooks, which are recommended for a more in-depth perspective of this topic: Baguelin *et al.* (1972), Briaud (1992), Clarke (1995) and Yu (2000). Extensive experience, gained over almost 50 years, has been summarized in five international symposia: Paris (1982), Texas (1986), Oxford (1990), Québec (1995) and Paris (2005). Pressuremeter technology, theoretical background, methods of interpretation, geotechnical applications and case studies on different geomaterials, both onshore and offshore, were covered in these venues. National Reports describing current practice are given in the 2005 Paris Symposium. National and regional codes of practice such as NFP 94-110-1 (2000), ASTM D 4719 (2000) and EN ISO 22476-4 give detailed recommendations for test procedures.

A brief review of pressuremeter technology and fundamental solutions for expansion and contraction cavities are presented in the following sections. Given the extent of the applications of pressuremeter data, the topics covered in this book are focused on theoretical methods of interpretation from which fundamental soil parameters are derived. The author considers the pressuremeter to be a more sophisticated instrument than the other in situ tests covered in the preceding chapters and, for this reason, that the instrument is most cost effective when rational methods are applied to the interpretation of testing data. The pressuremeter is intended for use in all soils and weak rocks and, although interpretation methods have been developed for sand and clay, they have been gradually extended to other geomaterials and environmental conditions, such as cemented, collapsible and expansible soils and unsaturated soil conditions, among others.

Types of instruments

Typically, the pressuremeter is composed of a probe and a control unit. Although the probe design varies significantly according to the method of installation, its wall is generally made of a flexible membrane protected by steel strips known as the 'Chinese Lantern'. The length of the flexible part has to be more than 6 diameters to conform to the requirements of interpretation methods which dictate that the expanding probe must be sufficiently long to be modelled as a cylindrical cavity. The capacity of the instrument ranges from 2.5 MN/m^2 up to 20 MN/m^2, this maximum pressure being suitable for use in hard soils and soft rocks.

The control unit consists of a pressure supply used to run the test, as well as displacement and pressure sensors designed to monitor the expanding cavity. The unit can be manually operated using visual displays to monitor pressure and displacements.[1] Furthermore, it can be assisted by sophisticated electronic servo-mechanisms, which control the rate of displacement or the rate of increase of pressure. In this last case, the system contains a power source, transducers and an analogue–digital converter that send signals directly to a computer for future interpretation.

Dependent on the method of installation into the ground, pressuremeters are generally classified into three groups: pre-bored, self-boring and push-in pressuremeters. The Ménard pressuremeter is a well-known example of a pre-bored probe, in which the device is lowered into a preformed hole. In a self-boring probe the device bores its own way into the ground with minimal disturbance, whereas in a push-in device a robust pressuremeter is simply pushed into the ground. Brief summaries of the different categories are given in the following sections.

Pre-bored pressuremeter (PBP)

In this group, the probe is simply lowered into a pre-bored hole to the test depth where the membrane is inflated. The Ménard-type pressuremeter (MPM) is the most widely used tool in this group, designed to be a simple and straightforward device to operate. Pressure applied by gas is recorded by a gauge placed at the surface, while displacements are determined by measuring the change in volume of a water-filled central cell. This central cell is protected by upper and lower guard cells that are inflated with gas to ensure that the central cell expands predominantly in the radial direction as a circular cylinder.

The basic characteristics of the Ménard pressuremeter are shown in Figures 5.1 and 5.2. The GC probe illustrated in these figures is 74 mm in diameter, and has a pressure capacity of 4 MN/m^2. The applied water pressure is set manually using a regulator and is measured by a Bourdon gauge. The unit has two gauges, the more sensitive to be used for lower pressures to obtain better resolution. The change in volume at each pressure increment is monitored by gas–water interface variations in a graduated tube. The control unit is connected to the pressuremeter probe by coaxial tubing: water circulates down the inner tube into the central cell and gas circulates down the outer tube into the guard cells. A set of rods are used to lower and raise the probe to each testing depth.

In addition, there are pre-bored tools on the market that utilize electronic sensors inside the membrane, allowing for high resolution of the measured radial displacement of the membrane (e.g. Hughes, 1982; Kratz de Oliveira *et al.*, 2001).

Self-boring pressuremeter (SBP)

The self-boring pressuremeter was developed simultaneously in France (Baguelin *et al.*, 1978) and in the UK (Wroth and Hughes, 1973) with the aim of reducing the inevitable soil disturbance caused by forming a borehole when inserting a cylindrical probe into the ground. In the self-boring technique, the pressuremeter

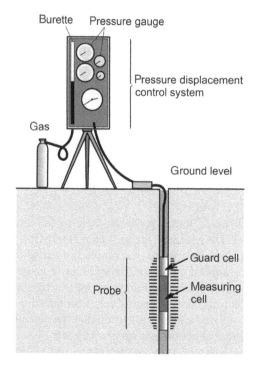

Figure 5.1 Schematic representation of the Ménard pressuremeter test.

is pushed hydraulically from the surface while the soil inside the sharp cutting edge of the tapered shoe is removed by a rotating cutting mechanism supplied with flush fluid that jets soil particles to the surface as the probe is advanced. In theory, the SBP offers the attractive possibility of performing tests on undisturbed soils, provided that high-quality installation is achieved.

There are two versions of the self-boring pressuremeter: the UK Cambridge Pressuremeter (Camkometer, after Wroth and Hughes, 1973) and the French Pressiomètre Autoforeur (PAF, after Baguelin *et al.*, 1972). For harder soils and rocks, a high pressure probe has been developed to enable much higher cavity pressures than the norm (for example, the Pressiomètre Autoforeur pour Sol Raide, PAFSOR – see Dalton, 2005). There are important differences between French and British self-boring devices. In the French PAF, the radial strains are inferred from measurements of volume change, while in the British device measurements are made using strain-gauged feeler arms. In terms of installation, the PAF prototype has cutters which are driven by downhole hydraulic motors, whereas the British cutter system is driven by rods extending to the ground surface.

Figure 5.2 Pre-bored Ménard pressuremeter.

The Cambridge self-boring pressuremeter, illustrated in Figures 5.3 and 5.4, is a hollow cylindrical probe, a little over 1 m in length and 83 mm in diameter. Three strain-gauged, full bridge cantilever springs and pivoted sensing arms positioned at 120° around the centre of the expanding length are used to measure the variations in diameter during the test. The gas pressure applied to expand the membrane and the soil porewater pressure generated during expansion are measured by electrical transducers mounted within the pressuremeter.

The boring technique should be consistent with achieving the minimum disturbance from the drilling to the surrounding soil, which is not always possible even with skilled operators. The technique requires the jet pressure to be adjusted to the total overburden pressure. The internal tapered shoe should be set near the edge of the shoe in stiff clay and dense sand. On the other hand, in soft clay and medium-dense sand the cutter should be placed at a distance equal to the radius of the SBP back from the leading edge (e.g.

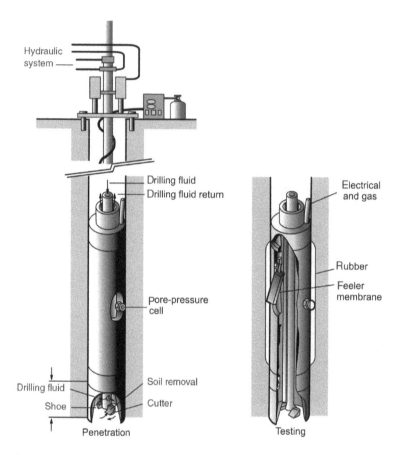

Figure 5.3 Details of a self-boring pressuremeter (adapted from Weltman and Head, 1983).

Clarke, 1995). The combined balance of thrust, speed of the cutter, pressure and flow rate of the drilling fluid is not easily achieved and therefore the use of a self-boring pressuremeter is time consuming and expensive.

Pressuremeter tests can be carried out under stress-controlled or strain-controlled conditions. It is common to adopt a stress-controlled approach in the early stages of the test, followed by a strain-controlled method after the onset of plastic strains. For clays, a strain rate of 1%/mm is recommended (Windle and Wroth, 1977). However, higher rates of strain may be required in order to ensure undrained shear. Small unload–reload loops are performed during the expansion phase to allow soil shear stiffness to be calculated.

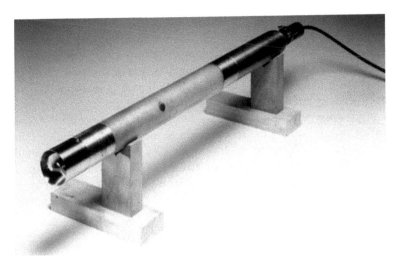

Figure 5.4 Cambridge self-boring pressuremeter (courtesy of Cambridge Insitu).

Push-in pressuremeter (PIP)

In terms of practical engineering applications, the push-in pressuremeter has many advantages over the self-boring pressuremeter. Installation is achieved by pushing the pressuremeter module from the bottom of a borehole or by inserting a pressuremeter module mounted behind a cone tip.

The cone pressuremeter prototype (CPM), previously shown in Figure 3.9 (Chapter 3), is pushed into the ground using a standard cone truck as part of the cone penetration test operation, with the cone driven into the soil at a constant penetration rate of 20 mm/s. Insertion is halted with the centre of the pressuremeter probe at the required depth to allow pressuremeter probe expansion. A continuous cone profile is obtained during penetration that is later used to estimate strength parameters in both sand and clay. The pressuremeter, in turn, enables soil stiffness and strength parameters to be assessed from the measured pressure–displacement curve.

Since the pressuremeter test is not carried out in undisturbed ground, the effects of installation have to be accounted for and large strain analysis is required. Nevertheless, the device has certain advantages, principally in the simplicity and economy of its installation, combining some of the merits of the pressuremeter with the operational convenience of the cone. A continuous profile of penetration resistance, pore pressures and friction ratio is obtained during penetration. This profile provides detailed and representative characterization of soil layers. Shear strength can be assessed from cone resistance and, in addition, stiffness and strength parameters are estimated from pressuremeter data. This technique is perceived as having a great potential that has not yet been fully realized in practice.

A robust driven pressuremeter (DPM) has recently been developed by Dalton (2005), in which the probe is hammered into hard soils and soft rocks. The DPM requires a conventional rotary coring or dynamic probing rig for installation to keep the costs down. This device is reported to give consistent profiles of strength and shear modulus.

Developments are currently under way to incorporate geophones into a pressuremeter rod, following the recent trend of measuring the small strain shear modulus G_0 (see Chapters 1, 2, 3 and 6). The combination of small strain measurements with the range of strains given by the SBP (about 0.01% to 1%) will help in building a shear modulus degradation curve, which may assist engineers in selecting operational stiffness for geotechnical design.

Calibration procedures

Calibration procedures are essential elements in obtaining the corrected pressure–expansion pressuremeter curve. Periodically, the device has to be calibrated to take account of errors arising from the instrument's accuracy, pressure losses and membrane corrections, as recommended by codes of practice and dedicated publications (e.g. Baguelin *et al.*, 1978; Mair and Wood, 1987; Clarke, 1995). According to Mair and Wood (1987) *pressuremeter testing is almost worthless if proper calibrations of the instrument are not carried out.*

There are three broad groups of calibrations, dealing with:

• absolute pressure and displacement transducers;
• compliance of measuring system;
• membrane stiffness.

Pressure and displacement transducers should be calibrated periodically. Repeatability, non-linearity and hysteresis should be evaluated and reported to check if the instrument accuracy is within the limits of established requirements.

Calibration for system compliance is necessary to evaluate the change in volume of the pressuremeter while the probe is inflated to its maximum pressure inside a thick-walled steel cylinder. For devices using volume measuring systems, calibration evaluates the compression of the probe, the lines connecting the probe to the surface and the control unit. For radial displacement type probes, the procedure is necessary to evaluate the change in thickness of the membrane due to pressure increase. During calibration the pressure is raised in increments until the maximum anticipated working pressure is reached.

System compliance has an appreciable influence on the measurement of the unload–reload shear modulus. In stiff soils, the strain amplitude observed during an unload–reload loop is small, and inaccuracies in the strain arm measuring system or the volume measuring system can result in large errors. The effect of pressuremeter compliance on the measured shear modulus has long been recognized (e.g. Mair and Wood, 1987; Fahey and Jewell, 1990; Houlsby and Schnaid, 1994). Fahey and Jewell (1990) suggested a calibration procedure to determine the compliance of the measuring system, which involves expanding the pressuremeter in the steel cylinder in a manner that mimics the procedure adopted in tests. Several unload–reload loops can be performed to estimate the system shear modulus G_{sys} at various levels of cavity pressure. The effect of compliance on the measured soil shear modulus can then be expressed as:

$$\frac{1}{G_{cor}} = \frac{1}{G_{mea}} - \frac{1}{G_{sys}} \tag{5.1}$$

where G_{cor} is the corrected modulus and G_{mea} the measured modulus. Houlsby and Carter (1993) suggest that another minor correction must be applied to account for the finite length of the pressuremeter probe.

Membrane stiffness is obtained by inflating the pressuremeter in air at an expansion rate similar to that used in strain-controlled field tests (stress-controlled tests are difficult to reproduce). This calibration has serious implications for the interpretation of tests in soft clay but is not critical in stiff clays and dense sands.

A block diagram shown in Figure 5.5 illustrates the correction procedure recommended to obtain the most accurate information from the pressure–expansion curve. Membrane stiffness is first subtracted from the inflation pressure recorded during the field test. Two other corrections are necessary for the shear modulus to account for compliance effects and the finite length of the pressuremeter probe.

Experimental results

Pressuremeter tests are carried out using stress-controlled or strain-controlled procedures. The Ménard test is a stress-controlled test in which the amount of expansion should ideally double the size of the initial cavity. This is achieved by applying equal increments of pressure at equal intervals of time. Each pressure increment is maintained for 60 s with readings of volume change measured at 15, 30 and 60 s after pressure application. The test starts immediately after the probe is in place at the prescribed depth with the membrane expanded to reach full contact with the borehole wall. The PBP is devised to provide information primarily for two parameters: the slope of the

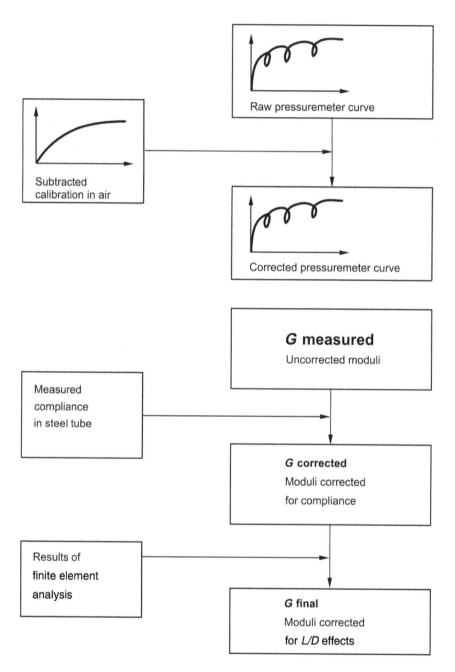

Figure 5.5 Corrections to the measured unload–reload shear modulus (Houlsby and Schnaid, 1994).

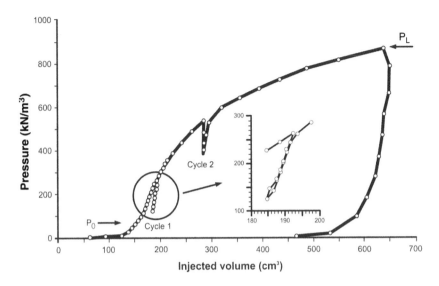

Figure 5.6 Typical Ménard pressuremeter test.

initial linear portion of the pressuremeter expansion curve and the limit pressure p_L, both identified in the typical test result shown in Figure 5.6.

For strain arm system devices, either a stress-controlled or a strain-controlled procedure can be adopted. The problem of adopting a fixed strain rate is that too few readings are recorded during the initial stiff elastic phase. On the contrary, the problem of adopting a fixed rate of applying stress increments is that the strain rate varies considerably once the onset of plastic behaviour is reached. A reasonable compromise is to carry out a stress-controlled test in the elastic phase and only when the soil begins to yield does the strain-controlled test take over. A typical SBP test is shown in Figure 5.7, in which inflation pressure is plotted against cavity strain. A highly non-linear pressure–expansion curve is observed until pressure approaches a maximum value close to its limit pressure. Three unload–reload loops were performed in this test to compute the soil shear modulus.

Analysis of pressuremeter tests

The pressuremeter test has long been recognized as having well-defined boundary conditions and therefore permits a more rigorous theoretical analysis than any other in situ test. Such analysis is based on the expansion of a cylindrical cavity, expressed by the numerous analytical solutions

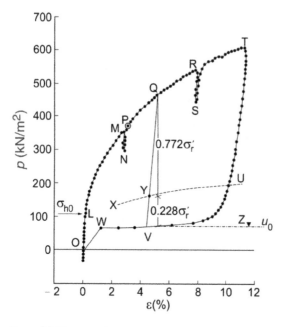

Figure 5.7 Typical self-boring pressuremeter test (Wroth, 1984).

developed on the basis of constitutive models of varying complexity. The intention in this section is not to provide a comprehensive overview of major developments in this subject, but to give a basic understanding of the mathematical framework of cavity expansion theory, which is achieved by describing the solution of cylindrical cavity expansion under the assumption of small strains.

The solution to the problem of expansion of a cylindrical cavity is based on the assumption that soil mass is a homogeneous, isotropic and continuous medium. Since the pressuremeter is idealized as an infinitely long cylinder, axial symmetry applies and all movements are in the radial direction. In the analysis, the convention that compression stresses and strains are positive is adopted, as is usual in soil mechanics. The solution is approached by using cylindrical coordinates so that equilibrium stress in an isotropic material is $\sigma_r = \sigma_\theta = \sigma_z$, σ_r and σ_θ being the normal stresses acting in the radial and tangential directions respectively and σ_z the vertical stress.

In an elastic medium, the solution of the cylindrical cavity problem is well known (e.g. Timoshenko and Goodier, 1970). Initially, the radius of the cavity is a_0 for a cavity volume V_0 and the internal pressure p_0 equal to the in situ total horizontal stress σ_{h0}, as illustrated in Figure 5.8. Everywhere in the surrounding material the radial and circumferential stress components are compressive and equal to p_0. The pressure inside the cavity is then increased

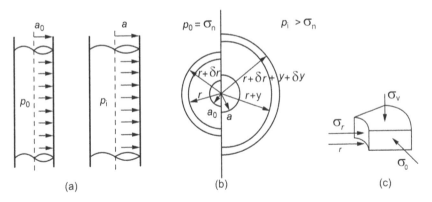

Figure 5.8 Definitions used in cavity expansion analysis: (a) expansion of cylindrical cavity, (b) expansion of an element at radius r and (c) stress action on an element (after Clarke, 1995).

to a value p_i and the cavity radius increases to a_i. A typical material point of the continuum now has a radial coordinate r, having moved to this position from its original position r_0. Using a cylindrical coordinate system, the axial, radial and circumferential strains are denoted by ε_z, ε_r and ε_θ.

Consider an element expanding to $r + y$ and the thickness increasing to $\delta r + \delta y$ as the pressure is increased from p_0 to p_i. In the initial part of loading it is assumed that soil behaves elastically and obeys Hooke's law until the onset of yielding. Using the small-strain theory for the initial phase of elastic loading, the definitions of strains are:

$$\varepsilon_r = \frac{\delta y}{\delta r} \qquad (5.2)$$

and

$$\varepsilon_\theta = -\frac{y}{r} \qquad (5.3)$$

r being the radial coordinate and y the radial displacement (assumed small compared to r) of a material point in the soil mass.

The only variables measured in a test are the applied cavity pressure p and the cavity radius a. The corresponding circumferential strain in the wall of the cavity is referred to as the cavity strain and is defined as:

$$\varepsilon_c = \frac{a - a_0}{a_0} \qquad (5.4)$$

These are definitions for small deformations; when deformations around the expanding cavity are large, care should be taken to use theoretically rigorous definitions of strains.

The equilibrium equation can be expressed in terms of total radial and tangential stresses as follows:

$$\frac{d\sigma_r}{dr} + \frac{\sigma_r - \sigma_\theta}{r} = 0 \tag{5.5}$$

The governing equations for both the elastic and plastic regions around an expanding cavity can now be introduced. This set of equations will be later used in the determination of soil properties from the measured cavity pressure and cavity strain.

Elastic soil

For a cylindrical cavity expanded in isotropic elastic material obeying Hooke's law, the principal stresses and principal strains are related as:

$$E \begin{bmatrix} \Delta\varepsilon_r \\ \Delta\varepsilon_\theta \\ \Delta\varepsilon_z \end{bmatrix} = \begin{bmatrix} ...1.. & -v.... & -v \\ -v.....1...... & -v \\ -v... & -v.......1 \end{bmatrix} \begin{bmatrix} \Delta\sigma_r \\ \Delta\sigma_\theta \\ \Delta\sigma_z \end{bmatrix} \tag{5.6}$$

where E is the modulus of elasticity and v is the Poisson's ratio. Since there is plane symmetry, the vertical strain $\delta\varepsilon_z = 0$ and therefore:

$$E \begin{bmatrix} \Delta\varepsilon_r \\ \Delta\varepsilon_\theta \end{bmatrix} = \begin{bmatrix} ..(1-v^2)....(-v-v^2) \\ (-v-v^2)......(1-v^2) \end{bmatrix} \begin{bmatrix} \Delta\sigma_r \\ \Delta\sigma_\theta \end{bmatrix} \tag{5.7}$$

For a cylindrical cavity expanded in isotropic elastic material, the radial displacement is inversely proportional to the radius:

$$y = \varepsilon_c \frac{a_0 a}{r} \tag{5.8}$$

and, consequently, the radial and circumferential strains are equal and opposite (equations 5.2 and 5.3) and since there are no vertical strains there must be no volume change in this region (i.e. the deformation within the elastic phase takes place at constant volume). Note that both ε_r and ε_θ vary inversely with the square of the radius.

The boundary conditions around the cavity are well-defined and can be expressed as $y = 0$ for $r = \infty$ and $y = (a - a_0)$ for $r = a$. From the equilibrium equation and these given boundary conditions, the stress distribution around the cavity can be expressed as:

$$\Delta\sigma_r = \sigma_r - \sigma_{h0} = 2G\varepsilon_c \frac{a_0 a}{r^2} \tag{5.9}$$

$$\Delta\sigma_\theta = \sigma_\theta - \sigma_{h0} = -2G\varepsilon_c \frac{a_0 a}{r^2} \tag{5.10}$$

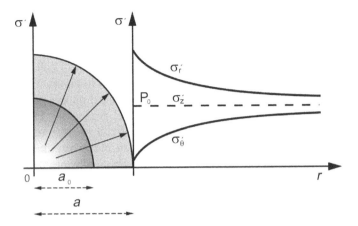

Figure 5.9 Stress around expanding cylindrical cavity in elastic soil.

where G is the shear modulus of the elastic material. Note that since radial and circumferential stresses are equal and opposite in respect of the in situ horizontal stress σ_{h0}, the mean stress level remains constant in the elastic region and no excess pore pressure is expected to develop during expansion (Figure 5.9).

At the cavity wall, $r = a$, and $\sigma_r = p$. As (a_0/a) is a small number, and therefore frequently omitted, equation (5.9) becomes:

$$p - \sigma_h = 2G\varepsilon_c \tag{5.11}$$

Thus, the in situ shear modulus of the soil can be determined simply by measuring displacement of the cavity wall as the cavity pressure increases above σ_{h0}:

$$G = \frac{1}{2}\frac{dp}{d\varepsilon_c} \tag{5.12}$$

The soil modulus can be also expressed in terms of volumetric strains as:

$$G = V_0 \frac{dp}{dV} \tag{5.13}$$

As pointed out by Mair and Wood (1987), the expansion of the cavity, which appears to be a compressive process, turns out to be an entirely shearing process. Properties deduced with reference to this analysis concern the shearing and not the compression of the surrounding soil.

Undrained analysis in elastic-plastic materials

Undrained analysis assumes no volume change in plane strain shearing with no strain in the direction parallel to the axis of the cylindrical cavity. All elements around the cavity, at all radii, are subjected to deformations which are similar in mode but of different magnitudes. The stress path of soil elements around an expanded cavity is schematically described in Figure 5.10, from which a cavity expansion solution can be deduced based on the stress-strain and strength behaviour of clays (e.g. Gibson and Anderson, 1961; Palmer, 1972). Point A, initially on the centre-line of the pressuremeter before installation, has been loaded elastically and then plastically during expansion. Elements between A and C have deformed plastically, point E is lying at the elastic-plastic boundary at this stage, while the material outside E is still elastic.

The exact solution for the shear stress τ at the cavity strain is (Baguelin et al., 1972; Ladanyi, 1972; Palmer, 1972):

$$\tau = \frac{1}{2}\varepsilon_c(1+\varepsilon_c)(2+\varepsilon_c)\frac{dp}{d\varepsilon_c} \tag{5.14}$$

implying that the shear strain in the wall of the cavity is approximately $2\varepsilon_c$. For small values of strains:

$$\tau \approx \varepsilon_c\frac{dp}{d\varepsilon_c} \tag{5.15}$$

Equation (5.15) can be conveniently written in terms of volumetric strains as:

$$\tau = \frac{dp}{d[\ln(\Delta V/V)]} \tag{5.16}$$

Usually referred to as Palmer's solution, these expressions allow for construction of the *sub-tangent* pressuremeter curve, given the fact that the shear stress is equal to the slope of the pressure–volumetric strain curve.

For an elastic-perfectly plastic soil, the value of τ is constant and equal to the undrained shear strength s_u (Figure 5.11). A solution for this condition was first presented by Gibson and Anderson (1961). During expansion, the soil responds as an elastic material until the onset of yielding at the wall of the cavity, which occurs when:

$$p = \sigma_{h0} + s_u \tag{5.17}$$

At this stage, the small elastic strain is:

$$\frac{\Delta V}{V} \approx \frac{\Delta V}{V_0} = \frac{s_u}{G} \tag{5.18}$$

As cavity pressure increases above the yield stress defined by equation (5.17), a plastic region develops around the cavity reaching the elastic-plastic boundary formed around the expanding probe. In this region the change in pressure can be obtained by integrating equation (5.16) with respect to $\ln(\Delta V/V)$ to give:

$$p = \sigma_{h0} + s_u \left[1 + \ln\left(\frac{G}{s_u}\right) + \ln\left(\frac{\Delta V}{V}\right) \right] \qquad (5.19)$$

In a plastic deforming material, the cavity pressure does not increase indefinitely and a limit pressure is gradually approached for $\Delta V/V = 1$. By substituting this limit value in equation (5.19) a limit pressure p_L is obtained:

$$p_L = \sigma_{h0} + s_u \left[1 + \ln\left(\frac{G}{s_u}\right) \right] \qquad (5.20)$$

This expression, proposed by Ménard (1957), demonstrates that the cavity limit pressure depends strongly on the undrained shear strength, as well as the shear stiffness of the soil.

Equation (5.19) applies between the limits of $s_u + \sigma_{h0} < p < p_L$, a region where the plastic response of the pressuremeter test can be most conveniently written as:

$$p = p_L + s_u \ln\left(\frac{\Delta V}{V}\right) \qquad (5.21)$$

This theoretical solution indicates that if the pressuremeter test results are plotted in terms of cavity pressure against the logarithm of the volume strain, the results of the plastic portion should lie on a straight line with a slope equal to the undrained shear strength of the soil s_u, as shown in Figure 5.11.

More recently, Jefferies (1988) and Houlsby and Withers (1988) independently extended the Gibson and Anderson method to include cavity unloading solutions. The solution derived by Jefferies (1988) was intended to be applied to SBPs and therefore small strain assumptions were adopted, whereas Houlsby and Withers (1988) is a large strain solution for interpretation of CPM tests.

Drained analysis in elastic-plastic materials

In a drained analysis, the soil is assumed to be an isotropic elastic-perfectly plastic material that behaves elastically and obeys Hooke's law until the onset of yielding, which is determined by the Mohr-Coulomb criterion. Since shear takes place under drained conditions with no excess pore pressures around the expanding cavity, volume changes will occur. Analysis of tests which fails to take the volume change into account is erroneous and cannot be used in the interpretation of experimental measured pressuremeter data.

Referring to Figure 5.12, as the pressure inside the cavity increases, the surrounding soil eventually yields and a plastic region will spread around the cavity. The size of the plastic region between points A to E can be determined,

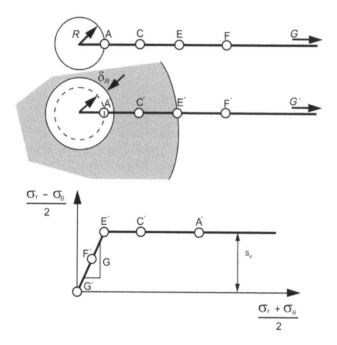

Figure 5.10 Undrained stress path around an expanding pressuremeter probe.

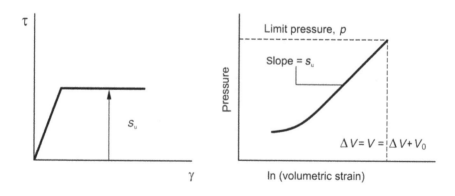

Figure 5.11 Undrained shear strength from pressuremeter test in clay.

with point E lying at the elastic-plastic boundary. Changes in volume within the sand are significant and several solutions have been developed incorporating the effects of shear dilation in drained expansion (Vesic, 1972; Hughes *et al.*, 1977; Manassero, 1989). The elastic deformation in the plastically deforming zone is small and is generally disregarded.

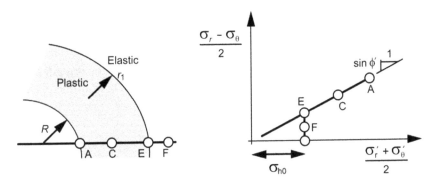

Figure 5.12 Drained stress path around an expanding pressuremeter probe.

Hughes *et al.* (1977) assumed that the rate of volume change is constant during cavity expansion and deduced the angles of soil friction and dilation from the pressuremeter loading test results. Assuming failure to be governed by a Mohr-Coulomb criterion:

$$\frac{\sigma'_r}{\sigma'_\theta} = \frac{1 - \sin\phi'}{1 + \sin\phi'} \tag{5.22}$$

Once a failure criterion is assumed, it is necessary to define the shape of the plastic potential, which is expressed in the form of a 'flow rule' relating directions of plastic strain increments to the current stress state. The use of Rowe's stress dilatancy law leads to the following expression:

$$\frac{1 + \sin\phi'}{1 - \sin\phi'} = \left(\frac{1 + \sin\psi}{1 - \sin\psi}\right)\left(\frac{1 + \sin\phi'_{cv}}{1 - \sin\phi'_{cv}}\right) \tag{5.23}$$

where ψ is the soil dilation angle and ϕ'_{cv} is the critical state friction angle. This follows the concept that when shearing medium dense and dense sands the material dilates, assigning to the soil a shear resistance given by the mobilized ϕ'_{cv} and mobilized angle of dilation ψ:

$$\sin\psi = -\frac{d\varepsilon_v}{d\gamma} \tag{5.24}$$

where $d\varepsilon_v$ is the change in volumetric strains. Principal stresses, volumetric strains and shear strains can now be combined as:

$$\frac{\sigma'_r}{\sigma'_\theta} = \left(\frac{1 + \sin\phi'_{cv}}{1 - \sin\phi'_{cv}}\right)\left(\frac{1 - d\varepsilon_v/d\gamma}{1 + dv/d\gamma}\right) \tag{5.25}$$

The onset of yielding at the wall of the cavity in drained expansion is:

$$p - u_0 = \sigma'_{h0}(1 + \sin \phi')$$ (5.26)

where u_0 is the in situ pore pressure and σ'_{h0} the horizontal in situ effective stress. The analytical solution for the cavity expansion curve in the plastic stage can then be approximated as (Hughes et al., 1977):

$$\ln(p - u_0) = S \ln \varepsilon_c + A$$ (5.27)

where A is a constant. Equation (5.27) indicates that pressuremeter results plotted as the effective cavity pressure versus the volumetric strains in logarithm scales should give a straight line with slope S expressed as:

$$S = \frac{(1 + \sin \psi) \sin \phi'}{1 + \sin \phi'}$$ (5.28)

The necessary link to obtain the values of ϕ' and ψ from the measured pressuremeter slope S and soil critical state friction angle is given by Rowe's stress dilatancy law in equation (5.23):

$$\sin \phi' = \frac{S}{1 + (S - 1) \sin \phi'_{cv}}$$ (5.29)

$$\sin \psi = S + (S - 1) \sin \phi'_{cv}$$ (5.30)

Pressuremeter tests in clay

The purpose of this section is to summarize key applications of cavity expansion theory to the interpretation of pressuremeter tests in clay, following the theoretical framework introduced in the preceding section. Pressuremeter tests are primarily used to estimate the shear modulus, in situ horizontal stress, undrained shear strength and coefficient of horizontal consolidation.

Shear modulus

One of the most common uses of the pressuremeter test is in the evaluation of soil moduli (Wroth, 1982). However, both the interpretation of soil moduli measured by the pressuremeter and the application of these moduli in engineering design are complex processes, as the moduli vary with both stress level and strain amplitude (e.g. Wroth et al., 1979; Jamiolkowski et al., 1985; Bellotti et al., 1986; Tatsuoka et al., 1997).

In the Ménard method an operational stiffness is obtained from the initial elastic segment of the pressure–expansion curve. This measured stiffness, which is obtained at relatively large shear strains and sensitive to disturbance produced by the installation technique, is often used as an input parameter in semi-empirical rules for shallow foundation design. The elastic region is defined within the limits determined by the point where the membrane comes into full contact with the sides of the borehole (Point P_0 in Figure 5.6) and the point marking the onset of plastic behaviour of the soil closest to the pressuremeter. Although these points are difficult to identify, the Ménard tangent modulus E_m is taken from the slope of this segment applying Lamé's solution:

$$E_m = 2(1+v)\left[V_0 + \left(\frac{V_B - V_A}{2}\right)\right]\frac{dp}{dV} \qquad (5.31)$$

where V_0 is the initial volume of the probe. In clay, plastic deformations begin in the early stages of the test and therefore E_m gives a low estimate of the in situ modulus.

In the previous discussion (under Analysis of pressuremeter tests), it is assumed that the pressuremeter is a long cylindrical cavity expanding radially under plane strain conditions in the axial direction. Therefore, if a pressuremeter is expanded in a linear isotropic elastic material, the shear modulus G is equal to half the slope of the pressure–expansion curve:

$$G = \frac{1}{2}\frac{dp}{d\varepsilon_c} \qquad (5.12)\text{idem}$$

This approach implies that any unloading of the expanding cavity brings the surrounding soil below the currently expanded yield surface into a zone where strains are small and, to a large extent, reversible, as indicated by the loading path represented by CDE in Figure 5.13. While carrying out these unload–reload loops, it is important to ensure that the loop remains in an elastic region. For an elastic-perfectly plastic Tresca soil, cavity expansion theory can be used to demonstrate that the magnitude of change in cavity pressure should not exceed (Wroth, 1982):

$$\Delta p_{max} = 2s_u \qquad (5.32)$$

where s_u is the undrained shear strength of the soil. Previous research with the SBP has shown that soil moduli measured from unload–reload loops, G_{ur}, are almost completely independent of the initial disturbance caused by the installation process (e.g. Hughes, 1982; Wroth, 1982).

Figure 5.14 shows a typical example of an unload–reload loop obtained from an SBP test measured by strain-gauged arms. The shear modulus varies with the strain range over which it is measured, due to the non-linear hysteretic response of the soil, even if reduction in pressure is not sufficient to cause failure in extension. Given the influence of non-linearity of a loop,

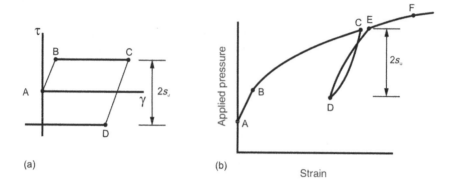

Figure 5.13 Range for elastic response in elastic-perfectly plastic clay: (a) shear stress path and (b) pressuremeter testing curve (Wroth, 1982).

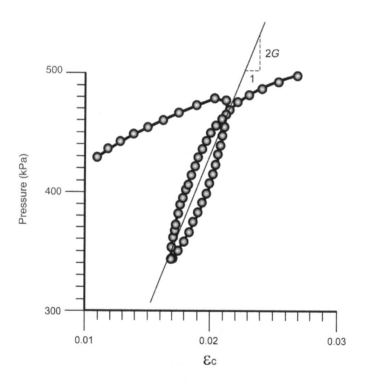

Figure 5.14 Expanded plot of a typical unload–reload loop calculated from the average readings of strain-gauged arms.

there is more than one procedure for calculating the average slope of measured data. It is usually recommended that a single line is drawn between the
two apexes of the loop or that the least squares fit for all points included
between these two apexes is calculated (Houlsby and Schnaid, 1992).

These interpretation methods allow soil stiffness to be assessed from the complete pressure–expansion curve or from interpretation of unload–reload loops.
Since the stress–strain relationships for soils are not linear, the challenge is to find
a means of characterizing and modelling non-linearity from pressuremeter data.
However, there remains the problem that the test does not measure 'element
stiffness' and interpretation of (non-linear) elemental stiffness characteristics
therefore relies on an appropriate numerical back-analysis method coupled with
a realistic soil constitutive model (e.g. Fahey and Carter, 1993; Bolton and
Whittle, 1999). The strains undergone by elements of soil at different distances
from the pressuremeter are inversely proportional to the square of the radius in
an undrained test and therefore a reference shear strain has to be arbitrarily
selected as representative of an unload–reload shear modulus, G_{ur}. This reference value is often taken as the strain applied at the pressuremeter surface.
Houlsby (1998) justifies the choice by demonstrating that the measured G_{ur} is
very strongly influenced by the stiffness of the soil close to the pressuremeter.

Muir Wood (1990) explores the variation of G_{ur} with shear strain amplitude in clays. In a pressuremeter it is possible to define a secant modulus
$G_s^p = (\psi - \sigma_{h0})/2\varepsilon$ and a tangent modulus $G_t^p = 1/2 \, d\psi/d\varepsilon$. This is similar to
laboratory tests in which it is also possible to define a secant $G_s = \tau/\gamma$, and
a tangent modulus $G_t = d\tau/d\gamma$, where τ is the shear stress. Following these
definitions it is straightforward to demonstrate that:

$$G_t = G_s + \gamma \frac{dG_s}{d\gamma} \tag{5.33}$$

$$G_t^p = G_s^p + 2\varepsilon \frac{dG_s^p}{2d\varepsilon} \tag{5.34}$$

and

$$G_s = G_t^p = G_s^p + 2\varepsilon \frac{dG_s^p}{2d\varepsilon} \tag{5.35}$$

Thus, in clay, according to Muir Wood (1990), the tangent modulus
measured from the pressuremeter curve is equal to the secant modulus
determined from a conventional laboratory test. This enables the results of
pressuremeter tests to be properly related to those of other tests by means of
degradation models designed to express the variation of G with γ. A range
of simple to fairly elaborate formulations have been proposed to express G
versus γ relationships in the form of a logarithm or a hyperbola. Regardless
of the model adopted in the analysis, the author would strongly advise that
combining seismic G_0 measurements and SBP tests incorporating multiple

unload–reload loops is currently the only accurate method of obtaining non-linear stiffness parameters from in situ tests (Schnaid, 2005).

In situ total horizontal stress

The SBP test appears to be the most appropriate technique for determining the in situ horizontal stress, as it offers the possibility of inserting a testing device into the ground with very little disturbance of the surrounding soil. We recall that when a pressuremeter is rapidly expanded in an elastic-perfectly plastic clay, the analysis of a cylindrical cavity takes as the reference state the size of the pressuremeter corresponding to zero cavity strain. At this initial state, the cavity pressure is equal to the horizontal stress, a condition that is rarely easily identifiable by a direct examination of data from pressuremeter tests due to the effects of drilling, inclination of the probe and system compliance. To overcome this uncertainty, several methods have been developed to assist in the selection of appropriate values of σ_{h0} (e.g. Briaud, 1992; Clarke, 1995).

The 'lift-off' approach is applicable to SBP tests where the insertion of the instrument into the soil does not substantially alter the geostatic stress prior to installation. At the initial loading stage, prior to lift-off, the membrane is unable to detach from the body of the probe. When the cavity pressure exceeds σ_{h0}, the cavity starts to expand and the movement of the membrane is recorded by the three feeler arms. This lift-off pressure corresponds to the in situ total horizontal stress of the soil. In practice, extreme care should be taken in order to obtain a reliable assessment of σ_{h0}. The example in Figure 5.15 offers a close examination of the initial stage of a test by plotting the p versus ε_c curve at a large scale. The output of each individual arm indicates lift-off at considerably different cavity pressures (in the range of 50 to 180 kPa) and so the choice of a unique representative value of σ_{h0} becomes subjective.

In a Ménard pressuremeter, the measurement that corresponds to the start of the linear elastic segment is an indication of the reference cavity pressure and one can argue that this reference value can be adopted to estimate σ_{h0}. Mair and Wood (1987) see no justification for this procedure, since the process of forming a borehole leaves the soil partially unsupported. An initial pressure measurement in a soil that has been stressed by unloading prior to an MPM test does not correspond to the undisturbed in situ stress state.

Marsland and Randolph (1977) proposed an alternative technique to estimate this initial reference pressure value, simultaneously based on the shear strength and yield stress of clays. During expansion, the yield stress at the wall of the cavity is equal to the sum of σ_{h0} and s_u and is identified by a pronounced change in slope on the test curve:

$$p = \sigma_{h0} + s_u$$

(5.17)idem

An iterative approach is then required to force consistency between the measured yield stress in a p versus ε_c curve and the sum of the initial estimate of horizontal stress and the undrained shear strength of the soil. The

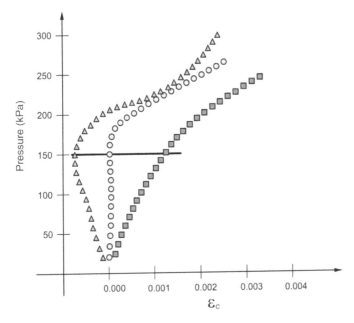

Figure 5.15 Close examination of the lift-off pressure in a p versus ε_c curve for determining σ_{h0}.

method only applies to clays which exhibit a significant elastic response and show a marked yield point in cavity expansion.

Undrained shear strength

There are several different methods for estimating strengths from pressuremeter tests in clay expressed in terms of total stress (Gibson and Anderson, 1961; Palmer, 1972; Houlsby and Withers, 1988; Jefferies, 1988) and effective stress analysis (Yu and Collins, 1998)[2] for both the loading and unloading segments of the pressuremeter curve.

For the loading analysis, the Gibson and Anderson method (1961) assumes the soil to behave as an elastic-perfectly plastic Tresca material. The analytical solution for the cavity expansion in the plastic stage of an infinitely long cylindrical cavity was derived above as:

$$p = \sigma_{h0} + s_u \left[1 + \ln\left(\frac{G}{s_u}\right) + \ln\left(\frac{\Delta V}{V}\right) \right] \qquad (5.19)\text{idem}$$

where p is the total pressuremeter pressure, σ_{h0} the in situ horizontal stress, s_u the undrained shear strength and G the shear modulus. The volumetric strain is:

$$\frac{\Delta V}{V} = \frac{a^2 - a_0^2}{a^2} \qquad (5.36)$$

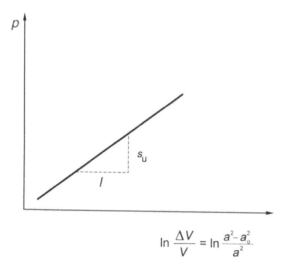

$$\ln \frac{\Delta V}{V} = \ln \frac{a^2 - a_0^2}{a^2}$$

Figure 5.16 Graphical representation of the loading analysis of pressuremeter tests in clay (after Gibson and Anderson, 1961).

Values a and a_0 are current and initial cavity radii. The analysis defined by equation (5.19) implies that results of an SBP test carried out in perfectly plastic clay should produce a straight line in a plot comparing cavity pressure to the logarithm of the volumetric strain (see Figure 5.16). In this space, the slope of the plastic portion is equal to the undrained shear strength of the soil. Figure 5.17 illustrates a result from an SBP test in clay, indicating a long straight plastic segment with the shear strength remaining approximately constant with continuing strain (Wroth, 1984).

For the unloading analysis, Jefferies (1988) represents the pressuremeter expansion by its maximum total cavity pressure p_{max}, followed by a slow contraction. Initially, the pressuremeter unloading curve is linear elastic and can be expressed accordingly:

$$p - p_{max} = 2G \left(\frac{a - a_{max}}{a_{max}} \right) \tag{5.37}$$

where a_{max} is the maximum displacement reached at the end of loading. The plastic unloading curve is then expressed as:

$$p - p_{max} = -2s_u \left[1 + \ln \left(\frac{G}{2s_u} \right) \right] - 2s_u \left(\frac{a_{max}}{a} - \frac{a}{a_{max}} \right) \tag{5.38}$$

assuming that the undrained strength on unloading is equal to that on loading. Regarding the Gibson and Anderson loading analysis method, the theoretical solution defined by equation (5.38) suggests that the pressuremeter

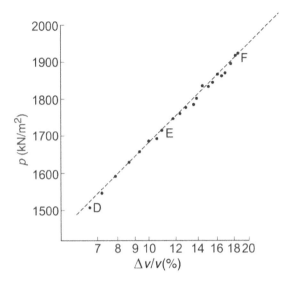

Figure 5.17 Interpretation of SBP test in clay (Wroth, 1984).

unloading results should be plotted in terms of cavity pressure p versus $-\ln(a_{max}/a - a/a_{max})$, as illustrated in Figure 5.18. In this space, the slope of the plastic unloading portion gives a straight line which is equal to twice the undrained shear strength of the soil $2s_u$.

A word of caution is necessary when predicting undrained shear strength values from SBP tests. There is enough evidence to suggest that strength values estimated from pressuremeter tests are higher than strength values estimated from other in situ and laboratory tests. This is partially due to end effects produced by the finite pressuremeter length, which introduces two-dimensional effects to an expanding cavity that have not been considered in analytical solutions. However, numerical studies carried out to quantify length to diameter ratio effects have produced results that do not completely eliminate the tendency to overestimate predicted s_u values (Yu, 1990; Houlsby and Carter, 1993; Charles *et al.*, 1999).

Coefficient of horizontal consolidation

Time-dependent parameters can be measured by pressuremeter probes instrumented with pressure transducers, allowing the changes in porewater pressure to be monitored while performing the so-called 'holding tests' (Clarke *et al.*, 1979). Conceptually, when a pressuremeter is expanded in clay under undrained conditions, the decay in excess pore pressure Δu generated by plastic expansion is measured, while the total pressure in the cavity or the deformation of the cavity is held constant. If the diameter of

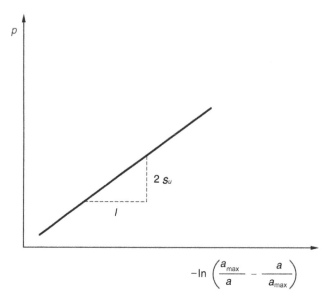

Figure 5.18 Graphical representation of the unloading analysis of pressuremeter tests in clay (after Jefferies, 1988).

the cavity is held constant, relaxation is observed in the decrease of the measured Δu, as well as the reduction in total cavity pressure. If the total pressure is held constant, relaxation is observed in the continuing decrease in measured Δu and the progressive increase in cavity diameter.

The increase in cavity pressure in undrained plastic expansion produces an increase in total cavity pressure that matches the increase in pore pressure in the wall of the cavity:

$$\Delta u = p - (\sigma_{h0} + s_u) \tag{5.39}$$

This excess pore pressure is linked to cavity volume strains by (Mair and Wood, 1987):

$$\Delta u = s_u \ln\left(\frac{G}{s_u}\right) + s_u \ln\left(\frac{\Delta V}{V}\right) \tag{5.40}$$

The horizontal coefficient of consolidation can be calculated from the measured decay in excess pore pressure using a dimensionless time factor, T_{50}:

$$T_{50} = \frac{C_h t_{50}}{a^2} \tag{5.41}$$

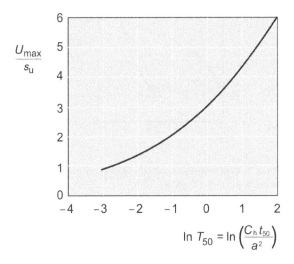

$$\ln T_{50} = \ln \left(\frac{C_h\, t_{50}}{a^2} \right)$$

Figure 5.19 Theoretical correlation between normalized maximum excess pore pressure and time factor (Randolph and Wroth, 1979).

where t_{50} is the time taken for the excess pore pressure to fall to half its maximum value. The relationship between dimensionless time factor T_{50} and normalized excess pore pressure $\Delta u_{max}/s_u$ is shown in Figure 5.19. The coefficient of horizontal consolidation C_h can be simply estimated from the combined values of t_{50} and $\Delta u_{max}/s_u$.

Pressuremeter tests in sand

Pressuremeter tests in sand can be used to estimate values of shear modulus, in situ horizontal stress and soil shear strength (angles of friction and dilation). Analytical solutions are based on small strain analysis, which suits the interpretation of SBP tests and assumes that expansion is carried out slowly under fully drained conditions.

Shear modulus

Pressuremeter results can be used to determine the initial modulus (typical of MPM – equation 5.31) and the unload–reload shear modulus (typical of SBP – equation 5.12). If a pressuremeter is expanded in a linear isotropic material, the relationship between the change in effective cavity pressure and the change in cavity strains is well-known; therefore, the shear modulus G can be determined from the slope of the pressure–expansion curve as:

$$G = \frac{1}{2}\frac{dp}{d\varepsilon_c}$$

(5.12)idem

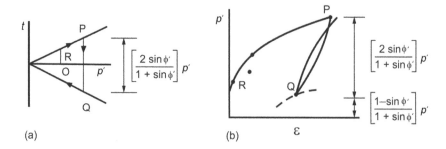

Figure 5.20 Range for elastic response in elastic-perfectly plastic sand: (a) shear stress path
and (b) pressuremeter testing curve (Wroth, 1982).

In a pressuremeter test in sand, the mean effective stress changes as
a function of both the applied cavity pressure and the distance from the
cavity wall. Since stiffness in sand depends on the stress level at which it is
measured, the modulus determined from successive unload–reload loops at
different load stages of a test might be expected to show a steady increase
in stiffness (for the same shear amplitude) as the test progresses.

As in the case of clay, it is important to ensure that any unloading of the
expanding cavity remains in the elastic region, where strains are small and,
to a large extent, reversible. Wroth (1982) suggested that the magnitude
of change in cavity effective stress during elastic unloading should not
exceed:

$$\Delta p_{max} = \frac{2 \sin \phi'}{1 + \sin \phi'} (p - u_0)_{max} \tag{5.42}$$

$(p - u_0)_{max}$ being the cavity effective stress at which unloading starts and ϕ' the
drained angle of friction (in plane strain). This limit state, represented by the
stress path PQ in Figure 5.20, indicates that the permissible extent of an unload
cycle depends on the stage of the tests. In other words, it is dependent on the
cavity effective stress at which unloading starts.

In situ total horizontal stress

Despite the various methods developed to predict the in situ state of stress,
it is still not possible to ensure that the value of the horizontal stress is
accurately determined in most natural sand deposits (see above under
Undrained analysis in elastic-plastic materials). Although the SBP appears
to be the most suitable technique for estimating σ_{h0}, numerous reports have

highlighted the uncertainties of evaluating the in situ horizontal stress in sand. Current experience is mainly from research efforts carried out in the 1980s (e.g. Windle, 1976; Fahey, 1980; Fahey and Randolph, 1984; Wroth, 1984). Problems associated with the self-boring technique lead to measured lift-off pressures which closely match the hydrostatic pore pressure (i.e. the lift-off approach tends to grossly underestimate σ_{h0} values).

Drained shear strength parameters

Earlier in this chapter the analytical solutions for small strain cavity expansion in isotropic elastic-perfectly plastic Mohr-Coulomb materials were introduced. These solutions enable the angles of shear resistance and dilation to be estimated.

Hughes et al. (1977) developed a method from which the cavity expansion curve in the plastic stage can be described by the following expression:

$$\ln(p - u_0) = S \ln \varepsilon_c + A \qquad\qquad (5.27)\text{idem}$$

where A is a constant and S is the slope of a graphical representation of the effective cavity pressure plotted against the volumetric strains in logarithm scales, expressed as:

$$S = \frac{(1 + \sin \psi) \sin \phi'}{1 + \sin \phi'} \qquad\qquad (5.28)\text{idem}$$

ϕ' being the friction angle and ψ the dilation angle. The necessary link to obtain the values of ϕ' and ψ from the measured pressuremeter slope S and soil critical state friction angle is given by Rowe's stress dilatancy law in equation (5.23):

$$\sin \phi' = \frac{S}{1 + (S - 1) \sin \phi'_{cv}} \qquad\qquad (5.29)\text{idem}$$

$$\sin \psi = S + (S - 1) \sin \phi'_{cv} \qquad\qquad (5.30)\text{idem}$$

This method is widely used in sand and requires the pressuremeter data to be plotted in terms of $\ln p$ versus $\ln \varepsilon_c$, as illustrated in Figure 5.21. In this plot, the slope of the plastic portion should produce a straight line that is equal to S, which can be used to determine the angles of friction and dilation from equations (5.29) and (5.30), provided that the critical friction angle is measured or estimated. Typical values of ϕ'_{cv} are shown in Table 5.1.

The accuracy of the predicted shear strength parameters using the method proposed by Hughes et al. depends on two aspects. First, the choice of the

Table 5.1 Typical values of ϕ'_{cv}

Material	Bolton (1979) ϕ'_{cv}	Robertson and Hughes (1986) ϕ'_{cv}
Dense, well-graded sand or gravel	35°	40°
Uniform medium-dense/coarse sand	32°	34–37°
Dense, sandy silt with some clay	32°	
Fine sand and sandy, silty clay	30°	30–34°
Clay-shale or partings	25°	
Clay (London)	15°	

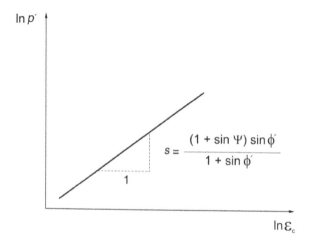

$$s = \frac{(1 + \sin \Psi) \sin \phi'}{1 + \sin \phi'}$$

Figure 5.21 Graphical representation of the loading analysis of pressuremeter tests in sand (after Hughes *et al.*, 1977).

correct reference strain is critical because of the extreme sensitivity of pressuremeter tests to small changes in cavity strain produced by installation disturbance (Fahey and Randolph, 1984). Second, the analysis assumes the relationship between volumetric and shear strains to be linear and elastic strains to be negligible. These assumptions are reasonable for dense sand; however, they are not necessarily valid for medium-dense and loose sands, producing low angles of shear resistance in these materials.

An alternative approach to interpreting pressuremeter loading tests has been presented by Manassero (1989) and Sousa Coutinho (1990). By assuming a plastic flow rule for sand, it is possible to deduce the complete pressuremeter stress–strain curve. No closed form solutions can be obtained and a finite difference method is required to determine the stress–strain relationship because of the effects of dilation. In the analysis, small strain is assumed, the elastic deformation in the plastic region is neglected and

Rowe's stress dilatancy relationship is adopted. Radial and circumferential strains are related by an unknown function f:

$$\varepsilon_r = f(\varepsilon_\theta) \qquad (5.43)$$

From strain compatibility and equilibrium, Manassero (1989) deduced a constitutive equation which links the changes in effective stress and development of strains as:

$$\frac{1}{\sigma_r}\frac{d\sigma_r}{d\varepsilon_\theta} = -\frac{1 + f'/K}{f - \varepsilon_\theta} \qquad (5.44)$$

where f' denotes the derivative of f with respect to ε_θ and K relates the ratio of the principal stresses to the ratio of the plastic strain increments. Unfortunately, equation (5.44) cannot be solved analytically and the finite difference method must be used to determine the relationship between ε_r and ε_θ.

Alternatively, Houlsby et al. (1986) developed a small strain analysis of the unloading section of the test using the same assumptions introduced by Hughes et al. (1977).[3] It is suggested that there are certain advantages in making use of the pressuremeter contraction data, as it is less sensitive to any initial disturbance caused by the installation of the probe. In this analysis it is necessary to account for the fact that unloading takes place from an initially non-homogeneous stress state, which has been set up by the preceding expansion phase. The approach is summarized in Figure 5.22, showing that the unloading slope S_d in a plot of $-\ln p$ against $\ln(\varepsilon_{max} - \varepsilon)$ is primarily controlled by the soil strength parameters and, to a small extent, by soil stiffness, where ε_{max} and ε denote the maximum and current cavity strain. The slope S_d can then be expressed as:

$$S_d = \frac{N - 1/N}{N + 1/n} \qquad (5.45)$$

where $N = (1 - \sin\phi')/(1 + \sin\phi')$ and $n = (1 - \sin\psi)/(1 + \sin\psi)$, in which ϕ' and ψ denote the plane strain friction angle and dilation angle, respectively. By making use of Rowe's stress dilatancy relationship (equation 5.23), it is possible to eliminate the dilation angle from equation (5.41), introducing instead the critical state friction angle ϕ'_{cv}. The result is:

$$\sin\phi' = \left(\sin\phi'_{cv} + \frac{1 + \sin\phi'_{cv}}{S_d}\right) - \sqrt{\left(\sin\phi'_{cv} + \frac{1 + \sin\phi'_{cv}}{S_d}\right)^2 - 1} \qquad (5.46)$$

Examination of pressuremeter data consists of plotting pressure contraction curves on logarithm scales to convert the slope of asymptotic lines to friction angles using equation (5.43) and comparing these values with data from other in situ tests or laboratory tests.

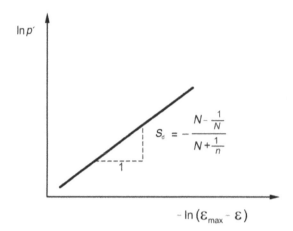

Figure 5.22 Graphical representation of the unloading analysis of pressuremeter tests in sand (after Houlsby et al., 1986).

State parameter

The state parameter introduced in Chapter 3 provides a simple and yet fundamental contribution to the characterization of granular geomaterials, by implicitly combining void ratio and compressibility. Attempts to estimate the state parameter from pressuremeter tests have been reported by Yu (1994, 1996). The approach is based on a critical state soil model with strain hardening and strain softening, for which there is no closed form solution. Finite element analysis, used to derive approximate relationships, can prove useful in practical geotechnical applications.

Numerical loading analysis demonstrates that the cavity expansion curve in a $\ln p$ against $\ln \varepsilon_c$ plot is approximately straight, having a slope S that is mainly controlled by the initial state parameter of the soil (Yu, 1994). This observation appears to be valid for different sands tested in large laboratory calibration chamber tests. Based on this information, the state parameter ψ was linearly related to S by the following correlation (Figure 5.23):

$$\Psi = 0.59 - 1.85S \qquad (5.47)$$

Soil friction angles can be directly estimated using an average correlation between the friction angle and the state parameter (Been *et al.*, 1987). This gives the following expression:

$$\phi'_{ps} = 0.6 + 107.8S \qquad (5.48)$$

where ϕ'_{ps} is the plane strain friction angle.

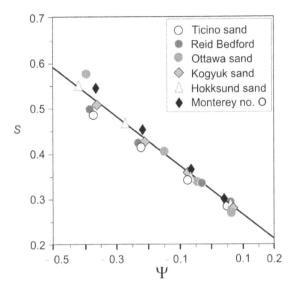

Figure 5.23 State parameter derived from the theoretical modelling of pressuremeter loading tests (after Yu, 1994).

Similarly, Yu (1996) developed a pressuremeter unloading analysis based on the same numerical approach described above. The work follows the experience gathered by analytical methods using the Mohr-Coulomb model, showing that the plastic unloading curve is straight in a plot of $\ln p$ against $\ln[(\varepsilon_c)_{max} - \varepsilon_c]$, from which a slope S_d is calculated. The average relationship between S_d and Ψ is shown in Figure 5.24 and can be approximated by the following linear relationship:

$$\Psi = -0.33S_d + 0.53 \tag{5.49}$$

which yields the following expression to derive the plane strain friction angle directly from the pressuremeter unloading slope (Figure 5.25):

$$\phi'_{ps} = 18.4S_d + 6.6 \tag{5.50}$$

Care should be taken when using the correlations given by equations (5.47) to (5.50) because they have not yet been systematically tested in practice.

Cone pressuremeter tests in clay and sand

The cone pressuremeter (CPM) is an in situ testing device that combines the 15 cm² cone with a pressuremeter module mounted behind the cone tip. As

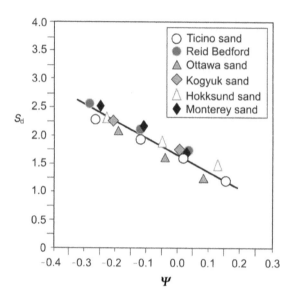

Figure 5.24 State parameter derived from the theoretical modelling of pressuremeter unloading tests (after Yu, 1996).

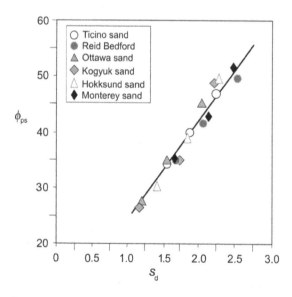

Figure 5.25 Soil friction angle derived from the theoretical modelling of pressuremeter unloading tests (after Yu, 1996).

the pressuremeter test is not carried out in undisturbed ground, the effects of installation have to be taken into account and large strain analysis is required. Since soil disturbance caused by installation is repeatable, the CPM test should be amenable to rational analysis.

Analysis in clay is achieved by a simple geometric construction of the pressure–expansion curve (Houlsby and Withers, 1988). Analysis in sand is significantly more complex and interpretation is based both on theory and on empirical methods derived from calibration chamber tests (Nutt and Houlsby, 1992; Schnaid and Houlsby, 1992; Yu et al., 1996). Research over the past 20 years has provided basic interpretation procedures from which engineering properties of soils may be assessed, including shear strength, relative density, state parameter, friction angle and in situ stress state.

Tests in clay

Interpretation of CPM tests in clay is based on the analysis developed by Houlsby and Withers (1988), from which undrained shear strength, shear modulus and in situ horizontal stress can be assessed simultaneously from the loading and unloading sections of the pressuremeter curve. The initial installation of the instrument into the soil is modelled theoretically as the expansion of a cylindrical cavity, whereas the expansion phase of the pressuremeter test is modelled as the continued expansion of the same cylindrical cavity, and the contraction phase as a cylindrical contraction. This is a large strain analysis that considers a cavity of zero initial diameter expanding to a finite size during installation, experiencing an additional 50% increase in radius during pressuremeter expansion.

The analysis is undrained with equations expressed in terms of total stresses, following the solutions developed above under the heading Analysis of pressuremeter tests. The cylindrical cavity limit pressure from the loading analysis is (Gibson and Anderson, 1961):

$$p_L = \sigma_{h0} + s_u\left[1 + \ln\left(\frac{G}{s_u}\right)\right] \qquad (5.20)\text{idem}$$

During the initial phase of cavity contraction, the whole of the soil behaves elastically before yielding in reverse plasticity when $\sigma_r - \sigma_\theta = -2s_u$ at the surface of the pressuremeter. The slope of the elastic section of the stress–strain curve is $2G$ calculated from $(p_L, \varepsilon_{cmax})$ and $(p_L, -2s_u)$, with ε_{cmax} being the maximum cavity strain at the start of the unloading phase.

After further unloading, a zone of plastic soil spreads outwards from the probe with the unloading pressure–displacement curve in contraction being defined as:

$$p = p_L - 2s_u\left[1 + \ln(\varepsilon_{cmax} - \varepsilon_c) - \ln\left(\frac{s_u}{G}\right)\right] \qquad (5.51)$$

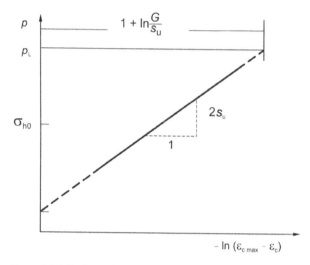

Figure 5.26 Analysis of cone pressuremeter test in clay (Houlsby and Withers, 1988).

Soil properties derived from this set of equations can be readily obtained by a simple geometrical construction based on the pressuremeter unloading curve, as represented in Figure 5.26 in a logarithm plot relating p to $-\ln(\varepsilon_{cmax} - \varepsilon_c)$. The expansion limit pressure can be readily obtained from this plot. The slope of the unloading curve is equal to twice the undrained shear strength of the soil. The abscissa of the intersection of the limit pressure and the plastic unloading line (point N) is used to determine the G/s_u ratio (and hence G) and the value of σ_{h0} is given, in theory, by the midpoint of segment KM.

Caution should be exercised when using this interpretation method because it has not yet been used extensively in practice. The horizontal stress implied in the analysis does not seem to give realist values. The shear modulus obtained from the G/s_u ratio does not refer to a particular shear strain amplitude and, therefore, does not eliminate the need to determine the shear modulus from unload–reload loops.

Tests in sand

Analytical and numerical methods of interpretation of CPM tests in sand proved to be significantly more complex than the equivalent analysis in clay. This is principally because of the difficulties involved in accounting for large strain analysis in frictional, dilative materials.

Under this condition, empirically based interpretation methods from calibration chamber tests give the appropriate option to calibrate and evaluate the CPM test in cohesionless soils. Experimental results of calibration chamber testing of the Fugro cone pressuremeter revealed that the ratio of cone tip resistance q_c to the pressuremeter limit pressure p_L correlates well with many soil properties, such as relative density and friction angle. It is worth noting that both cone resistance and pressuremeter limit pressure are dependent on the size of the calibration chamber used (Schnaid and Houlsby, 1991). The ratio of these two quantities is relatively unaffected by chamber size and, therefore, correlations established in the laboratory may be applied directly to field conditions. Approximate empirical expressions for relative density, D_r (Nutt and Houlsby, 1992; Schnaid and Houlsby, 1992), expressed as a percentage, are:

$$D_r = 9.6 \frac{q_c - \sigma_{h0}}{p_L - \sigma_{h0}} - 30 \tag{5.52}$$

Once the relative density is known, the soil friction angle can be estimated using Bolton's formula (1986). A relationship between $(q_c - \sigma_{h0})/(p_L - \sigma_{h0})$ ratio and triaxial peak friction angle for Leighton Buzzard sand gives (Schnaid and Houlsby, 1994):

$$\phi'_p = 1.45 \frac{q_c - \sigma_{h0}}{p_L - \sigma_{h0}} + 26.5 \tag{5.53}$$

Due to the limited practical experience available of using the CPM testing device, equation (5.53) is represented by a linear equation in natural scale.

Bearing in mind the fact that the cone resistance depends on the combined effects of horizontal stress and relative density, it is possible to make further use of cone data to obtain approximate correlations such as:

$$D_r = \frac{1}{3} \frac{q_c - \sigma_{h0}}{\sigma'_{h0}} + 10 \tag{5.54}$$

Equations (5.51) and (5.54) can be solved simultaneously to give a quadratic form of the horizontal stress σ'_{h0} as a function of q_c and p_L (Schnaid, 1990; Schnaid and Houlsby, 1994):

$$a(\sigma'_{h0})^2 + (bp_L + cq_c)\sigma'_{h0} + dp_L q_c = 0 \tag{5.55}$$

with $a = 92.0$, $b = -119.0$, $c = 28.0$ and $d = -1.0$. Although the potential application of this set of correlations has not yet been consistently validated by field tests, the good agreement between observed and measured values of relative density (Figure 5.27) and horizontal stress (Figure 5.28) strengthens the usefulness of the $(q_c - \sigma_{h0})/(p_L - \sigma_{h0})$ ratio in estimating soil constitutive parameters from CPM results.

Theoretical developments in interpretation of pressuremeter tests in sand have been supported by cavity expansion theory. Theoretical and experimental studies in sand clearly indicate that the pressuremeter limit pressure p_L

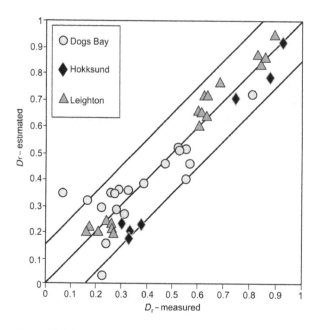

Figure 5.27 Estimates of relative density from CPM tests (after Houlsby, 1998).

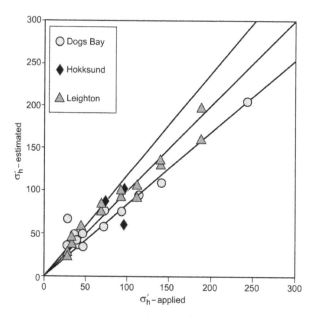

Figure 5.28 Estimates of horizontal stress from CPM tests (after Houlsby, 1998).

obtained from the CPM can be accurately predicted by the cylindrical cavity limit pressure σ_c (Schnaid, 1990; Schnaid and Houlsby, 1991, 1992):

$$p_L = \sigma_c \tag{5.56}$$

Previously, Ladanyi and Johnston (1974) proposed an analytical solution for the expansion of the spherical cavity σ_s, applied to determine cone resistance in sand:

$$q_c = \sigma_s \left(1 + \sqrt{3} \tan \phi'_{ps}\right) \tag{5.57}$$

where ϕ'_{ps} denotes the plane strain friction angle. By combining equations (5.53) and (5.54) the ratio of cone resistance to limit pressure can be expressed as:

$$\frac{q_c}{p_L} = \left(1 + \sqrt{3} \tan \phi'_{ps}\right) \frac{\sigma_s}{\sigma_c} \tag{5.58}$$

Using the Mohr-Coulomb yield criterion with a non-associated flow rule, Yu and Houlsby (1991) evaluated the variations of q_c/p_L with friction angle and shear modulus, from which the following expression is derived:

$$\phi'_{ps} = \frac{14.7}{\ln(G/p_0)} \frac{q_c}{p_L} + 22.7 \tag{5.59}$$

where ϕ'_{ps} is the (plane strain) friction angle (in degrees), G the shear modulus and p_0 the initial mean effective stress. Equation (5.59) suggests that the ratio of q_c/p_L is not only sensitive to the soil shear strength but also to shear stiffness. This can be conveniently shown in the $(q_c/p_L)/\ln(G/p_0)$ versus ϕ'_{ps} plot presented in Figure 5.29, in which these quantities can be correlated by a straight line for stiffness ratios G/p_0 ranging from 200 to 1,500.

An important contribution to the theoretical development of CPM tests occurred when Yu et al. (1996) published a theoretically based method for deriving soil properties in sand. The proposed analysis was supported by the experimental laboratory results reported by Schnaid (1990). In the theoretical development, the authors have assumed that both the cone resistance q_c and the pressuremeter limit pressure p_L are closely related to the limit pressure of spherical $(\sigma'_{s,lim})$ and cylindrical $(\sigma'_{c,lim})$ cavities respectively. The ratio of these two quantities can be estimated by the following equation:

$$\frac{\sigma'_{s,lim}}{\sigma'_{c,lim}} = C_1 (p'_0)^{C_2 + C_3(1+e_0)} \exp[C_4(1+e_0)] \tag{5.60}$$

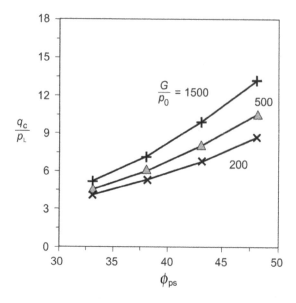

Figure 5.29 Theoretical correlation between the ratio of cone resistance to pressuremeter limit pressure with friction angle (Yu and Houlsby, 1991).

where C_1, C_2, C_3, C_4 are constants calibrated against six reference sands.
Solutions for cavity expansion in an elastic-perfectly plastic Mohr-Coulomb
soil have been used to correlate the ratio of q_c/p_L to the peak friction angle
of the soil. In addition, the limit pressure solutions for cavity expansion in
a strain hardening/softening soil using a state parameter-based soil model
are used to correlate q_c/p_L with the in situ sand state parameter. This
approach acknowledges the theory that, prior to the achievement of the
critical state, the behaviour of granular materials is largely controlled by the
state parameter Ψ, and that Ψ can be directly correlated with triaxial friction angles (Been and Jefferies, 1985). Figure 5.30 demonstrates that the
ratio q_c/p_L is mainly dependent on the initial state parameter of the soil and
that it can be conveniently expressed as:

$$\Psi = 0.46 - 0.3 \ln \frac{q_c}{p_L} \tag{5.61}$$

Although further verification of the proposed interpretation method from
field tests is still needed to bolster confidence in practical engineering applications, results shown in Figure 5.30 from Ghionna *et al.* (1995) and Wride
et al. (2000) support the method's validity.

   With regard to the self-boring pressuremeter, the evaluation of the shear
modulus can be estimated from unload–reload loops, provided that the
stress and strain levels associated with each individual loop are taken into

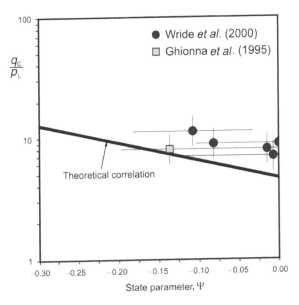

Figure 5.30 Theoretical ratio of cone resistance to pressuremeter limit pressure and *in situ* state parameter (after Yu *et al.*, 1996).

account (Schnaid, 1990; Schnaid and Houlsby, 1994). A method of calculating the mean stress level at the pressuremeter–soil interface was outlined by Houlsby and Schnaid (1994), in which a fully associated flow rule is used to compute the intermediate principal stress. This takes into account the fact that pressure levels around the cone pressuremeter are generally much higher than those surrounding the self-boring probe.

Some push-in devices make use of volume change measurements to compute the change in probe diameter during tests. This does not impose a severe restriction on the measured shear modulus from unload–reload loops, since good comparisons have been observed between the shear moduli calculated from strain arm measurement and volume change measurement (Houlsby and Schnaid, 1994). Careful calibrations are mandatory to obtain accurate and comparable values.

Pressuremeter tests in cohesive-frictional soils

Penetration tests, such as SPT and CPT, are extensively used in transported soils and also commonly adopted in the investigation of structured soils. Since interpretation in cemented geomaterials is imprecise, because of the difficulties involved in representing their behaviour using simple models with few constitutive parameters, empirical correlations for assessing soil properties are often the preferred option as they can be adapted according to specific site conditions to meet standards that reflect regional engineering practice (see Chapters 2 and 3). However, these empirical (or semi-empirical) methods do not provide satisfactory results in cohesive-frictional geomaterials, for the simple reason that two strength parameters (ϕ', c') cannot be determined from a single measurement (q_c or N_{60}).

The pressuremeter test emerges as the best option for checking the accuracy of a given set of design parameters in bonded soils. All the theories for the interpretation of pressuremeter tests in bonded soils make use of the in situ horizontal stress, soil stiffness and strength parameters: angle of internal friction, angle of dilation and cohesion intercept (which reduces with destructuration at high shear strains). The pressure expansion curve therefore represents a combination of all these parameters which cannot be assessed independently. Given this complexity, instead of attempting to derive a set of parameters from a single test in bonded soils, the pressuremeter should be viewed as a 'trial' boundary value problem, against which a theoretical pressure–expansion curve, predicted using a set of independently measured parameters, is compared to field pressuremeter tests. A good comparison between the measured and predicted curves provides reassurance in the process of selecting design parameters, whereas a poor comparison indicates that one or more of the constitutive parameters must be unrealistic (e.g. Schnaid, 2005).

Few methods have been developed for the interpretation of pressuremeter tests in cohesive-frictional materials, considering the effects of both internal frictional angle and cohesion intercept (Carter et al., 1986; Yu and Houlsby, 1991; Haberfield, 1997; Mantaras and Schnaid, 2002; Schnaid and Mantaras, 2003). Since none of the early work takes into account the effects of interparticle bonds on the rate of dilation, the solution conceived by Mantaras and Schnaid (2002) and Schnaid and Mantaras (2003) provides the most accurate means of describing the cavity expansion process in cohesive-frictional materials. This is because the authors include in their solution the concept that plastic dilatancy is inhibited by the presence of soil bonding.

A concept introduced by Rowe (1963), that plastic dilatancy is inhibited by the presence of soil bonding, was investigated by Mantaras and Schnaid (2002) and was used to describe the plastic components of the tangential and radial increments. The soil model and the stress and strain definitions used are essentially the same as those described by Yu and Houlsby (1991). For materials with no cohesion, the given equations are simplified to those adopted for describing a frictional material and the proposed solution approaches the Yu and Houlsby pressure–expansion relation.[4]

In the elastic-plastic analysis in a cohesive-frictional material, the shear stress is represented by two parameters – the internal friction angle of the material ϕ' and the cohesion intercept c', interpreted from the classical Mohr-Coulomb envelope. As the applied stress increases, initial yielding will eventually start at the cavity wall when the shear stress reaches the failure envelope:

$$\sigma_r = \frac{2c'\cos\phi'}{1-\sin\phi'} + \sigma_\theta \frac{1+\sin'}{1-\sin\phi'} \tag{5.62}$$

The non-associated flow rule can be expressed as $\dot\varepsilon_\theta^p/\dot\varepsilon_r^p = -\beta$, where $\dot\varepsilon_\theta^p$ and $\dot\varepsilon_r^p$ are the plastic components of the tangential and radial strain increments. The plastic strain increments are calculated by subtracting the elastic component from the total strain, leading to:

$$\beta\dot\varepsilon_r + \dot\varepsilon_\theta = \beta\dot\varepsilon_r^e + \dot\varepsilon_\theta^e \tag{5.63}$$

Substituting the values of the elastic strain components defined by Hooke's law leads to:

$$\beta\dot\varepsilon_r + \dot\varepsilon_\theta = \beta\frac{1-v^2}{E}\left[\Delta\sigma_r - \frac{v}{1-v}\Delta\sigma_\theta\right] + \frac{1-v^2}{E}\left[\Delta\sigma_\theta - \frac{v}{1-v}\Delta\sigma_r\right] \tag{5.64}$$

where v is Poisson's ratio. The general definition proposed by Rowe's law, based on the hypothesis of minimum absolute energy increment during shear, can be adapted not only by considering both the cohesive and frictional components, but also by allowing for degradation of cohesion during shear:

$$\beta = \frac{\sigma_r'}{\left[\tan\left(\frac{\pi}{4}+\frac{\phi_{cv}'}{2}\right)\right]^2\sigma_\theta + \frac{2c_0'}{[1+\varepsilon_r-\varepsilon_\theta-\gamma_p]^n}\tan\left(\frac{\pi}{4}+\frac{\phi_{cv}'}{2}\right)} \tag{5.65}$$

where ε_r is the radial strain, ε_θ the circumferential strain, γ_p the shear strain corresponding to the onset of yielding and c_0' the Mohr-Coulomb triaxial peak cohesion intercept. Note that the reduction in interparticle cohesion is expressed simply as a hyperbola, described as a function of shear strains and asymptotic to zero at large strains, so that:

$$c' = \frac{c_0'}{\left(1+\gamma-\gamma_p\right)^n} \tag{5.66}$$

n being the degradation index, calibrated using results from conventional laboratory tests, and γ_p the peak shear strain calculated using Hooke's law. Adopting equation (5.65) to describe the flow rule can eliminate one of the main limitations on classical cavity expansion formulations. There is no need to select a single constant value for the cohesion intercept to be representative of the complex stress variation with radii of material points within the soil at different loading stages of the test. A peak c_0' value is selected for the elastic region and in the plastic domain the analysis can keep careful track of the reduction in shear strength produced by increasing shear strain amplitudes around the cavity.

The solution for this set of equations is formulated within the framework of non-associated plasticity in which the Euler method is applied to solve two deferential equations simultaneously, which leads to the continuous variations of strains, stresses and volume changes produced by cavity expansion. It is not possible to establish beforehand the constitutive parameters at the cavity wall, as they have to be expressed as a function of stresses and shear strains around the expanding cavity. On the other hand, at the elastic-plastic interface, all variables are properly defined, which enables radial and circumferential stresses and strains to be calculated. Assuming an arbitrary position of the elastic-plastic interface b, the plastic region can be divided incrementally (Δr). Starting at $r = b$ it is possible to calculate β, then the deformation field around the plastic region and finally the stress distribution within the plastic region. The drawback is that the analysis is rather complex in mathematical terms and, given the fact that the set of equations cannot be integrated analytically, an iterative convergent numerical procedure is necessary to determine the complete pressure–expansion curve.

Figures 5.31 and 5.32 illustrate features of the behaviour described by the model by representing the ratios of $c'/c'_{(b)}$ and $\beta/\beta_{(b)}$ plotted against cavity radius, with $c'_{(b)}$ and $\beta_{(b)}$ being the cohesion intercept and dilation at the elastic-plastic interface, respectively. The variation in the cohesion ratio $c'/c'_{(b)}$ is presented in Figure 5.31, in which it is possible to observe the overall distribution of cohesion around the cavity with different values of the degradation index for a dilatant material expanded to one and a half times its initial diameter. Material points that lie at the elastic-plastic boundary do not exhibit any structural degradation and therefore cohesion values correspond to maximum peak values, so that the ratio $c'/c'_{(b)}$ is equal to unity. After expansion, the material between the cavity wall and the elastic-plastic boundary has been deformed plastically, with the shear strain amplitude reducing with increasing distance from the cavity wall. The ratio $c'/c'_{(b)}$ will therefore reduce with increasing γ and n values, this reduction being sensitive to the selected degradation index. Figure 5.32 shows the change in $\beta/\beta_{(b)}$ with radius for the same material and boundary conditions represented previously in Figure 5.31. An intact material is represented at $r = b$, which corresponds to $\beta/\beta_{(b)} = 1$. The ratio of $\beta/\beta_{(b)}$ gradually increases with reducing r values, the rate of increment being a function of the degradation index n.

Schnaid and Mantaras (2003) present results from a numerical analysis to illustrate the influence of structure degradation on a material that is at critical state ($\phi' = \phi'_{cv}$). The cohesion intercept is initially taken as 50 kPa, which is representative of a structured soil. Pressure expansion relationships are plotted in Figure 5.33 for a range of different n indexes. Results from the solutions proposed by Yu and Houlsby (1991) and Mantaras and Schnaid (2002) were also plotted in these figures to give a reference benchmark for cases in

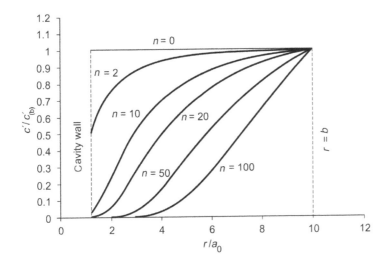

Figure 5.31 Variation in normalized cohesion within the plastic region (Schnaid and Mantaras, 2003).

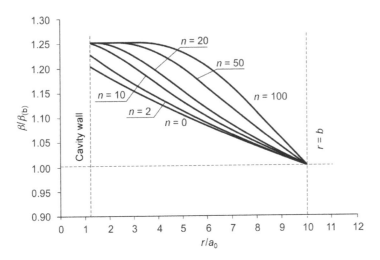

Figure 5.32 Normalized dilatation $\beta/\beta_{(b)}$ within the plastic region around an expanding cavity (Schnaid and Mantaras, 2003).

which strength parameters are considered as constants. The enlarged plot of initial expansion is presented in the figure to demonstrate the patterns of deformation for small strains. From the observed results, it is clear that limit pressure is strongly affected by the degradation of cohesion.

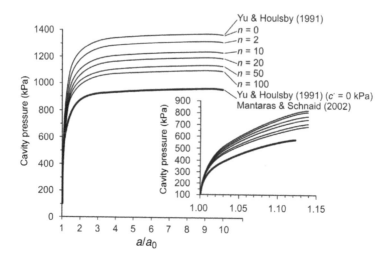

Figure 5.33 Typical pressure expansion results for different *n* values with $\phi' = \phi'_{cv}$ (Schnaid and Mantaras, 2003).

Case studies in the Hong Kong gneiss (Schnaid *et al.*, 2000), Porto granite (Mantaras, 2000) and São Paulo gneiss (Schnaid and Mantaras, 2003) have validated the application of the methodology described above. The trial fitting technique was used in all reported cases based on the recognition that strength, stiffness and in situ stresses interact to produce a particular pressuremeter expansion curve. The applicability of the proposed approach is illustrated in Figure 5.34, which shows a pressuremeter pressure–expansion curve taken as part of an extensive site investigation programme comprising laboratory triaxial tests, SPT and high-quality pressuremeter tests in residual gneiss. Interpretation of SBP data yielded ϕ'_{ps} from 27° to 31°, with considerable data scatter but within the range measured from laboratory testing data. The trial fitting applied to the loading portion of the SBP tests gave fairly consistent results, slightly above the assumed critical state values and compatible with laboratory data.

Pressuremeter tests in unsaturated soil conditions

In the interpretation of in situ tests, it is necessary to recognize that various geomaterials, such as hard soils and soft rocks, may not be saturated. In this case, the role of matrix suction and its effect on soil permeability has to be acknowledged and accounted for because it imparts a distinctive behaviour to a soil.

An important contribution to the analysis of unsaturated soils has been the extension of the elastic-plastic critical state concepts to unsaturated soil

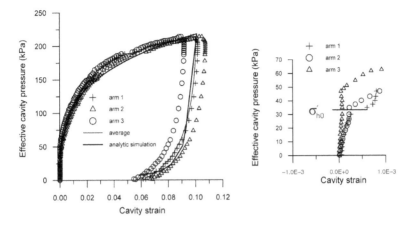

Figure 5.34 Comparisons between measured and predicted pressure–expansion response of a typical pressuremeter test carried out in the saprolite gneiss residual soil (data from Pinto and Abramento, 1997).

conditions by Alonso *et al.* (1990). In this method, the frame of reference is described by four variables: net mean stress $(p - u_a)$, deviator stress q, suction $s(u_a - u_w)$ and specific volume v, where u_a is the air pressure and u_w the porewater pressure. Several constitutive models have subsequently been proposed following these same concepts (e.g. Josa *et al.*, 1992; Wheeler and Sivakumar, 1995). These constitutive models allow derivation of the yield locus in the (p, q, s) space, an analysis that requires nine soil parameters. Model parameters are assessed from laboratory suction-controlled testing, such as isotropic compression tests and drained shear strength tests.

Isotropic conditions in unsaturated soils are characterized by a loading–collapse (LC) yield curve whose hardening laws are controlled by the total plastic volumetric deformation. The yield curve is expressed as a negative exponential function that obeys the irreversibility of Prager, as proposed by Balmaceda (1991):

$$p_0 = \left(p_0^* - p_c\right) + p_c[(1 - m)e^{-\alpha s} + m] \qquad (5.67)$$

where p_0 is the preconsolidation stress at a given suction stress, p_0^* the preconsolidation stress under soaked conditions, p_c a reference stress (smaller than p_0^*, but close to that value), α a non-dimensional shape factor and m a non-dimensional parameter, larger than unity, which controls the collapse evolution when p_0 increases.

A third state parameter has to be incorporated to take into account the effect of the shear stress q. The yield curve for a sample at a constant suction s is described by an ellipse, in which the isotropic preconsolidation stress is given by the previously defined p_0 value that lies on the loading–

collapse yield curve (Figure 5.35). The critical state line (CSL) for non-zero suction is assumed to result from an increase in cohesion, maintaining the slope M of the CSL for saturated conditions. If the increase in cohesion follows a linear relationship with suction, the ellipse intersects the p axis at a point for which $p = -p_s = -k_s$, where k is a constant. The ellipse can then be expressed as (Alonso et al., 1990):

$$q^2 = M^2(p + p_s)(p_0 - p) = 0 \qquad (5.68)$$

which is the yield function in a modified Cam clay critical state model (e.g. Muir Wood, 1990).

By combining critical state concepts with cavity expansion theory, Schnaid et al. (2004) developed the first interpretation method for deriving properties from pressuremeter penetration tests in unsaturated soils. The yield radial stress p represents the initial yielding at the cavity wall in the form (modified from equation 5.26):

$$p = \sigma_{h0}(1 + \sin \phi') + c \cos \phi' \qquad (5.69)$$

where σ_{h0} is the in situ horizontal stress, ϕ' the effective frictional angle of the soil and c the total cohesion. Unsaturated soil conditions require an expression that relates suction and cohesion:

$$c = c' + \frac{u_a - u_w}{a + b(u_a - u_w)} \qquad (5.70)$$

where c' is the effective cohesion and a and b are best-fit coefficients.

The mean stress is expressed in terms of octahedral stresses, which in cylindrical coordinates are defined by the two stress invariants, expressed in terms of the coefficient of earth pressure at rest, $K_0 = (\sigma_{h0} - u_{a0}) \sigma_v$:

$$p = \frac{\sigma_{h0}}{3}\left(2 + \frac{1}{K_0}\right) - u_a \qquad (5.71)$$

$$q = \sigma_{h0}\left[\frac{(4K_0^2 - 2K_0 + 1)}{K_0^2} - 3 \cos \phi'_p \left(\frac{P_0^2 \cos \phi'_p - 2P_0 c \sin \phi'_p - c^2 \cos \phi'_p}{P_0^2} \right) \right]^{1/2} \qquad (5.72)$$

where ϕ'_{ps} is the peak frictional angle of the soil. It is then possible to calculate the yield pressure at the isotropic stress state to each given suction level:

$$p_0 = \frac{q^2}{M^2(p + p_s)} + p \qquad (5.73)$$

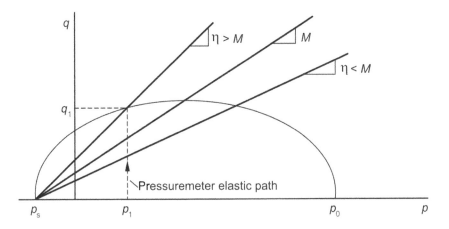

Figure 5.35 Stress state in p–q space for cylindrical expansion stress path.

where p_s and M are defined as:

$$-p_s = -c \cos \phi'_{cs} \tag{5.74}$$

$$M = \frac{q}{p} = \left[\frac{(4K_0^2 - 2K_0 + 1)}{K_0^2} - 3 \cos^2 \phi'_{cs} \right]^{1/2} \frac{3K_0}{2K_0 + 1} \tag{5.75}$$

ϕ'_{cs} being the critical state frictional angle of the soil. Note that for $K_0 = 1$, equation (5.75) is reduced to $M = \sqrt{3} \sin \phi'_{cs}$.

The assumption is made that the in situ suction remains constant during the expansion phase of a pressuremeter test, since the magnitude of shear strains is significant only at a very small radial distance from the pressuremeter cavity wall (Houlsby, 1998) and there are no important changes in the degree of saturation within the region affected by the cavity expansion. This assumption allows the same stress–strain relationships to be applied for drained analysis in saturated soils as in unsaturated soils, implying that unsaturated soils at constant suction behave qualitatively in the same way as soils in a saturated state. Based on this framework, the interpretation of pressuremeter tests can be used to assess the constitutive parameters that are necessary to describe the three-dimensional yield surfaces in a (p, q, s) space in unsaturated soils, as represented in Figure 5.36.

The recognition that matric suction produces an additional component of effective stress suggests the need to link the magnitude of in situ suction to the observed response of field tests. This led to the development of suction-monitored pressuremeter tests (SMPM), in which the in situ suction is monitored throughout the test by tensiometers positioned close to the pressuremeter probe. In these circumstances the standard self-boring technique cannot be applied to unsaturated soil conditions. The drilling technique,

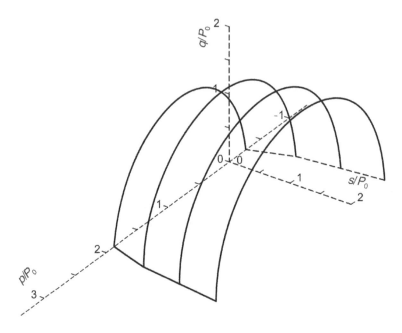

Figure 5.36 Three-dimensional view of yield surfaces in (p/P_0, q/P_0, s/P_0) space.

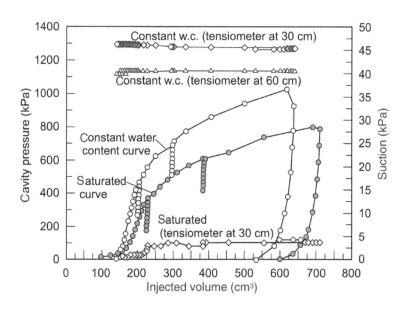

Figure 5.37 Typical SMPM tests (including suction measurements).

which uses either a flushing fluid or compressed air, would produce changes in the porewater pressure u_w or in the pore air pressure u_a, affecting the in situ soil suction $(u_a - u_w)$ in the vicinity of the pressuremeter probe. The pre-bored technique appears to be a viable option, despite its limitations.

Examples of typical pressuremeter curves in unsaturated residual granite soils are illustrated in Figure 5.37 (Schnaid et al., 2004). The first test was performed at an in situ suction of 40 kPa. After soaking the area, another test was carried out, producing a marked reduction in both pressuremeter initial stiffness response and cavity limit pressure. A straightforward assumption is that stiffness degradation with shear strain is likely to be shaped by changes in matric suction. The response of tensiometers installed at the same depth as the pressuremeter tests, at 30 and 60 cm from the centre of the SMPM borehole, is also shown in the figure. Suction measurements remained approximately constant throughout the expansion phase in the tests carried out in both soaked and unsaturated soil conditions. Clearly, the data suggest that shear strains induced by loading do not produce significant changes in matric suction, and this enables cavity expansion theory to be extended to accommodate the framework of unsaturated soil behaviour in the interpretation of pressuremeter tests (Gallipoli et al., 2001; Schnaid et al., 2004).

Design rules

In addition to cavity expansion theory, there is a set of design rules (known as the 'Ménard rules') that can support interpretation of Ménard pressuremeter tests, in which the results of tests are directly used to predict the performance of footings, piles, retaining walls, pavements and other soil–structure interaction problems. These are empirical correlations, supported by theory, which attempt to link the two Ménard pressuremeter parameters $(p_{LM}$ and $F_m)$ to the behaviour of full-scale structures.

The pressuremeter parameters are a function of the probe, the method of installation, the method of testing and the method of interpretation. In using the Ménard rules it is therefore essential to obtain these parameters according to standard procedures. The modulus E_m should be calculated according to equation (5.31). The limit pressure is defined as the pressure reached when the initial cavity has been inflated to twice its initial volume $(\Delta V/V_0 \approx 1)$. Since this degree of variation in volume is not often reached, some mathematical extrapolation is necessary (based on hyperbolic models). If various methods are used to define p_{LM}, the mean value should be selected as representative.

The ratio of these two quantities can serve as an indication of soil type. After Louis Ménard, the following relationships are adopted in cohesive soils:

$E_M/p_{LM} < 5$ remoulded clays
$5 < E_m/p_{LM} < 8$ underconsolidated clays
$8 < E_m/p_{LM} < 12$ normally consolidated clays
$12 < E_m/p_{LM} < 15$ slightly overconsolidated clays
$E_m/p_{LM} > 15$ highly overconsolidated clays

As a practical rule, $E_m/p_{LM} = 10$ gives an indication of normally consolidated clay. For cohesionless soils, the ratio of E_m/p_{LM} generally falls within the limits of 7 to 12.

Typical applications of design rules adopted in engineering practice are briefly outlined in the following sections. These rules form the basis of the French code for foundations (Fascicle 62V – M.E.L.T, 1983) and are in accordance to Eurocode 7. Step-by-step procedures of pressuremeter design methods have been detailed by Baguelin *et al.* (1985) and Briaud (1992).

Bearing capacity of shallow foundations

In this approach, the relationship between the unit end-bearing resistance q_b of a footing and the limit pressure p_{LM} established by an MPM test is linked to the ratio of the spherical and cylindrical cavities, respectively. From this conceptual background, and based on full-scale load tests, a bearing capacity factor k_p has been defined as:

$$k_p = \frac{q_b - \sigma_v}{p_{LM} - p_0} \tag{5.76}$$

where $(P_{LM} - P_0)$ is the equivalent net limit pressure within the zone of influence of the footing, σ_v is the total vertical stress at the foundation level and P_0 is the total horizontal stress estimated from the pressuremeter. The value of $(P_{LM} - P_0)$ is obtained as:

$$(p_{LM} - p_0)_e = \sqrt[n]{(p_{LM} - p_0)_1 (p_{LM} - p_0)_2 \ldots\ldots (p_{LM} - p_0)_n} \tag{5.77}$$

where n is the number of pressuremeter tests within $1.5B$ of the foundation level and B is the width of the footing. This approach applies to an equivalent footing embedment depth, calculated as:

$$D_e = \frac{1}{(p_{LM} - p_0)_e} \int_0^D (p_{LM} - p_0)(z) dz \tag{5.78}$$

z being the thickness of a layer for which the net limit pressure is $(p_{LM} - p_0)$. This definition allows stronger or weaker layers within the depth of influence of the footing to be taken into account. Using these input data, the bearing capacity factor k_p can be calculated from Table 5.2, or alternatively from Figure 5.38 for square and circular footings and from Figure 5.39 for strip footings.

A reduction factor $i_{\delta,\beta}$ should be applied to k_p equation (5.76) when calculating the end-bearing capacity for foundations subjected to inclined load or for footings adjacent to an excavation or slope:

$$i_{\delta,\beta} = \left(1 - \frac{A}{90}\right)^2 (1 - \lambda) + \left(1 - \frac{A}{20}\right)\lambda \tag{5.79}$$

Table 5.2 Bearing capacity factor k_p for shallow foundations (Frank, 1999)

Soil category	Class	Expression for k_p	k_p square and circular footings	k_p strip footing
Clay and silt Chalk	A	$0.8\left[1+0.25\left(0.6+0.4\dfrac{B}{L}\right)\dfrac{D_e}{B}\right]$	1.30	1.10
Clay and silt	B	$0.8\left[1+0.35\left(0.6+0.4\dfrac{B}{L}\right)\dfrac{D_e}{B}\right]$	1.50	1.22
Clay	C	$0.8\left[1+0.50\left(0.6+0.4\dfrac{B}{L}\right)\dfrac{D_e}{B}\right]$	1.80	1.40
Sand	A	$\left[1+0.35\left(0.6+0.4\dfrac{B}{L}\right)\dfrac{D_e}{B}\right]$	1.88	1.53
Sand and gravels	B	$\left[1+0.50\left(0.6+0.4\dfrac{B}{L}\right)\dfrac{D_e}{B}\right]$	2.25	1.75
Sand and gravels	C	$\left[1+0.80\left(0.6+0.4\dfrac{B}{L}\right)\dfrac{D_e}{B}\right]$	3.00	2.20
Chalk	B–C	$1.3\left[1+0.27\left(0.6+0.4\dfrac{B}{L}\right)\dfrac{D_e}{B}\right]$	2.18	1.83
Marl and weathered rock	–	$\left[1+0.27\left(0.6+0.4\dfrac{B}{L}\right)\dfrac{D_e}{B}\right]$	1.68	1.41

where A is either the inclination of the load δ or the slope of an excavation β. The value of λ is defined as:

$$\lambda = \lambda_D \lambda_M$$
$$\lambda_D = (1 - D/B)\text{.....................}0 < D/B < 1$$
$$\lambda_D = 0\text{...}D/B > 1 \qquad (5.80)$$
$$\lambda_M = (1 - M)\text{.........................}0 < M < 1$$
$$\lambda_M = 0\text{...}M > 1$$

where D is the depth of the footing and M is expressed as:

$$M = \frac{(p_{LM} - p_0)_{z=D}}{(p_{LM} - p_0)_{z=D+B}} \qquad (5.81)$$

If the load is inclined towards a slope, the factor is adjusted so that $A = (\delta + \beta')$. These correction coefficients are represented in Figure 5.40 for four different conditions: (a) inclined load on horizontal ground, (b) vertical load near a slope, with no embedment, (c) vertical load, near a slope, with embedment and (d) failure stress for inclined load, near a slope (after Frank, 1999).

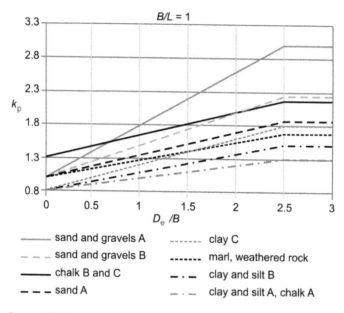

Figure 5.38 Bearing capacity factor k_p for square and circular footings (after Frank, 1999).

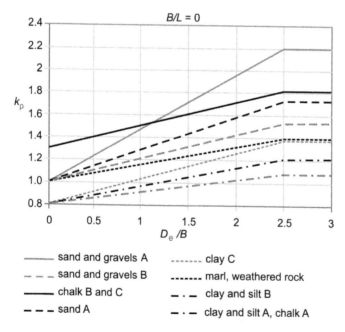

Figure 5.39 Bearing capacity factor k_p for strip footings (after Frank, 1999).

Settlement of a shallow foundation

As recommended by Ménard and Rousseau (1962), the settlement of a foundation is a function of both the isotropic stiffness E_c and the deviatoric stiffness E_d. Both quantities can be derived from pressuremeter data from which a ten-year settlement ρ of a rigid footing can be derived, provided that the embedment depth is greater than the width B:

$$\rho = \frac{2(q - \sigma_{v0})B_0}{9E_d}\left(\lambda_d \frac{B}{B_0}\right)^{\alpha} + \frac{\alpha\lambda_c(q - \sigma_{v0})B}{9E_c} \tag{5.82}$$

where B_0 is a reference width taken as 0.60 m, λ_c and λ_d are shape coefficients given in Table 5.3, α is a rheological factor given in Table 5.4 and q is the footing bearing pressure.

(a)

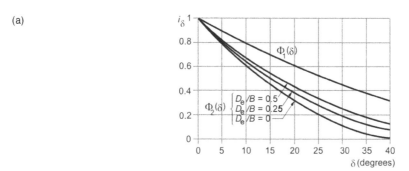

- Cohesive soils (clays, silts, marls), chalks, calcareous marls and weathered rocks:

$$i_\delta = \Phi_1(\delta)$$

- Granular soils (sands and gravels)

$$i_\delta = \Phi_2(\delta)$$

(b)

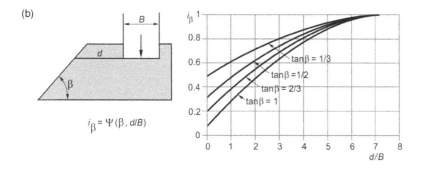

$$i_\beta = \Psi(\beta, d/B)$$

Figure 5.40 (Continued on next page)

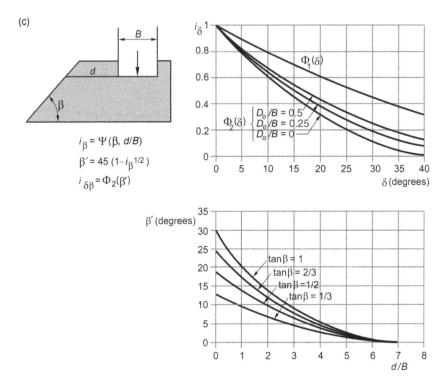

(c)

$i_\beta = \Psi (\beta, d/B)$

$\beta' = 45 (1 - i_\beta^{1/2})$

$i_{\delta\beta} = \Phi_2(\beta')$

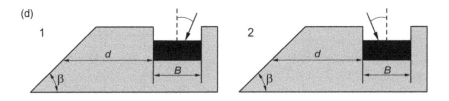

(d)

1. inclination towards the exterior of the slope
   $i_{\delta\beta} = \Phi_2(\delta + \beta)$

2. inclination towards the exterior of the slope
   $i_{\delta\beta} = \inf \{\Phi_1(\delta)\ \text{or}\ \Phi_2(\delta)\ ;\ \Phi_2\ (|\beta' - \delta|)\}$

Figure 5.40 Reduction of bearing capacity k_p for inclined load or adjacent excavation of slope for centred load: (a) inclined load on horizontal ground; (b) vertical load near a slope, with no embedment; (c) vertical load near a slope, with embedment; (d) failure stress for inclined load near a slope (see Frank, 1999).

Table 5.3 Shape coefficients for shallow foundations (Frank, 1999)

L/2B	Circle	Square	2	3	5	20
λ_c	1	1.12	1.53	1.78	2.14	2.65
λ_d	1	1.10	1.20	1.30	1.40	1.50

Table 5.4 Coefficient α for shallow foundations (adapted from Frank, 1999)

Ground type	Description	E_m/p_{LM}	α
Peat	Normally consolidated		1
Clay	Overconsolidated	> 16	1
	Normally consolidated	9–16	2/3
	Remoulded	7–9	1/2
Silt	Overconsolidated	> 14	2/3
	Normally consolidated	8–14	1/2
	Remoulded	5–8	1/2
Sand	Overconsolidated	> 12	1/2
	Normally consolidated	7–12	1/3
	Remoulded	5–7	1/3
Sand and gravel	Overconsolidated	> 10	1/3
	Normally consolidated	6–10	1/4
Rock	Low fractured		2/3
	Unaltered		1/2
	Extensively fractured		1/3
	Extensively weathered		2/3

A detailed procedure should be followed to calculate settlements from equation (5.82). The ground below the footing should be divided into 16 B/2 thick layers, as shown in Figure 5.41. Since the isotropic stiffness is at its maximum immediately beneath the footing, E_c is taken as equal to E_1. An equivalent modulus of the sixteenth layer, E_d, is calculated as:

$$\frac{4}{E_d} = \frac{1}{E_1} + \frac{1}{0.85E_2} + \frac{1}{E_{3,5}} + \frac{1}{2.5E_{6,8}} + \frac{1}{2.5E_{9,16}} \tag{5.83}$$

with a harmonic mean assumed for each layer ($E_{3,5}$, $E_{6,8}$, $E_{9,16}$) in equation (5.83).

Bearing capacity of vertically loaded piles

The method for obtaining the pile bearing capacity is similar to that outlined for shallow foundations and was originally proposed by Ménard (1963). The pile end-bearing capacity Q_b is expressed as:

$$\frac{Q_b}{A_b} = k_p(p_{LM} - P_0)_e + \sigma_v \tag{5.84}$$

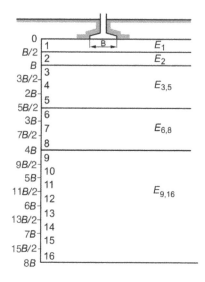

Figure 5.41 Discretization of the soil into layers for settlement analysis (see Frank, 1999).

where A is the pile base area, P_0 the total horizontal stress at rest and k_p the bearing capacity factor. The procedure for determining the equivalent net limit pressure is illustrated in Figure 5.42, from which $(p_{LM} - P_0)_e$ is calculated (e.g. Frank, 1999):

$$(p_{LM} - P_0)_e = \frac{1}{3a + b} \int_{D-b}^{D+3a} (p_{LM} - P_0)(z)dz \qquad (5.85)$$

The values of a and b are assumed to be:

$a = B/2$ if $B > 1$ m
$a = 0.5$ m   if $B < 1$ m
$b =$ minimum of a and h

where h is the embedment in the bearing layer. B is the pile equivalent diameter assumed to be $B = (4A/P)$, A and P being the area and the perimeter of the pile cross-section respectively.

Values of the bearing capacity factor k_p are listed in Table 5.5 for both displacement and non-displacement piles provided that equivalent embedment depth D_e is greater than $5B$. The equivalent embedment depth is the same given by equation (5.78). The value of k_p is reduced to k_e if $D_e < 5B$:

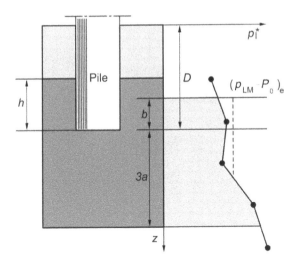

Figure 5.42 Estimation of the equivalent net limit pressure for bearing capacity calculations (see Frank, 1999).

Table 5.5 Bearing factor k_p for axially loaded piles (see Bustamante and Frank, 1999)

Soil type			Range of measurements		Non-displacement piles k_p	Displacement piles k_p
			p_{LM} (MN/m²)	q_c (MN/m²)		
Clay	A	soft	< 0.7	< 3	1.1	1.4
Silt	B	stiff	1.2–2	3–6	1.2	1.5
	C	hard (clay)	> 2.5	> 6	1.3	1.6
Sand	A	loose	< 0.5	< 5	1	4.2
Gravel	B	medium	1–2	8–15	1.1	3.7
	C	dense	> 2.5	> 20	1.2	3.2
Chalk	A	soft	< 0.7	< 5	1.1	1.6
	B	weathered	1–2.5	> 5	1.4	2.2
	C	dense	> 3	–	1.8	2.6
Marl	A	soft	1.5–4	–	1.8	2.6
Calcareous marl	B	dense	> 4.5	–	1.8	2.6
Rock	A	weathered*	2.5–4	–	1.1–1.8	1.8–3.2
	B	fragmented	4.5	–	–	–

Note
* Use the values of the most similar soil.

Table 5.6 Choice of limit unit skin friction curves Q_s (see Bustamante and Frank, 1999)

Soils	Clay and silt			Sand and gravel			Chalk			Marl		Rock
Type of piles	A	B	C	A	B	C	A	B	C	A	B	
Bored without casing	Q1	Q1 / Q2[a]	Q2 / Q3[a]		Q2 / Q1[b]	Q3 / Q2[b]	Q1	Q3	Q4 / Q5[a]	Q3	Q4 / Q5[a]	Q6
Bored under slurry	Q1	Q1 / Q2[a]		Q1	Q2 / Q1[b]	Q3	Q1	Q3	Q4 / Q5[a]	Q3	Q4 / Q5[a]	Q6
Bored with temporary casing	Q1	Q1 / Q2[c]			Q2 / Q1[b]	Q3 / Q2[b]	Q1	Q2	Q3 / Q4[c]	Q3	Q4	
Bored with permanent casing	Q1	Q1		Q1		Q2	[d]	Q2	Q3	Q2	Q3	
Piers[e]	Q1	Q2	Q3	Q1			Q1			Q2	Q3	Q6
Closed-ended steel tube	Q1		Q2	Q2		Q3	[d]			Q4	Q5	Q4
Driven precast concrete	Q1		Q2		Q3		[d]			Q3	Q4	Q4
Driven cast-in place	Q1		Q2	Q2		Q3	Q1	Q2	Q3	Q3	Q4	Q4
Concrete driven steel[h]	Q1		Q2	Q3		Q4	[d]			Q3	Q4	
Bored and low pressure grouted[f]		Q4		Q5			Q2	Q3	Q4		Q5	
Bored and high pressure grouted[f]			Q5	Q6				Q5	Q6		Q6	Q7[g]

Notes

a Prebored and grooved borehole wall.

b From long piles (longer than 30 m).

c Dry boring, no rotation casing.

d In chalk, below the water table, where Q_s can be significantly affected, a site-specific preliminary study is needed and full-scale test is recommended.

e Bored in dry soils or rocks above the water table without support.

f Low rate injection and repeated grouting at selected depths.

g Preliminary grouting of fissured or fractured surrounding ground and possible voids filling.

h A preformed steel pile or tubular H-section, with an enlarged shoe, is driven with simultaneous pumping of concrete (or mortar) into the annular space.

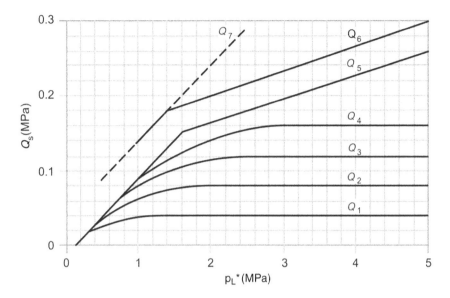

Figure 5.43 Unit skin friction for axially loaded piles (see Bustamante and Frank, 1999).

$$k_e = 0.8 + \left(\frac{k - 0.8}{25}\right) \frac{D_e}{B} \left(\frac{10 - D_e}{B}\right) \qquad (5.86)$$

The pile shaft friction capacity Q_s can be determined as:

$$Q_s = P \int_0^b q_s dz \qquad (5.87)$$

where q_s is the unit skin friction obtained from Table 5.6 in conjunction with Figure 5.43. In this approach, the shaft friction depends not only on the pressuremeter limit pressure, but also on the type of soil, type of pile and construction conditions of the pile.

Settlement of single piles

Settlements of single piles at half the ultimate load are taken as a percentage of the pile diameter, plus the sum of the elastic shortening of the free part pile (if any). The percentage is (see Frank, 1999):

$w = 0.6\%B$ for bored piles
$w = 0.9\%B$ for driven piles

where w is the pile settlement and B the pile diameter.

The complete load–settlement curves can be generated from pressuremeter data if both the unit end-bearing resistance–displacement curve $(q - w)$ and unit skin friction–displacement curve $(f - w)$ are obtained. Three procedures can be adopted to derive the $(q - w)$ and $(f - w)$ curves, as illustrated in Figure 5.44 (Gambin, 1963; Baguelin, 1982; Frank and Zhao, 1982; Frank, 1999). The only significant difference between these approaches is regarding recommendations for the slope of the elastic part of the curves. The method proposed by Gambin (1963) assumes a simple elastic-plastic model, whereas the method proposed by Frank and Zhao (1982) assumes a bilinear elastic-plastic model to match experimental data from full-scale tests on bored piles more closely.

Taking the example of Frank and Zhao's method, for fine-grained soils, the slope J, used to determine the base settlement, and the slope S, adopted to determine the shaft settlement, are, respectively:

$$J = \frac{q}{w_b} = 11.0\frac{E_m}{B} \quad \text{and} \quad S = \frac{f}{w_s} = 2.0\frac{E_m}{B} \tag{5.88}$$

For granular soils, the values of J and S are expressed as:

$$J = \frac{q}{w_b} = 4.8\frac{E_m}{B} \quad \text{and} \quad S = \frac{f}{w_s} = 0.8\frac{E_m}{B} \tag{5.89}$$

Alternatively, Baguelin et al. (1978) proposed the following equations for driven piles:

$$J = \frac{q}{w_b} = \frac{4E_m}{\pi(1 - v^2)B} \quad \text{and} \quad S = \frac{f}{w_s} = \frac{E_m}{(1 + v)(1 + \ln\frac{L}{B})B} \tag{5.90}$$

The initial phase of the load transfer curves can be derived from equations (5.88) to (5.90). The method prescribed to derive the limiting threshold $(q$ and $f)$ has already been outlined in the preceding section. It is therefore straightforward to determine the $(q - w)$ and $(f - w)$ curves for every material along the pile shaft.

Discretization of a pile into separate elements allows a numerical integration of forces and displacements along the pile shaft. This calculation is usually performed by a computer program, in which every element is associated with the corresponding $(q - w)$ and $(f - w)$ curves (e.g. Frank, 1999).

Horizontally loaded piles

A pressuremeter is the perfect model for a prototype pile subjected to horizontal load, given the similarities in shape and in the direction of load. This

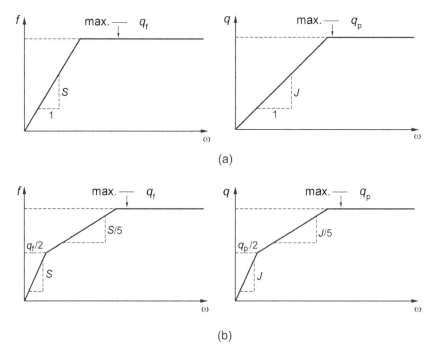

Figure 5.44 Elastic-plastic models to determine the settlement of axially loaded piles using a load transfer method: (a) Gambin (1979); (b) Frank and Zhao (1982).

similarity was recognized early on by Ménard *et al.* (1975) and Gambin (1979), giving rise to procedures designed for the assessment of the so-called 'p–y curve' from pressuremeter data (Matlock, 1970; Reese, 1977).

The p–y curve relates soil reaction to pile deflection as a means of representing the non-linear behaviour of the soil along the pile length. Existing expressions of p–y curves from MPM tests are summarized in Figure 5.45, in which the lateral load P is plotted against the horizontal movement y (after Frank, 1999). The reaction modulus E_s defined in this figure is expressed as:

$$E_s = \frac{P}{y} \tag{5.91}$$

E_s is known to be highly non-linear. In piles where the horizontal load is dominant over other conditions, this non-linear response is represented by a simple bilinear elastic-plastic model with the slope of the first linear portion of the curve being E_s and the second slope being half this value (= $E_s/2$).

A recommendation is made that the ultimate resistance at the surface is reduced to $\frac{1}{2}p_{LM}$. The effect of this reduction is illustrated in Figure 5.46

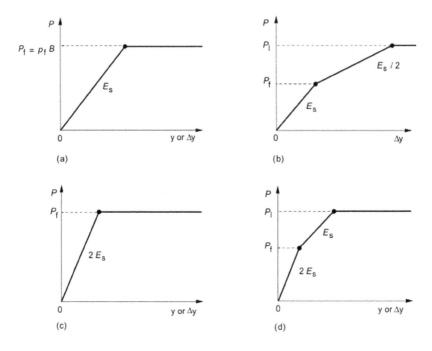

Figure 5.45 Reaction curves for an isolated pile subjected to horizontal loads: (a) permanent load at the head prevails; (b) lateral trusts along shaft prevails; (c) short-
term load at the head prevails; (d) accidental short-term load at the head
prevails (after Frank, 1999).

and is applied to soils above a critical depth, which is taken as $2B$ for clays
and $4B$ for sands.

The reaction modulus is calculated as (Ménard, 1975):

$$E_s = E_m \frac{18}{4\left(2.65\frac{B}{B_0}\right)^{\alpha}\frac{B}{B_0} + 3\alpha} \quad \text{for } B > 0.60\,\text{m} \tag{5.92}$$

and

$$E_s = E_m \frac{18}{4(2.65)^{\alpha} + 3\alpha} \quad \text{for } B \leq 0.60\,\text{m} \tag{5.93}$$

where B is the pile diameter, B_0 is the reference diameter ($= 0.60$ m) and α
is the coefficient introduced in Table 5.4. Representative p–y curves for

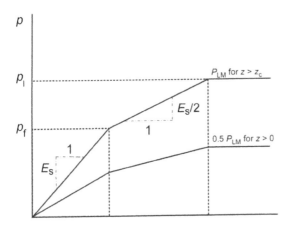

Figure 5.46 Reaction curves within and below the critical depth (after Baguelin *et al.*, 1978).

each soil layer can be derived from this set of equations, from which the distribution of pile deflection, bending moments and shear forces can be numerically determined.

Notes

1 A datalogger has recently been incorporated within the Ménard probe to conform to the current French Standard and to the forthcoming EN ISO Standard.
2 In 1998, Yu and Collins proposed a large strain analysis of SBP tests in overconsolidated clays, based on a critical state soil model that allows the dependency of soil strength on effective stresses to be properly taken into account. This approach fulfils the need for an interpretation method capable of modelling materials other than normally consolidated or lightly overconsolidated clays, where undrained cavity expansion theory expressed in terms of total stress produces reliable undrained shear strength predictions. The Yu and Collins (1998) method is described by Yu (2000).
3 Yu and Houlsby (1995) published a more rigorous unloading analysis that takes into account both large strains and elastic deformation in the plastic region. The analysis is complex, as reference should be made to the configuration at the end of the loading path, and involves a large number of variables that cannot be represented in a simple form.
4 Yu and Houlsby devised a complete large strain analysis, from which an explicit expression for the pressure–expansion relation is derived without imposing any restrictions on the magnitude of deformations.

Chapter 6

Flat dilatometer test (DMT)

In geotechnical design it has long been recognised that the assessment of soil properties is the most important single task. By comparison the details of numerical analyses are of much lesser consequence.

(N. Janbu, 1985)

General considerations

The dilatometer consists of a stainless steel blade with a circular, thin steel membrane mounted flat on one face. The blade is driven vertically into the soil using pushing rigs adapted from those used in the CPT. Penetration is halted every 20 cm and a test is performed by inflating the membrane and taking a series of pressure readings at prescribed displacements. The test is suitable for a wide variety of soils such as clay, sand, silt and hard formations. While the flat dilatometer test (DMT) is a commercial tool used worldwide in geotechnical investigation, only relatively recently have attempts been made to establish guidelines for good practice and to gain a more systematic understanding of interpretation and design applications. Understandably, this recent progress coincides with increasing acceptance of the DMT as a routine site investigation tool.

The DMT was developed in Italy by Silvano Marchetti in 1980. It is recognized here that Marchetti's efforts were of a pioneering nature and subsequent works on this subject followed his original basic conceptual framework. From a historical point of view, the DMT was conceived to establish a reliable operational modulus for the problem of laterally loaded piles (Marchetti, 2006). Later, the interpretation of test results was extended to provide measurements of in situ stiffness, strength and stress history of the soil. The equipment, test method and original correlations described by Marchetti (1980) were the essential initial steps towards establishing a procedure capable of supporting geotechnical applications.

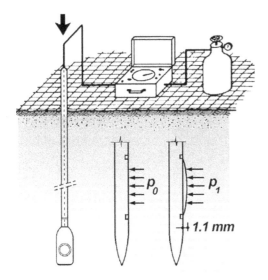

Figure 6.1 Schematic representation of the dilatometer test (Marchetti *et al.*, 2001).

Comprehensive assessments of the state of the art of DMT in research and practice were made by Lunne *et al.* (1989), Lutenegger (1988) and Marchetti (1997). In 2001, a report was issued under the auspices of the ISSMGE Technical Committee TC 16 (Ground Property Characterisation from *In Situ* Testing) by Marchetti *et al.* (2001). The increasing interest in the subject gave rise to a number of specific conferences – the 1st International Conference on the Flat Dilatometer (Canada, 1993), International Seminar on the Flat Dilatometer and its Applications to Geotechnical Design (Tokyo, 1999) and, most recently, the 2nd International Conference on Flat Dilatometer Testing (Washington, 2006). A research project sponsored by the Federal Highway Administration (FHWA) produced a practical manual designed to encourage the use of the DMT in the USA (Schmertmann, 1988). In addition to these contributions, a large number of key publications established a body of practical references on the DMT (e.g. Lacasse and Lunne, 1986; Robertson *et al.*, 1987; Powell and Uglow, 1988; Marchetti and Totani, 1989; Kamey and Iwasaki, 1995; Mayne *et al.*, 1999).

Equipment and procedures

The dilatometer is illustrated schematically in Figure 6.1, showing the control unit, dilatometer blade, push rods and gas tank. Figure 6.2 shows the DMT set up with a control unit that typically includes a pair of pressure

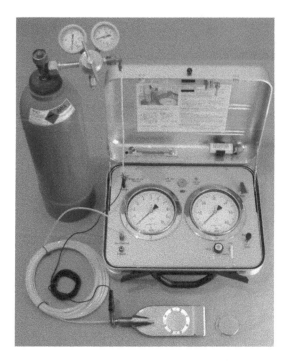

Figure 6.2 The flat dilatometer (courtesy of Marchetti).

gauges (for low-range (1 MPa) and high-range (6 MPa) measurements), valves for controlling gas flow, a galvanometer and audio buzzer signal.

The dilatometer steel blade front and side views are shown in Figure 6.3. Nominal dimensions of the blade are 95 mm width and 15 mm thickness, having a cutting edge angled between 24° and 32° to penetrate the soil. The blade is designed to safely withstand up to 250 kN pushing force. Also shown in Figure 6.3 is the 60 mm dia. and 0.20 mm thick circular steel membrane that is connected to the blade by a retaining ring. Figure 6.4 illustrates the design features of this membrane, which protects a sensing disc that is stationary at rest and kept in position press-fitted inside the insulating seat. The insulation prevents any electrical contact with the steel body of the dilatometer (Figure 6.5). The blade is connected to a control unit by a pneumatic-electrical tube capable of transmitting a combined gas pressure and electrical signal.

The test starts by driving the blade into the soil at a constant rate, generally between 10 and 20 mm/s. After penetration, the membrane is inflated through the control unit and a sequence of pressure readings is made at prescribed displacements, corresponding to:

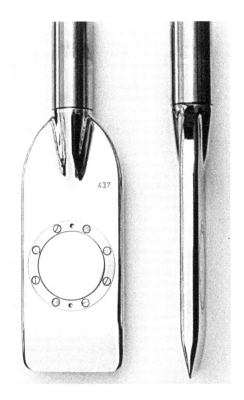

Figure 6.3 The flat dilatometer blades – front and side views (courtesy of Marchetti).

Figure 6.4 The flat dilatometer membrane (courtesy of Marchetti).

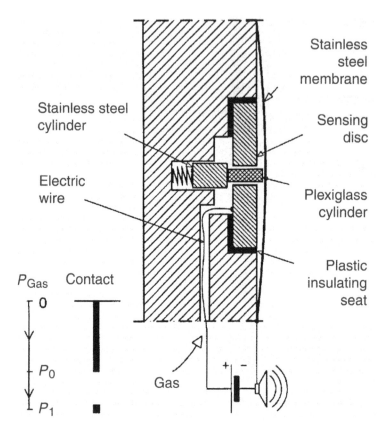

Figure 6.5 DMT working principle (Marchetti *et al.*, 2001).

1   the *A*-pressure at which the membrane starts to expand ('lift-off');
2   the *B*-pressure required to move the centre of the membrane by 1.1 mm against the soil;
3   the optional *C*-pressure, or closing pressure, taken by slowly deflating the membrane soon after the *B*-reading. The rationale behind this measurement is that, in sand, as the membrane deflates the closing pressure approximately equals the pressure of the water u_0.[1]

Before expansion, the sensing disc is grounded and the control unit emits a continual beeping sound. When the internal pressure slightly exceeds the external soil horizontal stress, the membrane starts to move, losing the electrical contact and cutting off the signal. This signal is reactivated when the centre of the membrane has moved 1.1 mm and the spring-loaded steel cylinder touches the inner face of the sensing disc. Deactivation and reactivation of the signal prompts the operator to take the A and B pressure readings respectively. Despite its simplicity, this working system has proved to be fairly accurate and reliable. Furthermore, the fact that results were highly reproducible was fully acknowledged by investigators and practitioners.

Once the process described above is complete and the membrane is fully deflated, the operator advances the blade to a new desired depth and repeats the test. The time delay between the end of the pushing stage and the start of inflation is no more than a few seconds. Eurocode 7 (1997) recommends that the rate of inflation shall be such that the A-reading is obtained within 20 s after reaching the test depth and the B-reading within 20 s after the A-reading. Each test sequence typically requires around 1 min. Tests carried out within the above limits give, essentially, a drained response in sand and an undrained response in clay. In intermediate permeability silty soils partial drainage conditions may occur which impact our ability to interpret test results.

As recognized by Marchetti and Crapps (1981) and Marchetti (1999), calibrations for membrane stiffness are an essential operation to obtain the corrected pressures during the expansion and contraction phases and they have an appreciable influence on the measured soil parameters derived from test results. Calibration consists of correcting the A and B readings for membrane stiffness to obtain:

- ΔA = external pressure that must be applied to the membrane in free air to guarantee a perfect contact against its seating (i.e. A-position);
- ΔB = internal pressure which in free air lifts the membrane centre 1.1 mm from its seating (i.e. B-position).

ΔA and ΔB can be measured by a simple procedure using the control unit and a syringe to generate vacuum or pressure. As a basic rule, this procedure must be carried out before and after each sounding, especially in soft soils where A and B are small numbers, of the order of magnitude of the membrane stiffness. These calibrations should fall within the tolerances given by Eurocode 7 (EN ISO 22476-5) which recommends that, before inserting the blade, the initial readings of ΔA must be within the range of 5 to 30 kPa and ΔB within the range of 5 to 80 kPa. Any membrane that fails to meet these recommendations should be replaced. In order to fall within the range of valid readings, the change of ΔA or ΔB must not exceed 25 kPa at the end of the sounding.

Values of ΔA and ΔB are used to correct the A and B readings respectively in order to determine the pressures p_0 and p_1 from the following expressions:

$$p_0 = 1.05(A - Z_m + \Delta A) - 0.05(B - Z_m - \Delta B) \tag{6.1}$$

$$p_1 = (B - Z_m - \Delta B) \tag{6.2}$$

where Z_m is the gauge zero offset when vented to atmospheric pressure.

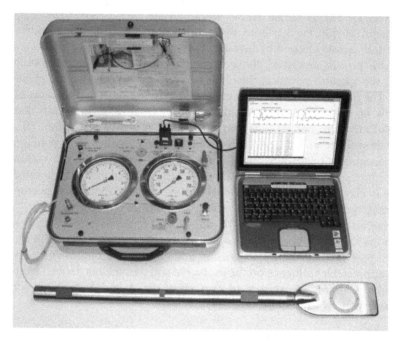

Figure 6.6 The seismic dilatometer (courtesy of Marchetti).

The calculated pressures p_0 and p_1 are subsequently used for interpretation of DMT results in the assessment of soil constitutive parameters. The p_0 pressure is inherently related to the in situ horizontal effective stress and therefore also related to the preconsolidation pressure and stress history. The difference between p_0 and p_1 forms the basis for evaluating soil compressibility.

A relatively recent development (Figure 6.6), the seismic dilatometer (SDMT) is a combination of the standard flat dilatometer (DMT) with a seismic module (Hepton, 1988; Martin and Mayne, 1997; Mayne *et al.*, 1999). This module houses two receivers spaced 0.5 m apart for measuring the shear wave velocity v_s from which the small strain shear modulus G_0 can be determined (see Chapter 3 under Seismic cone):

$$G_0 = \rho(v_s)^2$$

(3.1)idem

where ρ is the mass density. There are important benefits of including the v_s reading (and G_0) with a standard dilatometer. An independent measurement of soil stiffness enhances our ability to solve problems, such as the seismic behaviour of sandy soils during earthquakes and soil–structure interaction response under dynamic loads.

Intermediate DMT parameters

Interpretation of the DMT is based predominantly on empirical correlations related to three index parameters: material index, horizontal stress index and dilatometer modulus.

Material index I_D

The material index I_D is defined as:

$$I_D = \frac{p_1 - p_0}{p_0 - u_0} \tag{6.3}$$

where u_0 is the hydrostatic porewater pressure. The material index provides a reasonable estimate of soil type and was introduced by Marchetti (1980) after the observation that the difference between p_0 and p_1 is small for clay and large for sand.

Horizontal stress index K_D

The horizontal stress index K_D is expressed as:

$$K_D = \frac{p_0 - u_0}{\sigma'_{v0}} \tag{6.4}$$

where σ'_{v0} is the in situ vertical effective stress. K_D can be regarded as K_0 amplified by the penetration of the blade. Two experimental results are observed in clay deposits: in normally consolidated, uncemented clay the value of K_D is approximately 2 and the K_D profile is similar in shape to the OCR profile, giving an indication of the variation of the stress history of the soil with depth.

Dilatometer modulus E_D

The dilatometer modulus is obtained by relating the displacement s_0 to p_0 and p_1 using the theory of elasticity (Gravesen, 1960). The solution assumes that:

- the space surrounding the dilatometer is formed by two elastic half-spaces, in contact along the plane of symmetry of the blade; and
- zero settlement is computed externally to the loaded area:

$$s_0 = \frac{2D(p_1 - p_0)}{\pi} \frac{(1 - v^2)}{E} \tag{6.5}$$

where E is Young's modulus and v Poisson's ratio. For a membrane diameter D equal to 60 mm and a displacement s_0 equal to 1.1 mm, equation (6.5) approaches:

$$E_D = 34.7(p_1 - p_0) \tag{6.6}$$

Although E_D is inherently an operational Young's modulus (both E and E_D are calculated using elastic theory), it is recognized that the expansion of the membrane from p_0 and p_1 reflects the disturbed soil properties around the blade produced by DMT penetration.

Interpretation of test results

DMT interpretation methods are essentially based on correlations obtained by calibrating DMT pressure readings against thoroughly verified parameters (e.g. Marchetti, 1980, 1997; Lutenegger, 1988; Lunne et al., 1989). These correlations are supported by experience, which shows that DMT parameters can be related to soil type, soil unit weight, coefficient of earth pressure at rest (K_0), overconsolidation ratio (OCR), constrained modulus (M), undrained shear strength (s_u) and friction angle (ϕ'). Given the empirical nature of the interpretation methods, existing correlations are, at least in part, site-specific and some scatter is expected between estimated and predicted values.

Substantial effort has been expended on the validation of DMT correlations. Given the physical constraints of modelling a dilatometer, closed-form solutions have been obtained just under considerably simplified assumptions (e.g. Schmertmann, 1986; Roque et al., 1988; Kim and Paik, 2006; Lutenegger, 2006; Lutenegger and Adams, 2006). Finite element analysis should ideally model the penetration and the inflation of the membrane as a truly three-dimensional phenomenon, in contrast with the two-dimensional axisymmetric representation of cone penetration. Since this is not a simple task, only a limited number of reported numerical studies have tackled the problem of the numerical representation of boundary conditions around the blade (e.g. Baligh and Scott, 1975; Finno, 1993; Yu et al., 1993; Smith and Houlsby, 1995; Yu, 2004; Balachowski, 2006). Important contributions will be reviewed throughout this chapter.

Soil characterization

Recommendations and requirements for the graphical presentation of DMT data are fairly basic. The information to be presented combines two DMT intermediate parameters (I_D and K_D) with two derived soil parameters (the constrained modulus M and the undrained shear strength s_u) that are displayed side by side in four profiles. Note that I_D gives a clear indication of soil type, whereas K_D gives an assessment of the stress history of the deposit

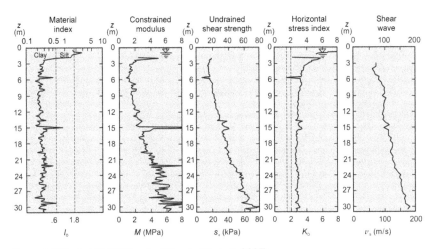

Figure 6.7 Typical SDMT in clay (Marchetti *et al.*, 2007).

by describing the shape of OCR with depth. For a seismic dilatometer, a fifth profile is added by displaying the shear wave velocity with depth.

Typical DMT profiles in clay and sand are shown in Figures 6.7 and 6.8, respectively. Figure 6.7 gives an example of the typical graphical representation of the SDMT-recorded output at the Futino clay research site (after Marchetti *et al.*, 2007). The output displays the profiles of the four basic DMT parameters, I_D (soil type), M, s_u and K_D (related to OCR), as well as the profile of v_s. Values of I_D fall consistently within the range of 0.1 to 0.5, which is characteristic of clay deposits. Both the constrained modulus and the undrained shear strength show a steady increase with increasing depth for values ranging from 2 to 6 MPa and 10 to 60 kPa, respectively. An interesting feature, which will be explored below, is the reduction of K_D from the surface to a relatively uniform value of $K_D = 2$. Since K_D profiles generally reflect OCR, this pattern would indicate a desiccated OC crust near the surface overlying a normally consolidated clay.

SDMT results at the Catania site in Italy are shown in Figure 6.8 to illustrate the DMT response in sand (Maugeri and Monaco, 2006). Values of I_D, M and K_D are fairly uniform at values of 2, 200 MPa and 5 respectively. The constrained modulus is much higher than values reported in the soft clay deposit at Futino, yielding a value of 200 MPa compared to the measured 5 MPa in clay. The shear velocity increases with depth, typically ranging from 200 m/s near the surface to about 300 m/s at 22 m depth.

Evidence of the consistency of test results has prompted the development of charts designed to identify soil type and estimate unit weight γ from I_D

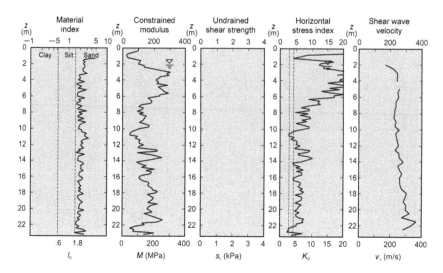

Figure 6.8 Typical SDMT in sand (Maugeri and Monaco, 2006).

and E_D (Marchetti and Crapps, 1981; Lacasse and Lunne, 1988), as illustrated in Figure 6.9. Charts of this type provide estimations of the sub-soil strata and the correspondent values of γ that are necessary to calculate σ'_{v0}.

Geotechnical parameters in clay

In clay, DMT data can provide direct assessment of the overconsolidation ratio, coefficient of earth pressure at rest, undrained shear strength, soil stiffness and coefficient of consolidation.

Overconsolidation ratio (OCR)

The similarity between the dilatometer K_D and OCR profiles was first pointed out by Marchetti (1980) and later confirmed by several authors (e.g. Jamiolkowski *et al.*, 1988; Powell and Uglow, 1988; Kamey and Iwasaki, 1995). For uncemented clays, OCR can be predicted to be:

$$OCR = (0.5K_D)^{1.56} \tag{6.7}$$

which is consistent with the experimental evidence that $K_D \approx 2$ for OCR = 1. The usefulness of this approach has been demonstrated experimentally by Kamey and Iwasaki (1995), as illustrated in Figure 6.10. Although a small variation in the coefficients in equation (6.7) may be acceptable to match local conditions, there is sufficient evidence to suggest that K_D profiles are

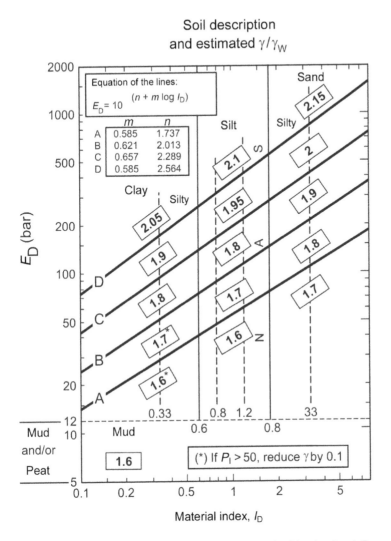

Figure 6.9 Chart for estimating soil type and unit weight (Marchetti and Crapps, 1981).

helpful in characterizing the stress history of the soil, allowing NC and OC clay deposits to be recognized and desiccated crusts to be identified.

Assuming that the installation of a flat dilatometer can be simulated by a flat cavity expansion process, Yu (2004) produced a sound numerical validation of the K_D and OCR relationship (Figure 6.11). The comparison between numerical simulations and the empirical correlation suggests that equation (6.7) can be used with reasonable confidence in most conditions, with the exception of heavily overconsolidated clays ($OCR > 8$).

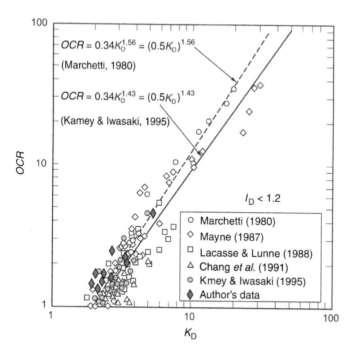

Figure 6.10 Correlations of K_D and OCR for cohesive soils (modified from Kamey and Iwasaki, 1995).

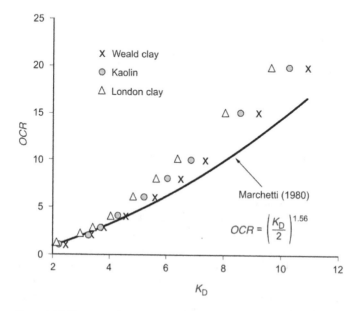

Figure 6.11 Theoretical correlation between K_D and OCR (Yu, 2004).

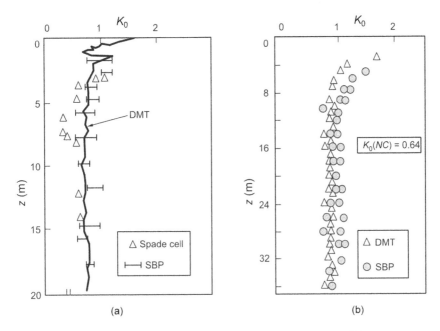

Figure 6.12 Comparison between K_0 values derived from DMT and self-boring pressure-meter (SBP) results at (a) Bothkennar research site (Nash et al., 1992) and (b) Fucino research site (Burghignoli et al., 1991).

As already recognized by Marchetti *et al.* (2001), equation (6.7) is not applicable to structured (cemented) materials. Soil structuration produces an increase in K_D, whereas soil destructuration, produced by driving the blade into the soil, would produce an opposite effect. The magnitude of these effects depends on the nature of the cementing agent and, for this reason, a standardized K_D–OCR correlation cannot be achieved.

Coefficient of earth pressure at rest

Several attempts have been made to correlate the coefficient of earth pressure at rest K_0 to the horizontal stress index K_D in uncemented clay (Marchetti, 1980; Lacasse and Lunne, 1988; Powell and Uglow, 1988; Kulhawy and Mayne, 1990). These correlations give the same broad information as that derived from Marchetti's original equation (1980):

$$K_0 = \left(\frac{K_D}{1.5}\right)^{0.47} - 0.6 \tag{6.8}$$

In Figure 6.12, it can be seen that, for the dilatometer, the deduced K_0 values are very close to K_0 values derived from the self-boring pressuremeter at the two research sites of Bothkennar (Nash et al., 1992) and Fucino (Burghignoli et al., 1991). Numerical work carried out by Yu (2004) shows that, although K_0–K_D relationships are sensitive to critical state parameters, equation (6.8) generally gives a reasonable estimation of K_0 in clay.

Undrained shear strength

The dependence of normalized undrained shear strength s_u/σ'_{v0} on OCR is well-known and has been extensively discussed throughout this book (see Chapters 2 and 3). The K_D versus OCR correlation has therefore prompted investigations into possible relations between K_D and s_u/σ'_{v0} and, consequently, the development of correlations to estimate s_u (Marchetti, 1980; Lacasse and Lunne, 1988; Powell and Uglow, 1988). The original correlation, as proposed by Marchetti (1980), is:

$$s_u = 0.22\sigma'_{v0}(0.5K_D)^{1.25} \tag{6.9}$$

Numerical analysis of the installation of flat dilatometers reported by Huang (1989), Finno (1993), Whittle and Aubeny (1993) and Yu (2004) generally supports the s_u–K_D correlation from equation (6.9) (e.g. Marchetti, 1980; Marchetti et al., 2001). Case studies comparing s_u values determined from the dilatometer with those obtained from other testing techniques have been presented by Lacasse and Lunne (1988), Powell and Uglow (1988), Burghignoli et al. (1991), Iwasaki et al. (1991), Nash et al. (1992), Coutinho et al. (1998) and others. The results of several reported cases are compiled in Figure 6.13.

Soil stiffness

Several attempts have been made to correlate the DMT intermediate parameters to soil stiffness, from which the constrained modulus, Young's modulus and small strain shear modulus can be assessed. Marchetti's original work in the 1980s has already explored this possibility by correlating the constrained modulus M and E_D in the form:

$$M_{DMT} = R_M E_D \tag{6.10}$$

where R_M is an empirical coefficient that ranges mostly between 1 and 3 and is known to be sensitive to I_D. The equations expressing R_M are:

$$
\begin{aligned}
&\text{if } I_D \leq 0.6 && R_M = 0.14 + 2.36 \log K_D \\
&\text{if } I_D \geq 0.3 && R_M = 0.5 + 2 \log K_D \\
&\text{if } 0.6 < I_D < 3 && R_M = R_{M0} + (2.5 - R_{M0})\log K_D \\
& && \text{with } R_{M0} = 0.14 + 0.15(I_D - 0.6) \\
&\text{if } K_D > 10 && R_M = 0.32 + 2.18 \log K_D \\
&\text{if } R_M < 0.85 && \text{set } R_M = 0.85
\end{aligned}
\tag{6.11}
$$

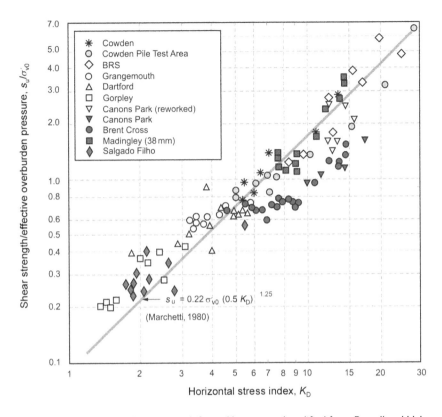

Figure 6.13 Undrained shear strength from dilatometer (modified from Powell and Uglow, 1988).

Given the nature of the correlated parameters, and because a vertical *drained* confined modulus cannot be expected to correlate to an *undrained* horizontal cavity expansion modulus, scatter in R_M is expected. In addition, although soil stiffness is known to be largely controlled by soil stress history, E_D does not reflect this dependency. Despite due recognition of these limitations, experience has generally shown that the DMT provides an indication of M (Figure 6.14) and consequently yields some useful predictions of settlements in clay deposits.

The drained Young's modulus can be derived directly from M_{DMT} using elastic theory:

$$E' = \frac{(1+v)(1-2v)}{(1-v)} M \tag{6.12}$$

bearing in mind that a Poisson's ratio of 0.25 gives $E' \approx 0.8 M_{DMT}$.

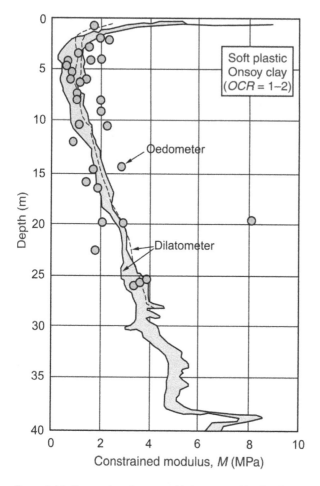

Figure 6.14 Comparison between M determined by DMT and by high-quality oedometer results (Lacasse, 1986).

Recently, attempts have been made to estimate the small strain shear modulus G_0 from E_D (Lunne *et al.*, 1989; Hryciw, 1990; Tanaka and Tanaka, 1998):

$$G_0/E_D = 7.5 \text{ for normally consolidated clays} \tag{6.13}$$

This equation is subject to the restrictions discussed throughout this book and, given the facilities available to measure G_0 from shear wave velocities, the use of a seismic dilatometer is recommended for this purpose.

Coefficient of consolidation

In low permeability soils, the excess pore pressure induced by the penetration of the blade dissipates much more slowly than the pressure induced by the DMT test. Since the dilatometer does not measure the environmental pore pressure, an assumption is made that the flow parameter can be assessed from the measured decay with time of a dilatometer reading (A or C). Three methods are currently used for dissipation tests: DMT-A method (Marchetti and Totani, 1989), DMT-A2 method (Schmertmann, 1988) and DMT-C method (Robertson *et al.*, 1988). Whereas the A and A2 methods monitor contact horizontal stress, the C method records the closing pressure (known to converge with the in situ equilibrium pore pressure).

   Interpretation procedures are fairly similar in all methods. For this reason, only the DMT-A method is detailed in this book (Marchetti *et al.*, 2001):

1   plot the A-log t DMT curve;
2   identify the contraflexure point in the curve (as represented in Figure 6.15) and the associated time (t_{flex});
3   obtain C_h from the following equation:

$$C_h(OC) = \frac{7cm^2}{t_{flex}} \qquad\qquad (6.14)$$

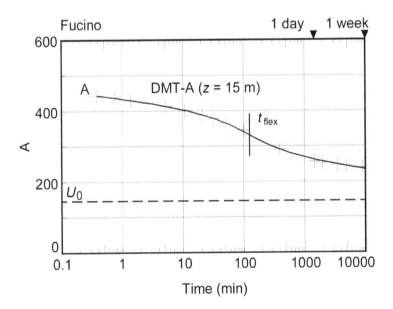

Figure 6.15 Coefficient of consolidation (Marchetti *et al.*, 2001).

Note that the value of C_h, estimated from equation (6.14), corresponds to soil behaviour in the overconsolidated range. As discussed in Chapter 3 under the heading Consolidation coefficients, a much lower C_h value should be used for estimating the rate of settlements in normally consolidated soils.

Geotechnical parameters in sand

In sand, both shear strength and stiffness can be assessed directly from measured DMT readings.

Friction angle

Prediction of the friction angle ϕ' in sand is primarily based on the evidence that the thrust necessary to push the dilatometer blade can be estimated directly from p_0 (and therefore from K_D, which is just a normalization of p_0 with respect to the vertical effective stress). Derived from the proportionality between K_D and penetration resistance (and then to tip cone resistance), Schmertmann (1982) proposed a method to estimate ϕ' using the bearing capacity theory of Durgunoglu and Mitchell (1975) already described in Chapter 3. This approach was later adapted by Marchetti (1985) in a graphical representation (Figure 6.16) that enables ϕ' to be estimated from K_0 and q_c or K_D. Since the value of K_0 is not easily determined, the author uses the values of the coefficients of earth pressure K_a and K_p and the previously mentioned Jacky's equation ($K_0 = 1 - \sin \phi'$) as reference to help in the decision-making process of selecting representative friction angles in sites where little geotechnical site information is available.

This basic empirical approach relies on the measured K_D from the DMT and requires an independent rough evaluation of K_0. A recent numerical analysis presented by Yu (2004) offered an instructive alternative approach to interpretation of DMTs in sand by demonstrating that, although K_D (or normalized K_D/K_0) increases with soil friction angle, the influence of the rigidity index (G_0/p_0) is also highly significant. In the numerical analysis, the DMT has been modelled as a flat cavity expansion process using linear-elastic, perfectly-plastic Mohr-Coulomb theory. The study clearly identifies the need to take the rigidity index into account when estimating ϕ' values. In addition, it appears that, in a stiff soil, the displacement that originates with K_D is sufficiently small to be regarded as elastic; however, in a softer soil plastic strains may occur.

More recently, Marchetti (2001) made full use of the relationship between q_c and K_D to propose a method of estimating ϕ' directly from DMT K_D values:

$$\phi'_{DMT} = 28° + 14.6°\log K_D - 2.1°\log^2 K_D \tag{6.15}$$

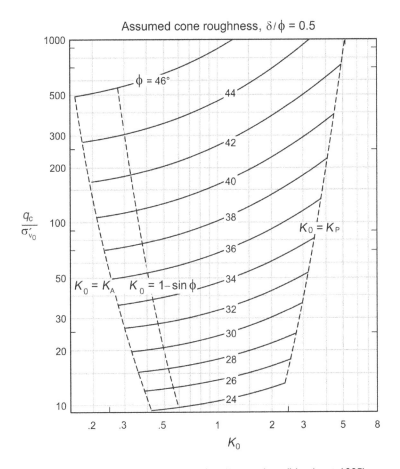

Figure 6.16 Graphical representation of q_c–K_0–ϕ' values (Marchetti, 1985).

Marchetti (2001) emphasizes that equation (6.15) is not intended to be an accurate estimation of ϕ', but a lower bound value that underestimates the in situ friction angle by between 2° and 4°.

State parameter

The state parameter concept is a key element for describing the behaviour of granular soils, following the recognition that it combines the effects of both relative density and stress level. Mobilized friction, dilatancy and liquefaction susceptibility are all governed in some way by the state parameter.

Recent research based on the state parameter model (Yu, 1998) supports the view of K_D as an index reflecting the in situ state parameter Ψ. Although

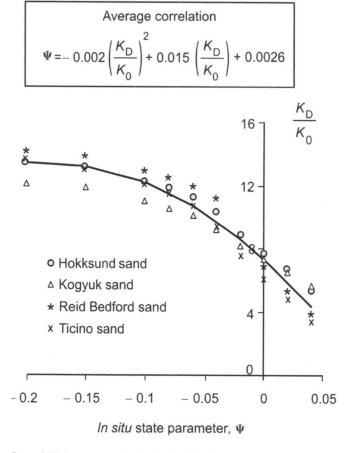

$$\boxed{\text{Average correlation} \\ \Psi = -0.002 \left(\frac{K_D}{K_0} \right)^2 + 0.015 \left(\frac{K_D}{K_0} \right) + 0.0026}$$

○ Hokksund sand
△ Kogyuk sand
* Reid Bedford sand
x Ticino sand

In situ state parameter, Ψ

Figure 6.17 Average correlation for deriving the *in situ* state parameter from DMT (Yu, 2004).

K_D–Ψ correlations reflect the Cam clay constitutive parameters (Figure 6.17), for practical purposes Ψ can be estimated from the normalized dilatometer horizontal index K_D/K_0 using the following equation:

$$\Psi = -0.002 \left(\frac{K_D}{K_0} \right)^2 + 0.015 \left(\frac{K_D}{K_0} \right) + 0.0026 \qquad (6.16)$$

Soil stiffness

The small strain stiffness can be computed from the measured shear wave velocity and subsequently associated with operative strains representative of static loading conditions (see under Coefficient of consolidation above).

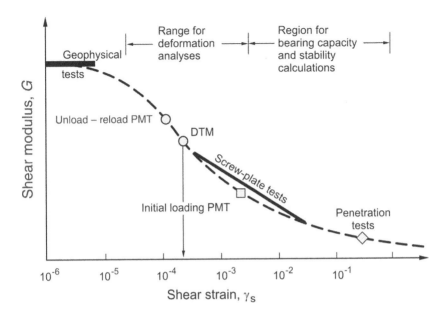

Figure 6.18 Decay of shear modulus with strain level and possible strain range of moduli from various *in situ* tests (Mayne, 2001).

This approach follows recent research efforts that are currently in progress to investigate the possible use of M_{DMT} for settlement predictions based on non-linear methods, taking into account the decay of soil stiffness with strain level using G–γ curves. Conceptually, the intention is to derive the in situ G–γ curve from two reference points: the initial shear modulus G_0 obtained from shear wave velocity v_s measurements and the M_{DMT} modulus at an operative strain. The M_{DMT} modulus is assumed to represent an inter-mediate strain level in the order of 0.05–0.1% (Mayne, 2001; Monaco *et al.*, 2006), as shown schematically in Figure 6.18.

In an alternative approach, research carried out in large laboratory cali-bration chamber tests has produced sound correlations for clean silica sand (Jamiolkowski *et al.*, 1988):

$$E'/E_D = 1.5 + 0.25 \text{ for normally consolidated} \tag{6.17}$$

and

$$E'/E_D = 3.66 + 0.80 \text{ for overconsolidated} \tag{6.18}$$

where the Young's modulus is defined at 0.1% axial strain.

Direct applications

Although geotechnical design should preferably be performed on the basis of sound analytical or numerical methods supported by constitutive parameters derived from both in situ and laboratory tests, DMT provides some useful additional applications for a wide range of practical problems, such as footing and pile design, anchored diaphragm walls, slope stability, liquefaction and sub-grade compaction control. Design of laterally loaded piles, establishing settlements of shallow foundations and assessing liquefaction are, in the author's view, the three most useful DMT applications.

Laterally loaded piles

The design of laterally loaded piles using the p–y approach was the original stimulus for the development of the flat dilatometer. In this approach, the p–y curve relates soil reaction to pile deflection as a means of representing the non-linear behaviour of the soil along the pile length. The two methods recommended for deriving the p–y curve are proposed by Robertson et al. (1987) and Marchetti et al. (1991).

Robertson et al. (1987) proposed a method that is an adaptation of early approaches, developed for estimating p–y curves from laboratory data (Matlock, 1970):

$$\frac{P}{P_u} = 0.5 \left(\frac{y}{y_c} \right)^{0.33} \tag{6.19}$$

where P/P_u = ratio of soil resistance and y/y_c = ratio of pile deflection.

Input parameters are derived directly from the DMT. In clays, the value of pile deflection y_c is a function of the undrained strength, soil stiffness and in situ effective stress (after Skempton, 1951) and can be conveniently expressed as:

$$y_c = \frac{23.67 s_u D^{0.5}}{F_c E_D} \tag{6.20}$$

where y_c is in cm, D is the pile diameter in cm and F_c is taken as equal to 10 as a first approximation.

The evaluation of the ultimate static lateral resistance P_u is given by Matlock (1970) as:

$$P_u = N_p s_u D \tag{6.21}$$

where N_p is a non-dimensional ultimate resistance coefficient that can be expressed as:

$$N_p = 3 + \frac{\sigma'_{v0}}{s_u} + \left[J \frac{x}{D} \right] \tag{6.22}$$

where

$N_p \leq 9$
x = depth
σ'_{v0} = effective vertical stress at a given depth x
J = empirical coefficient equal to 0.5 for soft clay and 0.25 for stiff clay (Matlock, 1970).

In cohesionless soils, the reference pile deflection is calculated as a function of internal friction angle, effective vertical stress and pile diameter, as well as the dilatometer modulus:

$$y_c = \frac{4.17 \sin \phi' \sigma'_{v0}}{E_D (1 - \sin \phi')} D \qquad (6.23)$$

where y_c and D are in cm. The ultimate lateral soil resistance P_u is determined by the following two equations (after Reese et al., 1974; Murchison and O'Neill, 1984):

$$P_u = \sigma'_{v0} \left[D(K_p - K_a) + x K_p \tan \phi' \tan \beta \right] \qquad (6.24)$$

$$P_u = \sigma'_{v0} D \left[K_p^3 + 2 K_0 K_p^2 \tan \phi' + \tan \phi' - K_a \right] \qquad (6.25)$$

where

σ'_{v0} = effective vertical stress at a given depth
D = pile diameter
ϕ' = internal friction angle
$K_a = \dfrac{1 - \sin \phi'}{1 + \sin \phi'}$ = Rankine active coefficient of earth pressure

$K_p = \dfrac{1 + \sin \phi'}{1 - \sin \phi'}$ = Rankine passive coefficient of earth pressure

K_0 = coefficient of earth pressure at rest
β = $45° + \phi'/2$.

Values of ϕ' and K_0 can be estimated directly from DMT data and other in situ or laboratory tests.

Alternatively, Marchetti et al. (1991) developed a simple and effective method for deriving p–y curves in clay, which is based on a non-dimensional hyperbolic equation (Figure 6.19):

$$\frac{P}{P_u} = \tanh \left(\frac{E_{si} y}{P_u} \right) \qquad (6.26)$$

with

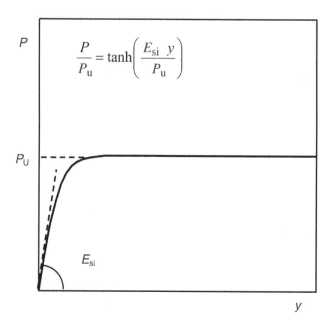

Figure 6.19 Hyperbolic representation of p–y curve (Marchetti et al., 1991).

$$P_u = \alpha K_1 (p_0 - u_0) D \qquad (6.27)$$

$$E_{si} = \alpha K_2 E_D \qquad (6.28)$$

$$\alpha = \frac{1}{3} + \frac{2}{3} \frac{z}{7D} \le 1 \qquad (6.29)$$

where

E_{si} = initial soil modulus
α = non-dimensional reduction factor for depth $z < 7D$ (α becomes 1 for $z = 7D$)
p_0 = corrected first DMT reading
u_0 = in situ pore pressure
z = depth
K_1 = empirical coefficient of soil resistance (assumed as 1.24)
K_2 = empirical coefficient of soil stiffness: $K_2 = 10(D/0.5m)^{0.5}$.

The two methods outlined above provide similar predictions and they are both fairly sensitive to changes in soil stiffness (Figure 6.20).

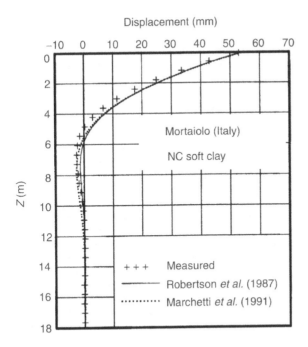

Figure 6.20 Comparison between measured and predicted displacement on laterally loaded pile (Marchetti *et al.*, 1991).

Settlement of shallow foundations

The DMT has proven to be a useful test in predicting settlements of shallow foundations in both sand and clay. Settlements are simply calculated as a one-dimensional problem (Figure 6.21):

$$S_{1DMT} = \sum \frac{\Delta \sigma_v}{M_{DMT}} \Delta z \qquad (6.30)$$

where $\Delta \sigma_v$ is the stress increment calculated by Boussinesq and M_{DMT} is the constrained modulus estimated from DMT data (see above under Intermediate DMT parameters). Although, in theory, settlements can be calculated using one-dimensional (rafts) or three-dimensional (small isolated footings) formulae, Marchetti *et al.* (1991) suggested that the one-dimensional formula be used in all cases for DMT interpretation. This recommendation seems to be based on two considerations: first, the one-dimensional and the three-dimensional calculations generally give similar answers in most cases and, second, the emphasis given to DMT interpretation should be on the accurate determination of simple parameters, such as the one-dimensional compressibility coupled with simple calculations.

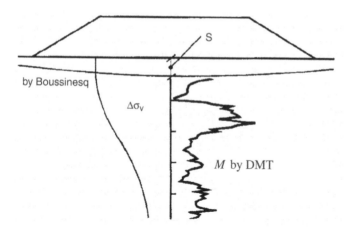

Figure 6.21 Recommended method for settlement calculation using DMT (Marchetti, 1997; Marchetti *et al.*, 2001).

This approach is particularly useful for clean uncemented sand where undisturbed samples cannot be retrieved. In addition, it can also be used as a first estimate of the primary settlement in clay with the M_{DMT} being treated as the average E_{oed} derived from the oedometer curve in the expected stress range.

Note that the settlement calculated from equation (6.30) is considered to be the settlement under *working conditions*, for a safety factor (SF) of about 3.0, since the M_{DMT} reflects an operative modulus that compares with moduli determined from back-calculations of shallow footings and rafts. A large number of documented case histories have confirmed this evidence by demonstrating that the elastic continuum solution from the dilatometer provides calculated settlements in line with the observed performance (Lacasse and Lunne, 1986; Schmertmann, 1986; Hayes, 1990; Woodward and McIntosh, 1993; Skiles and Townsend, 1994; Failmezger *et al.*, 1999; Pelnik *et al.*, 1999; Tice and Knott, 2000; Failmezger, 2001; Marchetti *et al.*, 2004; Mayne, 2005; Monaco *et al.*, 2006).

Schmertmann (1986) compiled measurements from 16 different locations and various soil types, including sand, silt, clay and organic soils with measured settlements ranging from 3 to 2,850 mm. The agreement observed between calculated and measured settlements is acceptable, with an average calculated/observed settlement ratio of 1.18, for ratios ranging from 0.7 to 1.3 and a standard deviation of 0.38 (Table 6.1). Monaco *et al.* (2006) updated Schmertmann's database to obtain a settlement ratio of ≈1.3 for footing sizes ranging from small footings to large rafts and embankments. Results summarized in Figure 6.22 demonstrate the generally satisfactory

Table 6.1 Comparison of DMT-calculated versus measured settlements from 16 case histories (Modified from Schmertmann, 1986)

Location	Structure	Compressible soil	Settlement (mm)		Ratio DMT/measured
			DMT	Measured	
Tampa	Bridge pier	Highly OC clay	25*	15	1.67
Jacksonville	Power plant (3 structures)	Compacted sand	15*	14	1.07
Lynn Haven	Factory	Peaty sand	188	185	1.02
British Columbia	Test embankment	Peat and organic soils	2,030	2,850	0.71
Fredricton	Surcharge	Sand	11*	15	0.73
”	3' plate load test	Sand	22*	28	0.79
”	Building (raft foundation)	Quick clayey silt	78*	35	2.23
Ontario	Road embankment	Peat	300*	275	1.09
”	Building	Peat	262*	270	0.97
Miami	4' plate load test	Peat	93	71	1.31
Peterborough	Apartment building	Sand and silt	58*	48	1.21
”	Factory	Sand and silt	20*	17	1.18
Peterborough	Water tank	Silty clay	30*	31	0.97
Linkoping	2 × 3 m plate	Silty sand	9*	6.7	1.34
”	1.1 × 1.3 m plate	Silty sand	4*	3	1.33
Sunne	House	Silt and sand	10**	8	1.25

* Ordinary method used (one-dimensional settlement, no adjustment of M for vertical effective stress during loading).

agreement between predicted and measured data, with the observed settlement falling within the range of ±50% from the DMT-predicted settlement.

Soil liquefaction

Procedures developed for estimating the liquefaction potential of soil deposits from the SPT and CPT have been reviewed in Chapters 2 and 3. These procedures are essentially supported by the early concepts developed by Seed and Idriss (1971), based on the cyclic shear stress ratios (CSR) induced by earthquake-generated ground motion (equation 2.68). The same concepts hold for

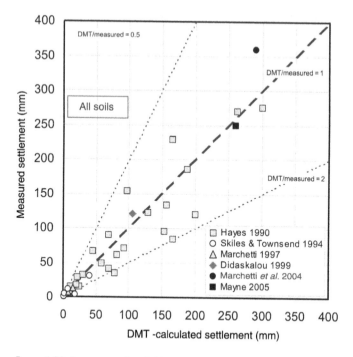

Figure 6.22 Summary of available comparisons of predicted versus observed settlements (Monaco *et al.*, 2006).

evaluation of seismic dilatometer liquefaction analysis methods, from which two parallel independent estimates of liquefaction resistance can be attained – one from K_D and one from v_s, using CRR–K_D and CRR–v_s correlations.

The use of v_s for evaluating CRR is acknowledged and has been described in Chapter 3 under the heading Liquefaction. Andrus and Stokoe (1997, 2000) present the derivation of a shear wave velocity dependent CRR for relatively small strains as a function of v_{s1} for magnitude $M_w = 7.5$ earthquakes.

The CRR–K_D correlations have been developed during the past two decades as an alternative to other in situ tests, prompted by the recognition of the sensitivity of K_D to stress history, aging, soil structure, cementation and relative density (Marchetti, 1982; Robertson and Campanella, 1986; Monaco *et al.*, 2005; Leon *et al.*, 2006; Maugeri and Monaco, 2006; Monaco and Marchetti, 2007; Monaco and Schmertmann, 2007). The ability of K_D to reflect the bonded structure in sand is a key element of the CRR–K_D correlation, since structure effect is considered to be a dominant factor controlling liquefaction behaviour.

The various methods developed to assess the liquefaction potential by differentiating between liquefiable and non-liquefiable soils are shown in

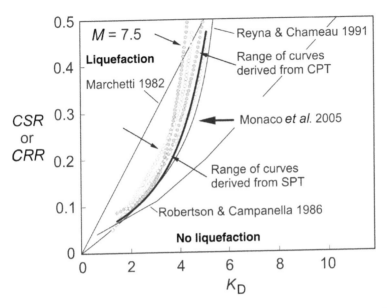

Figure 6.23 Tentative CRR–K_D correlation for evaluating liquefaction resistance from DMT (Monaco et al., 2005).

Figure 6.23. The Robertson and Campanella (1986) and the Marchetti (1982) methods show substantial discrepancies. Based on expanded data-bases and on the re-evaluation of CPT and SPT-based procedures, the Reyna and Chameau (1991) and Monaco et al. (2005) methods produced more consistent predictions. These methods exhibit little divergence and can be used to evaluate the liquefaction potential of magnitude $M=7.5$ earthquakes in clean sand. For magnitudes other than $M=7.5$, magnitude scaling factors should be applied following recommendations given in Chapter 2 under the heading Soil liquefaction (after Idriss, 1999; Youd and Idriss, 2001).

The Monaco et al. (2005) method is presented in Figure 6.23 where lique-faction potential is identified in the CRR–K_D space. CRR is calculated as:

$$CRR_{7.5} = 0.0107K_D^3 - 0.0741K_D^2 + 0.2169K_D - 0.1306 \qquad (6.31)$$

More data should be gathered before using equation (6.31) with confidence. However, Marchetti and his co-workers emphasize the potential use of the CRR–K_D approach by pointing out that K_D is more sensitive than v_s in factors such as stress history and aging. OCR crusts (which are very unlikely to liquefy) are accurately depicted by K_D but are almost unfelt by v_s, which is suggestive of the lesser ability of v_s to profile liquefiability. In addition, K_D is sensitive to both relative density and state parameter, which are known to greatly increase CRR.

Note

1   As discussed in Chapter 5, for a pressuremeter test in clean sand, the membrane deflates to its initial radius at approximately the equilibrium (hydrostatic) porewater pressure (e.g. Wroth, 1984). Campanella and Robertson (1991) reported data from a research dilatometer instrumented to measure pore pressure and the continuous deflection of the centre of the membrane. The authors observed remarkable similarities between the dilatometer and self-boring and full-displacement pressuremeter curves and concluded that, as the DMT membrane deflates, the sand arches and the C-reading closing pressure gives a measurement of u_0.

Chapter 7

Design parameters: guidelines

What can be called Modern Soil Mechanics was born at the turn of the nineteen-sixties, in particular with Roscoe, Schofield, Wroth and co-authors who integrated shear stress, mean effective stress, void ratio and shear strain in the same framework, often referred to as Critical State Soil Mechanics, CSSM. This work was reinforced with the stress path method (Lambe, 1967) and the concept of normalised behaviour (Ladd, 1969). It is also at that time that the influence of dilatancy was fully understood (Rowe, 1962; Lee and Seed, 1967). It was, however, in the nineteen-eighties that some major features of natural soils have been fully recognised: small strain stiffness (Jardine, 1985; Burland, 1989) and its measurement by geophysical methods (Stokoe II and Nazarian, 1985; Stokoe II et al., 1991); concept of collapse surface for explaining static liquefaction (Sladen et al., 1985); importance of effect of strain rate (Graham et al., 1983); influence of anisotropy and stress axis rotation (Hight and co-workers); extension of CSSM concepts to unsaturated soils (Alonso et al., 1990); the fact that most natural geomaterials are microstructured (Burland, 1990; Leroueil and Vaughan, 1990).

(S. Leroueil and D.W. Hight, 2003)

The assessment of reliable strength and stiffness parameters from in situ tests (as well as laboratory tests) is a necessary and fundamental step in every geotechnical design process. In this process, the degree of sophistication with which a geomaterial is characterized varies considerably according to the selected investigation technique – SPT, CPT, DMT, PMT, SBMT and VFT, used alone or in combination. The background information supporting the interpretation of test data comprises a large number of methods including empirical, semi-empirical, analytical and numerical approaches, from which constitutive parameters are determined. At this stage it is appropriate to review the accumulated experience and strategies regarding the characterization of soils, in order to bridge the gap between geotechnical engineers and the design needs that emerge from complex soil behaviour and lack of experience with natural soil deposits.

The possibility of characterizing soils from simple, straightforward index tests is attractive and offers a secure means of anticipating the possible magnitude of design parameters required for a given project. This chapter briefly summarizes existing approaches and correlations and provides examples of applications in clay, sand and residual soils. Although these approaches do not replace the need for direct measurements from laboratory and field tests, they help to build up local site experience at the early stages of design.

Clay

The tests for determining the liquid and plastic limits are by far the most widely used of the index tests and are often adopted to assess the mechanical behaviour of clay. Known as the Atterberg limits, standardized procedures are used to define the range of moisture contents within which the clay is of plastic consistency. The measure of the water content that must be mixed to the soil to make a slurry of viscous behaviour gives the liquid limit, whereas the gradual reduction of moisture to produce a sample of firm texture defines the plastic state. These index tests are performed in reconstituted samples that do not represent the mechanical response of natural soils, since the combined effects of fabric (arrangement of particles) and interparticle bonding cannot be reproduced. However, there is an extensive body of work demonstrating that compressibility and strength of reconstituted soils can be used as a frame of reference for interpreting the corresponding characteristics of natural sedimentary clays (e.g. Wroth and Wood, 1978; Burland, 1990; Leroueil and Vaughan, 1990; Leroueil and Hight, 2003). Both empirical approaches and critical state soil mechanics support correlations of index properties with basic engineering properties of soils (e.g. Wroth and Wood, 1978).

Estimates of undrained shear strength have been shown to be dependent on the liquidity index I_L, which scales the water content of soils at their liquid and plastic limits:

$$I_L = \frac{w - w_p}{w_L - w_p} = \frac{w - w_p}{I_p} \qquad (7.1)$$

where w is the water content and $(w_L - w_p)$ is the difference in water content between the liquid limit and the plastic limit, known as the plasticity index I_p. Figure 7.1 shows the variation of remoulded shear strength s_{ur} with liquidity index I_L for most reconstituted clays. Having assigned a strength of 2 kPa and 200 kPa to soils at their liquid and plastic limits respectively, Wroth and Wood (1978) estimated the reconstituted shear strength s_{ur} using the expression:

$$s_{ur}(kPa) = 2*100^{(1-I_L)} \qquad (7.2)$$

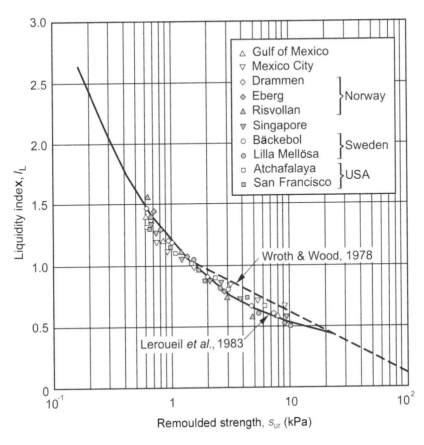

Figure 7.1 Relationship between liquidity index and remoulded shear strength (after Leroueil *et al.*, 1983).

For values of liquidity index between 0.4 and 3.0, Leroueil *et al.* (1983) proposed:

$$s_{ur}(kPa) = \frac{1}{(I_L - 0.21)^2} \tag{7.3}$$

It is worth recalling at this point that the strength of a soil is linked to its current in situ preconsolidation pressure σ'_p and that the normalized field shear strength s_u/σ'_p can be assumed to be:

$$\frac{s_u}{\sigma'_p} \approx 0.25 \tag{7.4}$$

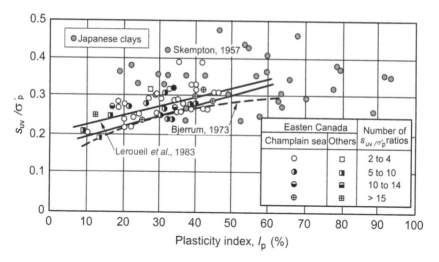

Figure 7.2 Variation of the s_u/σ'_p ratio with plasticity index (after Leroueil and Hight, 2003).

Equation (7.4) provides a reliable indication of the way in which the undrained shear strength mobilized in the field varies with the preconsolidation pressure. For normally consolidated clay deposits, Skempton (1957) proposed an expression that became widely used in engineering practice:

$$\frac{s_u}{\sigma'_{v0}} = 0.11 + 0.37 I_p$$

(7.5)

Variations of the s_u/σ'_p ratio with plasticity index for eastern Canada clays (Leroueil *et al.*, 1983) and Japanese clays (Tanaka, 2002) are summarized in Figure 7.2. For the normally consolidated clays shown in Figure 7.2, it is demonstrated that the correlations often adopted in engineering practice agree well with measured values obtained in a variety of different clays (Bjerrum and Simons, 1960; Bjerrum, 1973; Leroueil *et al.*, 1983). A compilation of published data from characterization studies of Brazilian clay sites is presented in Table 7.1. Sedimentary environments are encountered along the 9,000 km coastline of Brazil, exhibiting a wide range of clay fraction and organic matter content, containing more kaolin than other clay minerals. For the normally consolidated range, the undrained shear strength ranges from 5 to about 50 kPa, yielding a normalized field shear strength $s_u/\sigma'_p = s_u/\sigma'_{v0}$ typically in the range of 0.25 to 0.30.

Other empirical correlations have been found to provide satisfactory information based strictly on cumulative experience. For normally consolidated

Table 7.1 General characteristics of Brazilian soft clay deposits

Place	w (%)	I_L (%)	I_p (%)	CF^1 (%)	Activity	G_z	Organic matter	s_u (kPa)	s_t	References
P. Alegre, RS	47–140	80–130	30–57	37–70	0.9–1.7	2.54–2.59	0.4–6.3	10–32	2–7	Soares (1997)
Sarapuí, RJ	110–160	110–140	75–110	55–80	1.4–2.0	2.60–2.67	4.0–6.5	5–15	2–4	Costa Filho et al. (1985), Sayão (1980)
Santos, SP	100–140	80–150	30–90	30–80	1.0–2.2	2.60–2.69	4.0–6.0	10–60	4–5	Arabe (1995) Massad (1985)
Recife, PE	50–150	30–110	15–75	50–80	–	2.50–2.70	4.0–8.0	2–40	–	Ferreira et al. (1986) Coutinho and Ferreira (1988)
J. Pessoa, PB	35–150	30–60	15–30	30–80	–	2.50–2.65	–	13–40	2–3	Cavalcante (2002)
Jurtunaíba, RJ	40–400	50–390	30–280	–	–	2.10–2.60	7.0–70.0	5–37	3–20	Coutinho (1988)
Sergipe, SE	57–72	58–85	24–35	–	1.0–1.4	2.69	2.5–6.5	8–20	2–7	Brugger et al. (1994)
Rio Grande, RS	38–64	41–90	20–38	34–96	0.4–1.1	2.48–2.66	–	20–50	–	Dias and Bastos (1994)

Note:
1 Clay fraction.

Table 7.2 Internal friction angle of Brazilian clays

Place	ϕ'	References
Porto Alegre, RS	18.3–27.9	Soares (1997)
Rio Grande, RS	23–29	Dias and Bastos (1994)
Vale do Rio Quilombo, SP	19.5–31.6	Arabe (1995)
Vale do Rio Moji, SP	18–28	Arabe (1995)
Santos, SP	23–28	Samara *et al.* (1982)
		Arabe (1995)
		Massad (1988)
Sarapuí, RJ	23–26	Costa Filho *et al.* (1977)
Recife, PE	23–26	Coutinho *et al.* (1993)
João Pessoa, PB	18–21	Cavalcante (2002)

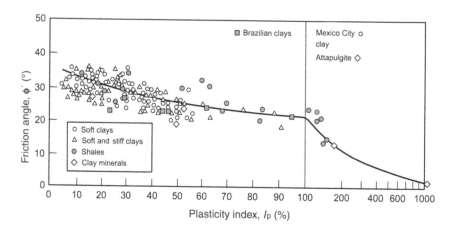

Figure 7.3 Variation of friction angle with plasticity index (Terzaghi *et al.*, 1996).

clay, the effective friction angle ϕ' is known to relate to the plasticity index (Mitchell, 1976; Terzaghi *et al.*, 1996), showing a general tendency of reducing with increasing plasticity. The variation of ϕ' with I_p for various normally consolidated (NC) clays is shown in Figure 7.3 and can be represented by one of several existing relationships – Bjerrum and Simons (1960), Bjerrum (1973) or Leroueil *et al.* (1983). A database compiled from Brazilian experience is summarized in Table 7.2 and Figure 7.3.

In addition, it is possible to relate the compression index of the remoulded soil to index properties. The remoulded (or intrinsic) compressibility C_c^* can be directly related to w_L:

$$C_c^* = \alpha(w_L - 10) \tag{7.6}$$

with $\alpha = 0.007$ or 0.009 according to Skempton (1944) and Terzaghi and Peck (1948) respectively. Wroth and Wood (1978) relate C_c to the plasticity index I_p:

$$C_c^* = \frac{I_p G_s}{2} \tag{7.7}$$

which, for a specific gravity $G_s = 2.7$, gives:

$$C_c^* = 1.35 I_p \tag{7.8}$$

Burland (1990) showed that the compressibility can be related to the void ratio at the liquid limit e_L as:

$$C_c^* = 0.256 e_L - 0.04 \tag{7.9}$$

Knowing that the compressibility of a clay reflects the initial structure, attempts have been made to correlate the compression index C_c and its initial void ratio e_0. Leroueil et al. (1983) suggest that C_c could be approximately defined as a function of void ratio and strength sensitivity, as indicated in Figure 7.4, for the range of consolidation pressures between σ'_p and $1.4 \, \sigma'_p$. Results from the moderated sensitivity clays from southern Brazil are shown in Figure 7.5 and yield the following C_c–e_0 relationship:

$$C_c = 0.69 e_0 - 0.32 \tag{7.10}$$

A database compiling the Brazilian experience is summarized in Table 7.3.

Similarly, the hydraulic conductivity and void ratio has been the subject of numerous investigations. Tavenas et al. (1983) showed that the e–$\log k_v$ relationships for clays are approximately linear for vertical strains up to about 20%, as shown in Figure 7.6. This type of correlation can be better characterized by the hydraulic conductivity change index C_k defined as (Tavenas et al., 1983):

$$C_k = \Delta e / \Delta \log k \tag{7.11}$$

which can be simply related to the initial void ratio e_0 of the clay (Figure 7.7):

$$C_k = 0.5 e_0 \tag{7.12}$$

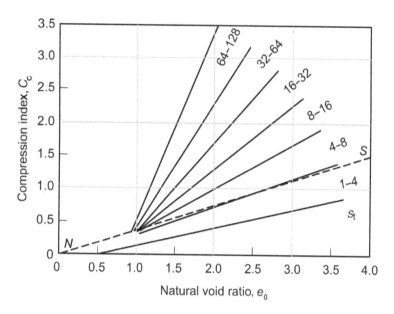

Figure 7.4 Relationship between compression index and initial void ratio for clay (Leroueil et al., 1983).

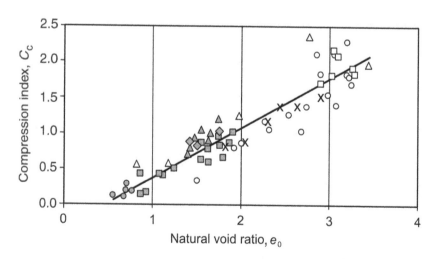

Figure 7.5 Relationship between compression index and initial void ratio for Brazilian clays.

Table 7.3 Typical values of C_c of Brazilian clays

Place	C_c	Reference
Ceasa, Porto Alegre, RS	0.34–2.27	Soares (1997)
Porto Alegre, RS	0.81–1.84	Soares (1997)
Tabaí-Canoas, RS	0.6–2.4	Dias and Gehling (1986)
Rio de Janeiro, RJ	0.5–1.8	Costa Filho et al. (1985)
	1.3–2.6	Ortigão (1980)
Sarapuí, RJ	1.35–1.86	Coutinho and Lacerda (1987)
Juturnaíba, RJ	0.29–3.75	Coutinho and Lacerda (1994)
Recife, PE	0.5–2.5	Coutinho et al. (1993, 1998)

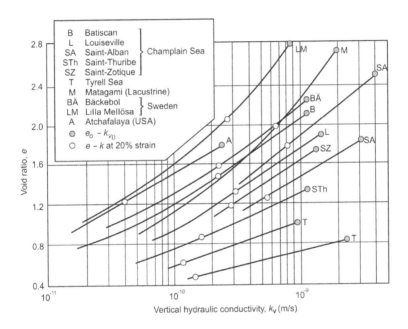

Figure 7.6 Variations of vertical hydraulic conductivity with void ratio (Tavenas et al., 1983).

The coefficient of consolidation can be expressed as a function of hydraulic conductivity and deformation modulus as:

$$C_v = \frac{kM}{\gamma_w} = \frac{k\sigma'_{v0}(1+e_0)}{0.434\gamma_w C_c} \qquad (7.13)$$

where M is the deformation modulus defined as $d\sigma'_v/d\varepsilon_v$ in one-dimensional compression. This expression is valid provided that Darcy's law applies and σ'_{v0}, M and k_v are constants. Table 7.4 summarizes the typical values of C_v of some Brazilian natural clay deposits. Much has been written on the variations between

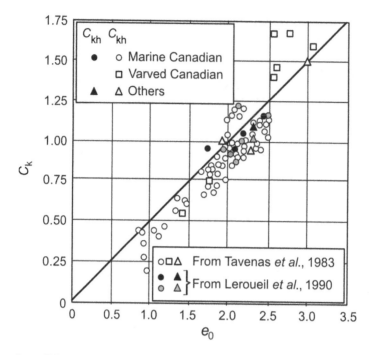

Figure 7.7 Relationship between the hydraulic conductivity change index and initial void ratio (Magnan and Tavenas, 1985).

the measured values of coefficients of consolidation. Figure 7.8 shows a typical case study at the Champlain Sea site (Leroueil *et al.*, 1995), where measurements of C_h range from 10^{-8} to 10^{-5} m²/s. Discrepancies between field and laboratory tests are due to one or more of the following main reasons (Simons, 1975; Hamouche *et al.*, 1995; Leroueil *et al.*, 1995, 2002; Leroueil and Hight, 2003):

- hydraulic conductivity measured in small laboratory specimens may underestimate the hydraulic conductivity of the soil mass, particularly in heterogeneous stratified deposits;
- inaccuracy of constitutive models and boundary conditions adopted in the interpretation of in situ tests;
- the two- or three-dimensional aspect of field conditions and of permeability anisotropy may be neglected in the interpretation of field problems.

Sand

The state of cohesionless soils is generally characterized by the relative density, which can be assessed from nearly every investigation tool used for field site characterization.

Table 7.4 Coefficients of consolidation of Brazilian clays

Place	C_v (cm²/s) * 10^{-4}	Test	Reference
Porto Alegre/ RS Ceasa	0.70–5.10	Vertical consolidation (NA)	Schnaid *et al.* (1997a, b)
	1.20–6.60	Radial consolidation (NA)	
	3.20–4.27	Piezocone (NA)	
Porto Alegre/ RS Airport	0.67–2.12	Vertical consolidation (NA)	Schnaid *et al.* (1997a, b)
	0.84–3.27	Piezocone (NA)	
	19.4–49.8	Piezocone (PA)	
Rio Grande, RS	1.00–5.00	Vertical consolidation (NA)	Dias and Bastos (1994)
Vale do Rio Quilombo, SP	4.00–8.90	Piezocone (NA)	Arabe (1995)
Vale do Rio Mogi, SP	4.00–8.90	Vertical consolidation (NA)	Massad (1985)
Baixada Santista, SP	0.001–0.10	Vertical consolidation (NA)	Souza Pinto and Massad (1978)
Sarapuí, RJ	1.40–4.40	Piezocone (NA)	Lacerda and Almeida (1995); Coutinho and Lacerda (1994); Rocha Filho (1987)
	24.0–102.0	Piezocone (PA)	
	1.0–10.0	Vertical consolidation (NA)	
Recife, PE	3.0–20.0	Vertical consolidation (NA)	Coutinho *et al.* (1993)
Salvador, BA	1.9–2.1	Vertical consolidation (NA)	Baptista and Sayão (1998)
	5.0–15.0	Piezocone (PA)	

Several attempts have been made to describe the mechanical behaviour of sands by the combined effects of relative density and stress level (e.g. Rowe, 1962; Lee and Seed, 1967), as indicated in Figure 7.9. This dependency can be conveniently expressed by a relative dilatancy index as (Bolton, 1986):

$$I_R = D_r(Q - \ln p') - R \tag{7.14}$$

where D_r is the relative density, p' the mean effective stress (in kPa), R is an empirical factor found equal to 1 as a first approximation and Q is a logarithm function of grain compressive strength, known to range from about 10 for silica sand to 7 for calcareous sand (see Chapters 2 and 3).

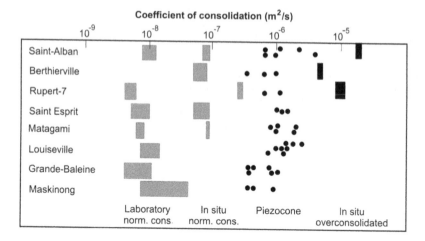

Figure 7.8 Range of variability of C_v for the Champlain site (Leroueil et al., 1995).

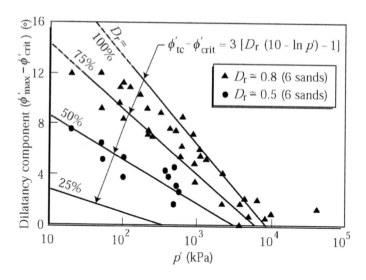

Figure 7.9 Dilatancy component for sand at various mean effective stress levels (Bolton, 1986).

For plane strain conditions:

$$\phi'_p - \phi'_{cv} = 5I_R \qquad (7.15)$$

For triaxial strain conditions:

$$\phi'_p - \phi'_{cv} = 3I_R \qquad (7.16)$$

Table 7.5 Properties of natural Japanese sand deposits (after Mimura, 2003)

	Yodo River sand	Yodo River sand	Yodo River sand	Natory River sand	Tone River sand	Edo River sand
G_z (g/cm3)	2.638	2.633	2.632	2.65	2.68	2.68
Gravel (%)	0.43	8.78	4.12	0.12	0.01	0.02
Sand (%)	1.11	91.01	93.78	99.65	96.21	99.56
Silt (%)	1.9	0.27	2.1	0.23	3.78	0.42
Clay (%)	0	0	0	0	0	0
U_c	98.46	3.3	3	2	2	2.2
D_{50} (mm)	0.32	0.82	0.62	0.22	0.18	0.29
e_{max}	1.054	0.883	0.921	1.167	1.33	1.227
e_{min}	0.665	0.569	0.567	0.765	0.775	0.812
Undisturbed frozen sand samples						
σ'_0 (kPa)	98	117.7	137.3	83.3	80.4	49
$\sigma'_1 - \sigma'_3$ (kPa)	405	396.1	469.7	388.3	355.9	103.4
E_{50} (MPa)	30.3	26.2	34.7	35.5	27.3	19.2
ϕ' (degrees)	42.2	38.4	39.1	40.9	41.7	39.7
ψ (degrees)	23.7	17.5	13	22.7	29.3	21.4
Reconstituted sand samples						
σ'_0 (kPa)	98	117.7	137.3	83.3	80.4	49
$\sigma'_1 - \sigma'_3$ (kPa)	352	316.7	405	308.9	398.1	191.2
E_{50} (MPa)	19.8	13.7	17.9	13.5	15.7	10.9
ϕ' (degrees)	39.5	34	36.6	39.5	42.5	41.9
ψ (degrees)	19.5	4.5	7.5	16.4	32	23.9

Basic physical and strength characteristics of four Japanese sands are summarized in Table 7.5 to illustrate some aspects of the behaviour of cohesionless soils (after Mimura, 2003). The grain size distribution and grain shapes of these sands are similar, although the mineral composition is different. Particles are predominantly angular with low values of roundness and sphericity. Results of the internal friction angle ϕ' and dilation angle Ψ measured from consolidated drained compression triaxial tests in both undisturbed frozen samples and reconstituted samples have been reported. Values of ϕ' and ψ for the undisturbed and reconstituted samples are generally similar, with ϕ' (*undisturbed*) slightly greater than ϕ' (*reconstituted*), indicating that effects of aging and cementation do not have such a significant effect on shear strength in these reported studies. However, these geomaterials show remarkable differences in the liquefaction resistance during undrained cyclic triaxial tests, due to the mode of particle interlocking.

Clearly, distinct behaviour emerges from geological environment, diagenesis and deposition processes of sand deposits. It is now widely recognized that mechanical behaviour of cohesionless geomaterials is influenced by

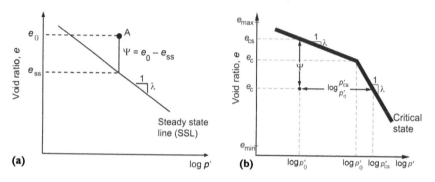

Figure 7.10 (a) Definition of the state parameter (after Been and Jefferies, 1985), and (b) non-linear characteristic of the critical state line (after Konrad, 1998).

sand gradation (determined by the grain size distribution), shape and mineralogy, as well as by the state of the soil, which is dictated by stress history, anisotropy, cementation and aging.

Although none of the Japanese sands shown in Table 7.5 is particularly sensitive to soil crushing, there is now an increasing consensus that, unlike the case for clays, the critical state line on sands is non-linear in the $\upsilon - \ln p'$ space (e.g. Konrad, 1998; Jefferies and Been, 2000), exhibiting a steeper slope at high stresses due to changes in gradation and grain shape induced by grain crushing, as shown in Figure 7.10. The amount of particle crushing that occurs in an element of soil under stress depends on particle size distribution, particle shape, mean effective stress, effective stress path, void ratio and particle bonding (e.g. Hardin, 1978). Triaxial tests in samples of equal relative density have conclusively shown that the dilatant component reduces with increasing mean effective stress, since at high stresses, crushing eliminates dilation (e.g. Bolton, 1986).

It is therefore advisable to make extensive use of relative density to characterize conhesionless natural soil deposits, provided that engineering judgement is exercised when using correlations established on the basis of tests carried out on fresh, uncemented laboratory samples characterized by their relative density alone.

Residual soils

Residual soils formed by in situ weathering exhibit behaviour which is distinct from transported soils, given the complex spatial structure arrangement that emerges from both soil particles and 'bonding' between particles, which can be progressively destroyed during plastic straining. Although most geomaterials are recognized as being structured according to the conceptual framework proposed by Leroueil and Vaughan (1990), the natural structure of bonded soils has a dominant effect on their mechanical

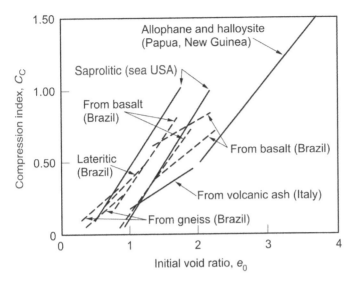

Figure 7.11 Relationship between compression index and initial void ratio for residual soils (Leroueil and Vaughan, 1990).

response, since the cohesion/cementation component can dominate soil shear strength during engineering applications involving low stress levels.

As with clay, attempts have been made to establish C_c–e_0 relationships for residual soils, as shown in Figure 7.11 (after Leroueil and Vaughan, 1990), but considerable scatter is observed. Atterberg limits on the plasticity chart are purely indicative of the clay mineral composition of the soil and do not represent mechanical behaviour. Characterization should necessarily be linked to geological, pedological and engineering features of the deposit and, at present, there are no suitable index tests for residual soils.

Given the bonded structure and unsaturated conditions of many residual soil deposits, shear strength has to be expressed in terms of both the internal friction angle and cohesion intercept. Database values from Singapore soils are summarized in Table 7.6 to illustrate the large variation in measured effective friction angles that can be attributed to variable parent rock types and the consequent changes in soil mineralogy and structure. Values of cohesion will vary over an even larger range, given the changes in suction and cementation.

Stiffness

Recent developments in the characterization of prefailure deformation properties indicate that stiffness response of geomaterials is complex, being a function of a

Table 7.6 Effective shear strength parameters of Singapore residual soils (after Leong et al., 2003)

Residual soils	References	Range of c′ (kPa)	Average c′ (kPa)	Range of φ′ (°)	Average φ′ (°)
Bukit Timah granite	Dames and Moore (1983)	0–125	0	13–36	30
	Poh et al. (1985)	0–42	–	20–35	–
	Yang and Tang (1997)	5–10	–	35–40	–
	Tan et al. (1987)	0–40	15	30–35	32
	Rahardjo (2000)	12–50	26	29–33	30
		0–14	9	27–31	29
	KarWinn et al. (2001)	–	–	20–40	–
	Zhou (2001)	–	7	–	32
Jurong Formation	Dames and Moore (1983)	5–100	–	17–46	28
		10–65	–	17–36	–
	Young et al. (1985)	–	12	13–40	35
		–	17	–	28
	Lo et al. (1988)	–	6	–	32
	Lim (1995)	19–50	31	24–40	27
	Gasmo (1997)	15–22	20	–	27
	Hritzuk (1997)	–	95	–	35
	Rahardjo (2000)	5–9	7	29–32	30.5
	Zhu (2000)	10–30	–	24–40	–
	Seah et al. (2001)	0–40	–	24–40	33
	Orihara et al. (2001)	–	–	–	

number of factors such as kinematic yielding, stress history, anisotropy, structuration and destructuration, non-linearity of strain and pressure, etc. (e.g. Jardine, 1985; Stokoe II et al., 1994; Jamiolkowski et al., 1995b; Tatsuoka et al., 1997; Lo Presti et al., 1999). As stated by Tatsuoka et al. (1997), uncemented and highly cemented geomaterials, and unstructured and structured soils, show a wide range of small strain Young's modulus (typically between 10 and 10^4 MPa) and non-linear deformation characteristics, as shown in Figure 7.12. Generally, well-cemented soils exhibit a greater number of linear deformation characteristics and a stiffer degradation due to destructuration. Anticipating realistic deformation properties and representing the full range of behaviour by a single model are among the most difficult tasks in geotechnical design.

In conclusion, the information presented in this chapter is useful in indicating the general range of values for typical geotechnical properties. A database compiling the experiences of studies worldwide will clearly be of direct benefit to students and professionals who are willing to explore the field of soil mechanics in depth. The reader should bear in mind that no attempt has been made to compile and present these reported data in a unified way. Because of

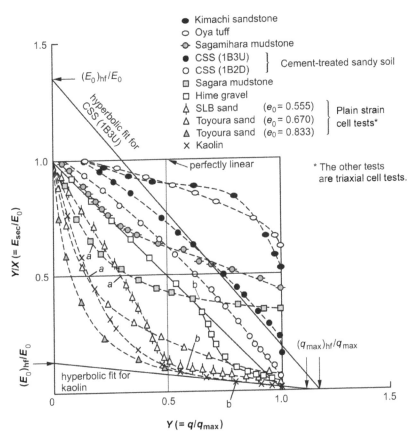

Figure 7.12 Non-linear deformation characteristics of various geomaterials (Tatsuoka *et al.*, 1997).

the extremely variable nature of geologic materials, the range of values presented in the tables and figures in this chapter should be considered generic, but not necessarily representative. It cannot be overstated that generalized knowledge provided by the compilation of published information is no substitute for site-specific laboratory and field investigation.

References and further reading

Aas, G. 1965. A study of the effect of vane shape and rate of strain in the measured values of in situ shear strength of clays. *Proc. 6th Int. Conf. Soil Mech. Found. Engng*, 1:141–145.

Aas, G. 1967. Vane tests for investigation of undrained shear strength of clays. *Proc. Geotech. Conf.*, Oslo, 1:3–8.

Aas, G., Lacasse, S., Lunne, T. and Hoeg, K. 1986. Use of in situ test for foundation design on clay. *Proc. ASCE Specialty Conf. In Situ '86: Use of In Situ Tests in Geotech. Engng*, Blacksburg, VA, 1–30.

ABNT MB-3406 1991. *Cone penetration test: CPT*, Brazilian Standard.

Abu-Farsakh, M., Tumay, M. and Voyiadjis, G. 2003. Numerical parametric study of piezocone penetration test in clays. *Int. J. Geomech.*, ASCE, 3(2):170–181.

AFNOR 2000. *Ménard pressuremeter test. Part 1*. La Plaine Saint-Denis, France. French Standard, in French.

Ajalloeian, R. and Yu, H.S. 1988. Chamber studies of the effects of pressuremeter geometry on test results in sand. *Géotechnique*, 48:621–636.

Almeida, M.S.S. and Marques, M.E.S. 2003. The behaviour of Sarapui soft clay. *Charact. and Engng Properties of Natural Soils*, AA Balkema Publishers, The Netherlands, 1:477–504.

Alonso, E.E., Gens, A. and Josa, A. 1990. A constitutive model for partially saturated soils. *Géotechnique*, 40(3):405–430.

Anderson, W.F., Pyrah, I.C. and Haji-Ali, F. 1987. Rate effects in pressuremeter tests in clay. *J. Geotech. Engng*, ASCE, 113(11):1344–1358.

Andresen, A., Berre, T., Kleven, A. and Lunne, T. 1979. Procedures used to obtain soil parameters for foundation engineering in the North Sea. *Marine Geotech.*, 3(3):201–266.

Andrus, R.D. and Stokoe II, K.H. 1997. Liquefaction resistance based on shear wave velocity. Proc. NCEER Workshop on Evaluation of Liquefaction Resistance of Soils, Technical Report NCEER-97-0022. T.L. Youd and I.M. Idriss (eds), National Center for Earthquake Engineering Research, Buffalo, NY, 89–128.

Andrus, R.D. and Stokoe II, K.H. 1999. Liquefaction evaluation procedure based on shear wave velocity. Wind and seismic effects. *U.S./Japan Natural Resources Development Program (UJNR). Joint Meeting*, 31st Technical Memorandum of PWRI 3653. Proceedings. 11–14 May 1999, Tsukuba, Japan, 469–474.

Andrus, R.D. and Stokoe II, K.H. 2000. Liquefaction resistance of soils from shear-wave velocity. *J. Geotech. and Envir. Engng*, ASCE, 126(11):1015–1025.

Andrus, R.D., Stokoe, K.H., II and Chung, R.M. 1999. Draft guidelines for evaluating liquefaction resistance using shear wave velocity measurements and simplified procedures. *Report*, Building and Fire Research Laboratory, National Institute of Standards and Technology, Gaithersburg, MD.

Andrus, R.D., Stokoe, K.H., II and Juang, C.H. 2004. Guide for shear-wave-based liquefaction potential evaluation. *Earthquake Spectra*, 20(2):285–308.

Aoki, N. and Velloso, D.A. 1975. An approximate method to estimate the bearing capacity of piles. *Proc. 5th Pan Am. Conf. Soil Mech. Found. Engng*, Buenos Aires, HUELLA Estudio Grafico, 1:367–376.

Aoki, N. and Cintra, J.C.A. 2000. The application of energy conservation Hamilton's principle to the determination of energy efficiency in SPT test. *Proc. 6th Int. Conf. on the Application of Stress-Wave Theory to Piles*, São Paulo, 457–460.

Arabe, L.C.G. 1995. Geotechnical properties of the experimental testing site of Baixada Santista. *Proc. Brazilian Conf. Soil Mech. Found. Engng*, 7(5):25–47, in Portuguese.

Aragão, C.J.G. 1975. *Geotechnical properties of Grande Rio soft clay deposits.* MSc thesis, PUC, Rio de Janeiro, 251pp, in Portuguese.

Archie, G.E. 1942. The electrical resistivity as an aid in determining some reservoir characteristics. *Trans. American Institute Min. Engng*, 146:54–62.

ASTM D 1586-99 *Standard test method for penetration test and split-barrel sampling of soils.* American Society for Testing and Materials, 1999.

ASTM D 2573 *Standard test method for field vane test in cohesive soils.* American Society for Testing and Materials, 2007.

ASTM D 4719 *Standard test method for pre-bored pressuremeter testing in soils*, Volume 04.08. American Society for Testing and Materials Standard, 2000.

ASTM D 4719 *Standard test method for pre-bored pressuremeter testing in soils.* American Society for Testing and Materials, 2007.

ASTM D 5778 *Standard test method for performing electric friction cone and piezocone penetration testing of soils.* American Society for Testing and Materials, 2000.

ASTM D 6635 *Standard test method for performing the flat plate dilatometer.* American Society for Testing and Materials, 2007.

Azzouz, A.S., Baligh, M.M. and Ladd, C.C. 1983. Corrected field vane strength for embankment design. *J. Geotech. Engng Div.*, ASCE, 109(5):730–734.

Baguelin, F. 1982. Rules of foundation design using self-boring pressuremeter results. *Symp. Pressuremeter and its Marine Applications*, Editions Technip, Paris.

Baguelin, F., Jezequel, J.F., Lemee, E. and Mehause, A. 1972. Expansion of cylindrical probes in cohesive soils. *J. Soil Mech. Found. Div.*, ASCE, 98 (SM11):1129–1142.

Baguelin, F., Jezequel, J.F. and Shields, D.H. 1978. *The pressuremeter and foundation engineering*. Trans. Tech. Publications, Clausthall, Germany, 278pp.

Balachowski, L. 2006. Analysis of dilatometer test in calibration chamber. *Proc. 2nd Int. Flat Dilatometer Conf.*, Washington DC.

Baldi, G., Bellotti, R., Ghionna, V.N., Jamiolkowski, M. and Pasqualini, E. 1981. Cone resistance of a dry medium sand. *Proc. 10th Int. Conf. on Soil Mech. and Found. Engng*, Stockholm, AA Balkema Publishers, The Netherlands, 2:455–458.

Baldi, G., Bellotti, R., Ghionna, V.N., Jamiolkowski, M. and Pasqualini, E. 1982. Design parameters for sand from CPT. *Proc. 2nd European Symp. on Penetration Testing*, ESOPT-II, Amsterdam, 2:425–432.

Baligh, M.M. 1975. *Theory of deep site static cone penetration resistance.* MIT, Cambridge, MA, Research Report R75-56.

Baligh, M.M. 1985. Strain path method. *J. Soil Mech. Found. Engng Div.*, ASCE, 11(7):1108–1136.

Baligh, M.M. 1986. Undrained deep penetration. I: Shear stresses. *Géotechnique*, 36(4):471–485.

Baligh, M.M. and Levadoux, J.N. 1986. Consolidation after undrained piezocone penetration. II: Interpretation. *J. Soil Mech. Found. Engng Div.*, ASCE, 112(7): 727–745.

Baligh, M.M. and Scott, J.N. 1975. Quasi static deep penetration in clays. *J. Soil Mech. Found. Engng Div.*, ASCE, 101(11):1119–1133.

Balmaceda, A.R. 1991. *Compacted soils: an experimental and theoretical study.* PhD thesis, Universitat Politècnica de Catalunya, Barcelona, in Spanish, 434.

Baptista, H.M. and Sayão, A.S.F.J. 1998. Geotechnical properties of the Enseada do Cabrito soft clay deposit. *Proc. 11th Brazilian Cong. on Soil Mech. Geotech. Engng*, Brasília, 2:911–916, in Portuguese.

Barden, L., Ismail, H. and Tong, P. 1969. Plane strain deformation of granular material at low and high pressures. *Géotechnique*, 19(4):441–452.

Barentsen, P. 1936. Short description of a field testing method with cone-shaped sounding apparatus. *Proc. 1st Int. Soil Mech. Found. Engng*, Cambridge, MA, 1, B/3, 6–10.

Barksdale, R.D. and Blight, G.E. 1997. Compressibility and settlement of a residual soil. *Mech. of Residual Soils*, Blight (ed.), AA Blakema Publishers, The Netherlands, 99–154.

Barros, J.M.C. 1997. *Dynamic shear modulus in tropical soils*, PhD thesis, São Paulo University, in Portuguese.

Battaglio, M. and Maniscalco, R. 1983. *Il piezocono: Esecuzione ed interpretazione*, Politechnico di Torino, Report No. 607, in Italian.

Battaglio, M., Bruzzi, D., Jamiolkowski, M. and Lancellotta, R. 1986. Interpretation of CPTs and CPTUs. *Proc. 4th Int. Geotech. Seminar*, Singapore, 129–143.

Battaglio, M., Jamiolkowski, M., Lancellotta, R. and Maniscalco, R. 1981. Piezometer probe test in cohesive deposits. Cone penetration testing and experience. *Proc. Session ASCE National Convention*, St Louis, MO, 264–302.

Becker, D.E. 2001. *Site Characterisation Geotechnical and Geo-environmental Engineering Handbook.* Kluwer Academic Publishing, Norwell, MA, 69–105.

Been, K. and Jefferies, M.G. 1985. A state parameter for sands. *Géotechnique*, 35(2):99–112.

Been, K., Jefferies, M.G., Crooks, J.H.A. and Hotemburg, L. 1987. The cone penetration test in sands. Part 2: General influence of state. *Géotechnique*, 37(3):285–299.

Begemann, H.K.S. 1953. Improved method of determining resistance to adhesion by sounding through a loose sleeve behind a cone. *Proc. 3rd Int. Conf. Soil Mech. Found. Engng*, Zurich, 1:213–217.

Begemann, H.K.S. 1963. The use of the static soil penetrometer in Holland. *New Zealand Engng*, Wellington, 18(2):41–49.

Begemann, H.K.S. 1965. The friction jacket cone as an aid in determining the soil profile. *Proc. 6th Int. Conf. Soil Mech. Found. Engng*, Montreal, 1:17–20.

Belicanta, A. 1985. *SPT dynamic energy from a theoretical-experimental investigation.* MSc thesis, Polytechnic University – USP, São Paulo, Brazil, in Portuguese.

Belicanta, A. 1998. *Evaluation of factors affecting the SPT penetration resistance.* PhD Thesis, São Paulo University, in Portuguese.

Bellotti, R., Ghionna, V., Jamiolkowski, M., Lancellotta, R. and Manfredini, G. 1986. Deformation characteristics of cohesionless soils from in-situ tests. *In-situ '86 – Use of In-situ Tests in Geotech. Engng*, ASCE, Blacksburg, 47–73.

Bellotti, R., Ghionna, V.N., Jamiolkowski, M., Robertson, P.K. and Peterson, R.W. 1989. Interpretation of moduli from self-boring pressuremeter tests in sand, *Géotechnique*, 39(2):269–292.

Bellotti, R., Jamiolkowski, M., Lo Presti, D.C.F. and O'Neill, D.A. 1996. Anisotropy of small strain stiffness in Ticino sand. *Géotechnique*, 46(1):115–131.

Bemben, S.M. and Myers, H.J. 1974. The influence of rate on static cone resistance in Connecticut River Valley varved clay. *Proc. European Symp. on Penetration Testing*, ESOPT, Stockholm, 2:33–34.

Berezantzev, U.G. 1961. Load bearing capacity and deformation of piled foundations. *5th Int. Conf. Soil Mech. Found. Engng*, 2:11–12.

Biscontin, G. and Pestana, J.M. 2001. Influence of peripheral velocity on vane shear strength of an artificial clay. *ASTM Geotech. Testing J.*, 24(4):423–429.

Bishop, A.W. 1971. Shear strength parameters for undisturbed and remoulded soil specimens. Stress-strain behaviour of soils. *Proc. Roscoe Memorial Symp.*, Cambridge, 3–39.

Bjerrum, L. 1972. Embankments on soft ground. *Proc. ASCE Speciality Conf. on Performance of Earth and Earth-Supported Structures*, Purdue University, USA, 2:1–54.

Bjerrum, L. 1973. Problems of soil mechanics and construction on soft clays. *Proc. 8th Int. Conf. Soil Mech. Found. Engng*, Moscow, 3:111–159.

Bjerrum, L. and Simons, N.E. 1960. Comparison of shear strength characteristics of normally consolidated clays. *Proc. ASCE Research Conf. on Shear Strength of Cohesive Soils*, Boulder, CO, 711–726.

Blight, G.E. 1968. A note on field vane testing of silty soils. *Can. Geotech. J.*, 5(3):142–149.

Blight, G.E. 1997. Mechanics of residual soils. *Tech. Committee 25, Int. Society Soil Mech. Found. Engng*, AA Balkema Publishers, The Netherlands, 237pp.

Bolton, M.D. 1979. *A guide to soil mechanics.* Macmillian Press, London, 439pp.

Bolton, M.D. 1986. The strength and dilatancy of sands. *Géotechnique*, 36(1):65–78.

Bolton, M.D. and Whittle, R.W. 1999. A non-linear elastic/perfectly plastic analysis for plane strain undrained expansion tests. *Géotechnique*, 49(1):133–141.

Bolton, M.D., Gui, M.W., Garnier, J., Corte, J.F., Bagge, G., Laue, J. and Renzi, R. 1999. Centrifuge cone penetration tests in sands. *Géotechnique*, 49(4):543–552.

Briaud, J.-L. 1992. *The pressuremeter.* AA Balkema Publishers, The Netherlands, 192pp.

Brinch Hansen, J. 1961. The ultimate resistance of rigid piles against transversal forces. *Danish Geotechnical Institute Bulletin*, 12:5–9.

Brown, D.N. and Mayne, P.W. 1993. Stress history profiling of marine clays by piezocone. *Proc. 4th Canadian Conf. Marine Geotech. Engng*, Memorial University, Newfoundland, 1:1305–1312.

Brugger, P.J., Almeida, M.S.S., Sandroni, S.S., Brant, J.R., Lacerda, W.A. and Danziger, F.A.B. 1994. Geotechnical parameters of the Sergipe soft clay. *Proc. 10th Brazilian Conf. Soil Mech. Found. Engng*, 1:539–546.

BS 1377-90 1990. *Methods for test for soils for civil engineering purposes. In-situ tests*. British Standard.

Burbidge, M.C. 1982. *A case study review of settlements on granular soils*. MSc. dissertation, Imperial College, London.

Burghignoli, A., Cavalera, L., Chieppa, V., Jamiolkowski, M., Mancuso, C., Marchetti, S., Pane, V., Paoliani, P., Silvestri, F., Vinale, F. and Vittori, E. 1991. Geotechnical characterization of Fucino clay. *Proc. Xth European Conf. Soil Mech. Found. Engng*, Florence, 1:27–40.

Burland, J.B. 1989. Small is beautiful: the stiffness of soils at small strains. *Can. Geotech. J.*, 26:499–516.

Burland, J.B. 1990. On the compressibility and shear strength of natural clays. *Géotechnique*, 40(3): 329–378.

Burland, J.B. and Burbidge, M.C. 1985. Settlement of foundations on sand and gravel. In: *Proc. ICE*, 78:1, London.

Burland, J.B., Broms, B.B. and de Mello, V.F.B. 1977. Behaviour of foundations and structures. *State-of-the-Art Review*, 9th Int. Conf. Soil Mech. Found. Engng, Tokyo, 3:495–546.

Burmister, D.M. 1948. The importance and practical use of relative density in soil mechanics. *Symp. ASTM 48*, 1249–1268.

Burns, S.E. and Mayne, P.W. 1998. Monotonic and dilatory pore-pressure decay during piezocone tests in clay. *Can. Geotech. J.*, 35(6):1063–1073.

Bustamante, M. and Gianeselli, L. 1982. Pile bearing capacity prediction by means of static penetrometer CPT. *Proc. 2nd European Symp. on Geotech. Engng of Hard Soils–Soft Rocks*, Athens, Greece, AA Balkema Publishers, The Netherlands, 405–416.

Bustamante, M. and Frank, R. 1999. Current French practice for axially loaded piles. *Ground Engng*, March, 38–44.

Butler, F.G. 1975. Heavily overconsolidated clays. General report and state-of-the-art review for session. *Proc. 3rd Conf. on Settlement of Structures*. Pentech Press, London.

Butler, J.J., Caliendro, J.A. and Goble, G.G. 1998. Comparison of SPT energy measurements methods. *Int. Conf. in Site Characterisation, ISC '98*, Atlanta, 2:901–906.

Cadling, L. and Odenstad, S. 1948. *The vane borer*. Royal Swedish Geotechnical Institute, Stockholm.

Campanella, R.G. 1999. Geo-environmental site characterization of soils using in-situ testing methods. *Asian Institute of Technology 40th Year Conf., New Frontiers and Challenges*.

Campanella, R.G. and Robertson, P.K. 1986. Research and development of the UBC cone pressuremeter. *Proc. 3rd Canadian Conf. Marine Geotech. Engng*, St John's, Newfoundland, I, 205–214.

Campanella, R.G. and Robertson, P.K. 1988. Current status of the piezocone test. *Proc. Int. Symp. on Penetration Testing, ISOPT-1*, Orlando, 1, AA Balkema Publishers, The Netherlands, 93–116.

Campanella, R.G. and Weemees, I. 1990. Development and use of an electrical resistivity cone for groundwater contamination studies. *Canadian Geotech. J.*, 27(5):557–567.

Campanella, R.G. and Robertson, P.K. 1991. Use and interpretation of a research dilatometer. *Canadian Geotech. J.*, 28:113–126.

Campanella, R.G., Gillespie, D. and Robertson, P.K. 1982. Pore pressure during cone penetration testing. *Proc. 2nd European Symp. Penetration Testing, ESOPT-II*, Amsterdam, AA Balkema Publishers, The Netherlands, 1:507–512.

Campanella, R.G., Robertson, P.K. and Gillespie, D. 1986. Seismic cone penetration test. *Proc. ASCE Spec. Conf. In Situ '86 on Use of In Situ Tests in Geotech. Engng*, Blacksburg, VA, 116–130.

Caquot, A. and Kerisel, J. 1956. *Traité de méchanique des sols*. Gauthier-Villars, Paris.

Carlsson, L. 1948. Determination in-situ of the shear strength of undisturbed clay by means of a rotating auger. *Proc. 2nd Int. Conf. on Soil Mech. Found. Engng*, Rotterdam, 1:265–270.

Carter, J.P. 1978. CAMFE, a computer program for the analysis of cavity expansion in soil. *Cambridge Univ. Press Dept. of Engng*. Report CUEDIC–Soils TR52.

Carter, J.P. and Kulhawy, F.H. 1992. Analysis of laterally loaded shafts in rocks. *J. Geotech. Engng*, ASCE, 118(6):839–855.

Carter, J.P., Randolph, M.F. and Wroth, C.P. 1979. Stresses and pore pressure changes around a driven pile. *Int. J. Numer. Analyt. Meth.Geomech.*, 3:305–322.

Carter, J.P., Booker, J.R. and Yeung, S.K. 1986. Cavity expansion in cohesive frictional soils. *Géotechnique*, 36:349–358.

Cavalcante, E.M. 2002. *Experimental and theoretical investigations based on the SPT*. PhD thesis, COPPE/UFRJ, in Portuguese.

Cecconi, M., Viggiani, G. and Rampello, S. 1988. An experimental investigation on the mechanical behaviour of pyroclastic soft rock. *Proc. 2nd Int. Symp. on Geotech. Engng of Hard Soils–Soft Rocks*, Naples, 1:473–482.

Cecconi, M., Viggiani, G. and Rampello, S. 1998. An experimental investigation of the mechanical behaviour of a pyroclastic soft rock. *The Geotechnics of Hard Soils–Soft Rocks*, AA Balkema Publishers, The Netherlands, pp. 473–482.

Cerato, A.B. and Lutenegger, A.J. 2004. Disturbance effects of field vane tests in a varved clay. *Proc. 2nd Int. Symp. on Geotech. Site Characterization*, Porto, Portugal, 1:861–867.

Cetin, K.O., Seed, R.B., Der Kiureghian, A., Tokimatsu, K., Harder, Jr. L.F., Kayen, R.E. and Moss, R.E.S. 2004. SPT-based probabilistic and deterministic assessment of seismic soil liquefaction potential. *J. Geotech. Geoenvironm. Engng*, ASCE, 130(12):1314–1340.

Chandler, R.J. 1988. The in-situ measurement of the undrained shear strength of clays using the field vane. *Vane Shear Strength Testing in Soils. Field and Laboratory Studies*, A.F. Richards (ed.) ASTM STP 1014, Philadelphia, 13–44.

Chang, C.S., Misra, A. and Sundaram, S.S. 1991. Properties of granular packings under low amplitude cyclic loading. *Soil Dyn. Earthquake Eng.*, 10(4):201–211.

Chapman, G.A. and Donald, I.B. 1981. Interpretation of static penetration tests in sand. *Proc. 10th Int. Conf. on Soil Mech. Found. Engng*, Stockholm, AA Balkema Publishers, The Netherlands, 2:455–458.

Charles, M., Yu, H.S. and Sheng, D. 1999. Finite element analysis of pressuremeter tests using critical state soil models. *Proc. 7th Int. Symp. on Num. Models in Geomech.*, Graz, 645–650.

Chen, B.S.-Y. and Mayne, P.W. 1994. Profiling the overconsolidation ratio of clays by piezocone tests. *Report GIT-CEE/GEO-94-1*, National Science Foundation, Georgia Institute of Technology, 280pp.

Chen, B.S.-Y. and Mayne, P.W. 1996. Statistical relationships between piezocone measurements and stress history of clays. *Canadian Geotech. J.*, 33(3):488–498.

Clarke, B.G. 1995. *Pressuremeter in Geotechnical Design*. Chapman & Hall, London.

Clarke, B.G., Carter, J.P. and Wroth, C.P. 1979. In situ determination of the consolidation characteristics of saturated clays. *Proc. 7th European Conf. Soil Mech. Found. Engng*, Brighton, 2:207–213.

Clayton, C.R.I. 1993. *The standard penetration test (SPT) – methods and use*. Construction Industry Research and Information Association, Funder Report/CP/7, CIRIA, London, 129pp.

Clayton, C.R.I. 1995. *The standard penetration test (SPT): methods and use*. Report 143. CIRIA, London.

Clayton, C.R.I. and Hababa, M.B. 1985. Dynamic penetration resistance, and the prediction of the compressibility of a fine grained sand – a laboratory study. *Géotechnique*, 25(1):19–31.

Clayton, C.R.I. and Serratrice, J.F. 1993. General report session 2: the mechanical properties and behavior of hard soils and soft rocks. *Proc. 1st Int. Symp. on the Geotech. of Hard Soils–Soft Rocks*, Athens, 3:1839–1877.

Clayton, C.R.I. and Heymann, C. 2001. The stiffness of geomaterials at very small strains. *Géotechnique*, 51(3):245–256.

Clayton, C.R.I., Simons, N.E. and Matthews, M.C. 1982. *Site Investigation*. Granada Publishing, London, 424pp.

Clayton, C.R.I., Matthews, M.C. and Simons, N.E. 1995b. *Site investigation*. Blackwell Science, 584 pp.

Clough, R.W. and Penzien, J. 1975. *Dynamics of Structures*, McGraw-Hill, 629pp.

Cole, K.W. and Stroud, M.A. 1976. Rock socket piles at Coventry Point, Marketway. Coventry Symposium on Piles in Weak Rock. *Géotechnique*, 26(1):47–62.

Collins, I.F., Pender, M.J. and Wang, Y. 1997. Cavity expansion in sands under drained loading conditions. *Int. J. Num. Anal. Methods in Geomech.* 16(1):3–23.

Consoli, N.C., Prietto, P.D.M., Carraro, J.A.H. and Heineck, K.S. 2001. Behavior of soil-fly ash-carbide lime compacted mixtures. *J. Geotech. Geoenvironm. Engng*, 127(9):774–782.

Consoli, N.C., Rotta, G.V. and Prietto, P.D.M. 2000. The influence of curing under stress on the triaxial response of cemented soils. *Géotechnique*, 50(1):99–105.

Consoli, N.C., Ulbrich, L.A. and Prietto, P.D.M. 1998. Influence of fiber and cement addition on behavior of sandy soil. *J. of Geotech. Geoenvironm. Engng*, ASCE, 12(124):1211–1214.

Coop, M.R. and Atkinson, J. 1993. The mechanics of cemented carbonate sands. *Géotechnique*, 43(1):53–67.

Coop, M.R. and Airey, D.W. 2003. Carbonate sands. *Charact. and Engng Properties of Natural Soils, Singapore*. AA Balkema Publishers, The Netherlands, 2:1049–1086.

Cornforth, D.A. 1961. *Plane strain failure characteristics of a saturated sand*. PhD thesis, University of London.

Correia, G., Viana da Fonseca, A. and Gambin, M. 2004. Routine and advanced analysis of mechanical in situ tests. *Proc. 2nd Int. Conf. on Site Charact.*, Millpress, Porto, 1:74–88.

Costa Filho, L.M. and Vargas, Jr. E. 1985. *Hydraulic properties. Peculiarities of geotechnical behaviour of tropical lateritic and saprolitic soils.* Progress Report (1982–1985). Brazilian Society of Soil Mechanics, 67–84.

Costa Filho, L.M., Werneck, M.G.L. and Collet, H.B. 1977. The undrained strength of a very soft clay. In: *Proc. 9th Int. Conf. Soil Mech. Found. Engng*, Tokyo, Vol. 1, pp.79–82.

Costa Filho, L.M., Aragão, C.J.G. and Velloso, P.P.C. 1985. Geotechnical characteristics of soft clay deposits from Rio de Janeiro. *Soils and Rocks*, 8(1):3–13, in Portuguese.

Cotecchia, F. and Chandler, R.J. 1997. The influence of structure on the pre-failure behaviour of a natural clay. *Géotechnique*, 47(3):523–544.

Coutinho, R.Q. 1988. Stress-strain and strength response of the Jurtunaíba soft clay. *Proc. Symp. New Concepts on Laboratory and Field Tests*, Brazil, 2:709–726, in Portuguese.

Coutinho, R.Q. and Lacerda, W.A. 1976. *Compacted fills on compressible soils: vertical and radial consolidation of the Rio de Janeiro soft clay.* Instituto de Pesquisa Rodoviária, Rio de Janeiro, in Portuguese.

Coutinho, R.Q. and Lacerda, W.A. 1987. Characterization and consolidation of Jurtunaíba organic clays. *Proc. Int. Symp. on Geotech. Engng on Soft Clays*, Mexico, 1:17–24.

Coutinho, R.Q. and Ferreira, S.R.M. 1988. Organic clays from Recife: compressibility of six deposits. *Sidequa, Proc. Symp. from Brazilian Quaternary Near-shore Deposits.* Rio de Janeiro, 3:35–54, in Portuguese.

Coutinho, R.Q. and Lacerda, W.A. 1994. Characterization and consolidation characteristics of Juturnaíba organic clay. *Soils and Rocks*, ABMS, São Paulo, 17(2):145–154, in Portuguese.

Coutinho, R.Q., Oliveira, J.T.R. and Danziger, F.A.B. 1993. Geotechnical characterization of the Recife soft clay. *Soils and Rocks*, ABMS, São Paulo, 16(4):255–266, in Portuguese.

Coutinho, R.Q., Oliveira, J.T.R. and Danzinger, F.A.B. 1995. Geotechnical characterization of a soft clay deposit from Recife. *COPPE Symp., COPPEGEO '95*, Rio de Janeiro, 1:41–54.

Coutinho, R.Q., Oliveira, J.T.R., França, A.E. and Danziger, F.B. 1998. Piezocone tests in the soft clay deposit of Ibura, Recife, PE. *Proc. 11th Brazilian Cong. Soil Mech. Geotech. Engng*, 2:957–966, in Portuguese.

Crespellani, T. and Vannucchi, G. 1991. Dynamic properties of soils. In: *Seismic hazard and site effects in the Florence area.* G. Vannucchi (ed.), Assoc. Geotech. Italiana, Rome, 71–80.

Crooks, J.H.A., Been, K., Becker, D.E. and Jefferies, M.G. 1988. CPT interpretation in clays. *Proc. 1st Int. Symp. on Penetration Testing, ISOPT-1*, Orlando, AA Balkema Publishers, The Netherlands, 2:715–722.

Crova, R., Jamiolkowski, M., Lancellotta, R. and Lo Presti, D.C.F. 1993. Geotechnical characterisation of gravelly soils at Messina site: selected topics, *Wroth Memorial Symp. Predictive Soil Mech.*, Oxford, Thomas Telford, London, 199–218.

Cubrinovski, M. and Ishihara, K. 1999. Empirical correlations between SPT N-value and relative density for sandy soils. *Soils and Found.*, 39(5):61–72.

Cuccovillo, T. and Coop, M.R. 1997. Yielding and pre-failure deformation of structured sands. *Géotechnique*, 47(3):491–508.

Cuccovillo, T. and Coop, M. 1999. On the mechanics of structured sands. *Géotechnique*, 49(4):349–358.

Cundall, P.A. and Streck, O.D.L. 1979. A discrete numerical model for granular assemblies. *Géotechnique*, 29(1):47–65.

Dalton, J.C.P. 2005. United Kingdom experience with pressuremeter 1982–2005. *Proc. Int. Symp. 50 Years of Pressuremeter*, Presses de L'Ecole Nationale des Ponts et Chaussées, Paris, 2:201–210.

Dames and Moore. 1983. *Singapore Mass Rapid Transit System – Detailed Geotechnical Study Interpretative Report*. Prepared for Provisional Mass Rapid Transit Authority, Singapore.

Daniel, C.R. 2000. *Split spoon penetration testing in gravels*. MASc. thesis, University of British Columbia, Vancouver, BC.

Daniel, C.R., Howie, J.A., Campanella, R.G. and Sy, A. 2004. Characterisation of SPT grain size effects in gravels. *Proc. 2nd Int. Conf. on Site Charact.*, Milpress, Porto, 1:299–306.

Danziger, F.A.B. 1990. *Development of piezocone testing equipment for use in soft clays*. COPPE/UFRJ Technical Publication, in Portuguese.

Danziger, F.A.B. and Lunne, T. 1997. Rate effects in cone penetration testing. *Géotechnique*, 47(5):901–914.

Danziger, F.A.B., Soares, M.S.S. and Sills, G.C. 1996. The significance of the strain path analysis in the interpretation of piezocone dissipation data. *Géotechnique*, 46(2):143–154.

Dauncey, P.C. and Woodland, A.R. 1988. Bored cast in situ pile foundations in Keuper Marl for the Birmingham International Arena. *Proc. Conf. on Piling and Ground Treatment*, Thomas Telford, London.

de Beer, E.E.W. 1977. Static cone penetration testing in clay and loam. *Sondeer Symp.*, Utrecht.

de Mello, V.F.B. 1971. The standard penetration test. *Proc. 4th Pan Am. Conf. Soil Mech. Found. Engng*, Porto Rico, 1:1–87.

de Mello, V.B.F. 1977. Reflections on design decisions of practical significance to embankment dams – 17th Rankine lecture. *Géotechnique*, 27(3):281–354.

Decourt, L. 1982. Prediction of the bearing capacity of piles based exclusively on N values of the SPT. *Proc. 2nd European Symp. on Penetration Testing, ESOPT-II*, Amsterdam, 1:29–34.

Decourt, L. 1989. The standard penetration test. State-of-the-art report. *Proc. 2nd Int. Conf. Soil Mech. Found. Engng*, Rio de Janerio, 4, AA Balkema Publishers, The Netherlands.

Decourt, L. and Quaresma, A.R. 1978. Pile bearing capacity from SPT penetration. *6th Brazilian Congress Soil Mech. Found. Engng*, CBMSEF, Rio de Janeiro, ABMS/ABEF, in Portuguese.

Decourt, L., Muromachi, T., Nixon, I.K., Schmertmann, J.H., Thorburn, C.S. and Zolkov, E. 1988. Standard penetration test (SPT): international reference test procedure. *Proc. Int. Symp. on Penetration Testing, ISOPT-1*, Orlando, AA Balkema Publishers, The Netherlands, 1:37–40.

Decourt, L., Quaresma, A.R., Almeida, M.S.S. and Danziger, F. 1996. Geotechnical investigation. In: *Foundations: from theory to practice*, ABMS/ABEF, Pini Publishers, São Paulo, Brazil, in Portuguese.

Deere, D.V. and Patton, F.D. 1971. Slope stability in residual soils. *Proc. 4th Pan Am. Conf. Soil Mech. Found. Engng*, Puerto Rico, 1:87–170.

Dias, C.R.D. and Bastos, C.A.B. 1994. Geotechnical properties of the Rio Grande marine soft clay: geological interpretation. *Proc. 10th Brazilian Cong. Soil Mech. Geotech. Engng*, Foz do Iguaçu, 2:555–562, in Portuguese.

Dias, R.D. and Gehling, W.Y.Y. 1986. Shear strength and compressibility of the Porto Alegre clay deposit. *Proc. 11th Brazilian Cong. Soil Mech. Geotech. Engng*, Porto Alegre, 1:107–111, in Portuguese.

Diaz-Rodrigues, J.A. 2003. Characterisation and Engineering Properties of Mexico City Lacustrine Soils. *Charact. and Engng Properties of Natural Soils*, AA Balkema Publishers, The Netherlands, 1:725–756.

Diaz-Rodriguez, J.A., Leroueil, S. and Aleman, J.D. 1992. Yielding of Mexico City clay and other natural clays. *J. Geotech. Div.*, ASCE, 118(7):981–995.

Didaskalou, G. 1999. Comparison Measured vs DMT-predicted Settlements Hyatt Hotel Thessaloniki, personal communication.

Dobry, R., Ladd, R.S., Yokel, F.Y., Chung, R.M. and Powell, D. 1982. *Prediction of Pore Water Pressure Build Up and Liquefaction of Sands During Earthquakes by the Cyclic Strain Method*. NBS Building Science Series, US, 138.

Donald, I., Jordan, D.O., Parker, R.J. and Teh, C.T. 1977. The vane test – a critical appraisal. *Proc. 9th Int. Conf. Soil Mech. Found. Engng*, Tokyo, 1:81–88.

Douglas, B.J. and Olsen, R.S. 1981. Soil classification using electric cone penetrometer. Cone penetration testing and experience, *ASCE National Convention*, St. Louis, 209–227.

Duncan, J.M. and Buchignani, A.L. 1976. An engineering manual for settlement studies. *Department of Civil Engineering*, University of California, Berkeley, CA.

Durgunoglu, H.T. and Mitchell, J.K. 1975. Static penetration resistance of soils. *1st Proc. of the ASCE Specialty Conf. on In Situ Measurement of Soil Properties*, Raleigh, 1:151–189.

Einav, I. and Randolph, M.F. 2005. Combining upper bound and strain path methods for evaluating penetration resistance. *Int. J. on Num. Methods in Geotech. Engng*, 63(14):1991–2016.

EN ISO 22476: 2005 Geotechnical Investigation and Testing: Field Testing

Eslaamizaad, S. and Robertson, P.K. 1997. A framework for in-situ determination of sand compressibility. *49th Can. Geotech. Conf.*, St John's, Newfoundland.

Eurocode 7: Geotechnical Design

Fahey, M. 1980. *A Study of the Pressuremeter Test in Dense Sand*. PhD thesis, Cambridge University.

Fahey, M. 1998. Deformation and in situ stress measurement. *Geotech. Site Charact., ISC '98*, AA Balkema Publishing, The Netherlands, 1:49–68.

Fahey, M. and Randolph, M.F. 1984. Effect of disturbance on parameters derived from self-boring pressuremeter tests in sand. *Géotechnique*, March, 34(1):81–97.

Fahey, M. and Jewell, R.J. 1990. Effect of pressuremeter compliance on measurement of shear modulus. *Proc. 3rd Int. Symp. on Pressuremeters*, British Geotechnical Society, Oxford, 115–124.

Fahey, M. and Carter, J.P. 1993. A finite element study of the pressuremeter test in sand using a non-linear elastic plastic model, *Can. Geotech. J.*, 30:348–362.

Failmezger, R.A. 2001. Discussion to Duncan, J.M. 2000. Factor of safety and reliability in geotechnical engineering, in: *J. GGE*, ASCE, 126(4). *J. GGE*, ASCE, 127(8):703–704.

Failmezger, R.A., Rom, D. and Ziegler, S.B. 1999. Behavioral characteristics of residual soils. SPT? – a better approach to site characterization of residual soils using other in-situ tests. *ASCE Geotech. Special Publ.* No. 92, Edelen, B. (ed.), ASCE, Reston, VA, 158–175.

Fernandez, A.L. and Santamarina, J.C. 2001. Effect of cementation on the small-strain parameters of sands. *Can. Geotech. J.*, 38(1):191–199.

Ferreira, R.S. and Monteiro, L.B. 1985. Identification and evaluation of colluvial soils that occur in the São Paulo State. *1st Int. Conf. on Geomech. in Tropical Lateritic and Saprolitic Soils*, Brasília, 1:269–280.

Ferreira, R.S. and Robertson, P.K. 1992. Interpretation of undrained self-boring pressuremeter test results incorporating unloading. *Can. Geotech. J.*, 29:918–928.

Ferreira, S.R.M., Amorin Jr, W.M. and Coutinho, R.Q. 1986. *Proc. 7th Brazilian Cong. Soil Mech. Geotech. Engng*, 1, 183–197, in Portuguese.

Findlay, J.D. 1984. Discussion. *Piling and Ground Treatment*, Thomas Telford, London, 190–198.

Finno, R.J. 1993. Analytical interpretation of dilatometer penetration through saturated cohesive soils. *Géotechnique*, 43(2):241–254.

Fioravante, V., Jamiolkowski, M. and Lancellotta, R. 1994. An analysis of pressuremeter holding tests. *Géotechnique*, 44(2):227–238.

Flaate, K. 1968. Baereevne av Friksjonspeler I Leire, *om Beregning av Baereevne pa grunnlag av Geotekniske undersokelser*, Oslo, Veglabolatoriet, 38.

Fletcher, M.S. and Mizon, D.H. 1984. Piles in chalk for Orwell bridge. *Piling and Ground Treatment*, Thomas Telford, London, 203–209.

Flodin, N. and Broms, B.B. 1981. History of civil engineering of soft clays. *Soft Clay Engineering*. Elsevier, Amsterdam, 27–156.

Fonseca, A.P.V. 1996. *Mechanical characterisation of the residual granite from Porto: design of shallow foundations*. PhD thesis, Porto University, Portugal, in Portuguese.

Foray, P. 1991. Scale and boundary effects on calibration chamber pile tests. *Proc. 1st Int. Symp. on Calibration Chamber Testing*, Potsdam, New York, 147–160.

Frank, R. 1999. Calculated response of shallow and deep foundations. *Techniques de l'Ingénieur (TI) et Presses de l'Ecole nationale des ponts et chaussées*, 2ème trimestre, 139pp, in French.

Frank, R. and Zhao, M. 1982. Prediction of parameters from pressuremeter tests and their applications to pile design in fine soils. *Bulletin des Laboratoires des Ponts et Chaussées*, Paris, 119:17–24, in French.

Fukue, M., Minato, T., Matsumoto, M., Horibe, H. and Tayan, N. 2001. Use of a resistivity cone for detecting contaminated soil layers. *Engineering Geology*, 60:361–369.

Futai, M.M., Almeida, M.S.S. and Lacerda, W.A. 2007. Yield, strength and critical state of a tropical saturated soil. *J. Geotech. Geoenvironm. Engng*, 130(11): 1169–1179.

Gallipoli, D., Karstunen, M. and Wheeler, S.J. 2001. Numerical modelling of pressuremeter tests in unsaturated soil. *Proc. 10th Int. Conf. on Computer Methods and Advances in Geomech.*, 7–12 January, Tuscon, Arizona, USA, AA Balkema Publishers, The Netherlands, 807–812.

Gambin, M. 1963. Prediction of pile settlement from pressuremeter tests. *Sols–Soils*, 7.

Gambin, M. 1979. Design of foundations subjected to horizontal forces using MPM data. *Sols–Soils*, 30/31.

Garga, V.K. and Blight, G.E. 1997. Permeability. *Mechanics of Residual Soils*. AA Balkema Publishers, The Netherlands, Chapter 6, 79–94.

Gasmo, J.M. 1997. *Stability of unsaturated residual soil slope as affected by rainfall*. MEng thesis, School of Civil and Structural Engineering, Nanyang Technological University, Singapore.

Gaudin, C., Schnaid, F. and Garnier, J. 2005. Sand characterization by combined centrifuge and laboratory tests. *Int. J. of Physical Modelling Geotech.*, 1:42–56.

Gens, A. and Nova, R. 1993. Conceptual bases for a constitutive model for bonded soils and weak rocks. *Proc. 1st Int. Symp. on the Geotech. of Hard Soils–Soft Rocks*, Athens, 1:485–494.

Ghionna, V.N. and Jamiolkowski, M. 1991. A critical appraisal of calibration chamber testing of sands. *Proc. 1st Int. Symp. on Calibration Chamber Testing*, Potsdam, New York, 13–39.

Ghionna, V.N., Jamiolkowski, M., Pedroni, S. and Piccoli, S. 1995. Cone penetrometer tests in Po River sand. *The Pressuremeter and its Marine Applications*, Gerard Ballivy (ed.), 471–480.

Giacheti, H.L. 1991. Personal communication.

Gibbs, H.J. and Holtz, W.G. 1957. Research on determining the density of sands by spoon penetration testing. *4th Int. Conf. Soil Mech. Found. Engng*, London, 1:35–39.

Gibson, R.E. 1950. Correspondence. *J. Instn Civ. Engrs*, 34:382–383.

Gibson, R.E. and Anderson, W.F. 1961. In situ measurement of soil properties with the pressuremeter. *Civil Engng and Public Works Review*, 56:615–618.

Gomes Correia, A., Viana da Fonseca, A. and Gambim, M. 2004. Routine and advanced analysis of mechanical in situ tests. Keynote Lecture. *2nd Int. Conf. Geotech. Site Characterization*, Millpress, 1:75–96.

Graham, J. and Houlsby, G.T. 1983. Anisotropic elasticity of a natural clay. *Géotechnique*, 33(2):165–180.

Graham, J., Crooks, J.H.A. and Bell, A.L. 1983. Time effects on the stress-strain behaviour of soft natural clays. *Géotechnique*, 33(3):327–340.

Gravesen, S. 1960. *Elastic semi-infinite medium bounded by a rigid wall with a circular hole*. Laboratoriet for Bygninsteknik, Danmarks Tekniske Højskole, Meddelelse No. 10, Copenhagen.

Gui, M.W., Bolton, M.D., Garnier, J., Corte, J.-F., Bagge, G., Laue, J. and Renzi, R. 1998. Guidelines for cone penetration tests in sand. *Proc. Int. Conf. Centrifuge '98*, Tokyo, AA Balkema Publishers, The Netherlands, 1:155–160.

Haberfield, C.M. 1997. Pressuremeter testing in weak rock and cemented sand. *Proc. Instn Civ. Engrs, Geotech. Engng*, 125:168–178.

Halwachs, J.E. 1972. *Analysis of Sediment Shear Strength at Varying Rates of Shear*, US Naval Academy, Trident Scholar Project Report, TSPR 28, 24pp.

Hamouche, K., Roy, M. and Leroueil, S. 1995. A pressuremeter study of the sensitive Louiseville clay. *Proc. 4th Int. Symp. on Pressuremeter*, Sherbrooke, 361–366.

Hanzawa, H. and Tanakaz, H. 1992. Numerical undrained strength of clay in normally consolidated state and in the field. *Soils and Found.*, 32(1):132–148.

Harder, L.F. Jr. and Seed, H.B. 1986. Determination of penetration resistance for coarse-grained soils using the Becker hammer drill. Rep. UCB/EERC-86/06, *Earthquake Engrg Res. Ctr.*, University of California at Berkeley.

Hardin, B.O. 1978. The nature of stress-strain behavior for soils. *Proc. ASCE Geotech. Div. Specialty Conf. on Earthquake Engng and Soil Dynamics*, Pasadena, 1:3–90.

Hardin, B.O. and Black, W.L. 1968. Vibration modulus of normally consolidated clay. *J. Soil Mech. Found. Div.*, ASCE, 95(6):1531–1537.

Hatanaka, M. and Uchida, A. 1996. Empirical correlation between penetration resistance and effective friction angle for sandy soils. *Soils Found.*, 36(4):1–9.

Hayes, J.A. 1990. The Marchetti dilatometer and compressibility. *Seminar on In Situ Testing and Monitoring*, Southern Ontario Section of Canadian Geotechnical Society, September, 21pp.

Hepton, P. 1988. Shear wave velocity measurements during penetration testing. *Penetration Testing in the UK*, Thomas Telford, London, 275–278.

Hight, D.W. 1998. Soil characterisation: the importance of structure and anisotropy. *38th Rankine Lecture*, London.

Hight, D.W. and Leroueil, S. 2003. Characterisation of soils for engineering purposes. *Charact. and Engng Properties of Natural Soils*, AA Balkema Publishers, The Netherlands, 1:255–362.

Hight, D.W., Georgiannou, V.N. and Ford, C.J. 1994. Characterization of clayey sand. *Proc. 7th Int. Conf. of Offshore Structures*, USA, 1:321–340.

Hight, D.W., Bennell, J.D., Chana, B., Davis, P.D., Jardine, R.J. and Porovic, E. 1997. Wave velocity and stiffness measurements at Sizewell. *Proc. Symp. on Pre-Failure Deformation Behaviour of Geomaterials*, London, 65–88.

Hight, D.W., Paul, M.A., Barras, B.F., Powell, J.M., Nash, D.F.T., Smith, P.R., Jardine, R.J. and Edwards, D.H. 2003. The characterization of the Bothkennar Clay. *Charact. and Engng Properties of Natural Soils*, AA Balkema Publishers, The Netherlands, 1:543–598.

Hird, C.C. and Hassona, F. 1986. Discussion on 'A state parameter for sands'. *Géotechnique*, 36(1):124–127.

Hobbs, N.B. 1977. Behaviour and design of piles in chalk – an introduction to the discussion of the papers on chalk. *Piles in Weak Rock*, Thomas Telford, London, 149–175.

Holden, J.C. 1976. *The Calibration of Electrical Penetrometers in Sand*. Final Report, Norwegian Council for Scientific and Industrial Research (NTNF), reprinted in Norwegian Geotechnical Institute Internal Report 52108-2, January 1977.

Hornsnell, M.R. 1988. The use of cone penetration testing to obtain environmental data. *Proc. Geotech. Conf.: Penetration Testing in the UK*, Birmingham, Thomas Telford, London, 289–295.

Houlsby, G.T. 1988. Discussions: Sessions 3 (Piezocone penetration test) and 5 (pressuremeter, dilatometer and other developments). *Proc. Conf. on Penetration Testing in the UK*, Birmingham, Thomas Telford, London, 180–187 and 318–319.

Houlsby, G.T. 1998. Advanced interpretation of field tests, *Geotech. Site Charact.*, ISC '98, AA Balkema Publishers, The Netherlands, 1:99–112.

Houlsby, G.T. and Hitchman, R. 1988. Calibration chamber tests of a cone penetrometer in sand. *Géotechnique*, 38:575–587.

Houlsby, G.T. and Teh, C.I. 1988. Analysis of the piezocone in clay. *Proc. Int. Symp. on Penetration Testing, ISOPT-1*, Orlando, AA Balkema Publishers, The Netherlands, 2:777–783.

Houlsby, G.T. and Withers, N.J. 1988. Analysis of the cone pressuremeter test in clays. *Géotechnique*, 38(4):575–587.

Houlsby, G.T. and Schnaid, F. 1992. Measurement of the properties of sand in a calibration chamber by the cone pressuremeter test. *Géotechnique*, 42(4):587–601.

Houlsby, G.T. and Carter, J.P. 1993. The effects of pressuremeter geometry on the results of tests in clays. *Géotechnique*, 43:567–576.

Houlsby, G.T. and Schnaid, F. 1994. Interpretation of shear moduli from cone penetration tests in sand. *Géotechnique*, 44(1):147–164.

Houlsby, G.T., Wroth, C.P. and Clarke, B.G. 1986. Analysis of the unloading of a pressuremeter in sand. *Proc. 2nd Int. Symp. on Pressuremeter and its Marine Applications*, ASTM, SPT950, 245–262.

Hritzuk, K.J. 1997. *Effectiveness of drainage systems in maintaining soil suctions.* MEng thesis, School of Civil and Structural Engineering, Nanyang Technological University, Singapore.

Hryciw, R.D. 1990. Small-strain-shear modulus of soil by dilatometer. *J. Soil Mech. Found. Div.*, ASCE, 116(11):700–716.

Huang, A.B. 1989. Strain path analysis for arbitrary three dimensional penetrometers. *Int. J. Num. and Anal. Methods in Geomech.*, 13:551–564.

Huang, A.B. and Ma, M.Y. 1994. An analytical study of cone penetration tests in granular material. *Can. Geotech. J.*, 31:91–103.

Huang, A.B. and Hsu, H.H. 2004. Advanced calibration chamber for cone penetration testing in cohesionless soils. *Proc. 2nd Int. Conf. on Site Charact.*, Milpress, Porto, 1:147–166.

Hughes, J.M.O. 1982. Interpretation of pressuremeter tests for the determination of elastic shear modulus. *Proc. Engng Found. Conf. on Updating Subsurface Sampling of Soils and Rocks and Their In-Situ Testing*, Santa Barbara, CA, 279–289. Available from American Society of Civil Engineers, 345 East 47th Street, New York, NY 10017.

Hughes, J.M.O. and Ervin, M.C. 1980. Development of a high pressure pressuremeter for determining the engineering properties of soft to medium strength rocks. *Proc. 3rd Aus.–NZ Conf. Geomechanics*, Brisbane, 292–296.

Hughes, J.M.O., Wroth, C.P. and Windler, D. 1977. Pressuremeter tests in sands. *Géotechnique*, 27(4):455–477.

Idriss, I.M. and Boulanger, R.W. 2004. Semi-empirical procedures for evaluating liquefaction potential during earthquakes. *11th Int. Conf. on Soil Dynamics & Earthquake Engng (ICSDEE)*, Berkeley, CA, 32–56.

Idriss, M. 1999. An update of the Seed–Idriss simplified procedure for evaluating liquefaction potential. *Proc. TRB Workshop on New Approaches to Liquefaction.* Publication No. FHWA-RD-99-165, Federal Highway Admin., USA.

Imai, T. and Yokota, K. 1982. Relationships between N value and dynamic soil properties. *Proc. 2nd European Symp. on Penetration Testing, ESOPT-II*, Amsterdam, 1:73–78.

Ireland, H.O., Moretto, O. and Vargas, M. 1970. The dynamic penetration test: a standard that is not standardised. *Géotechnique*, 20(2):185–192.

IRTP/ISSMGE 1988. International reference testing procedure for cone penetration tests (CPT). *Report of the ISSMGE Technical Committee on Penetration Testing of Soils: TC-16, with reference to Test Procedures*, Swedish Geotech. Inst. Linkoping, Information 7, 6–16.

Ishihara, K. 1982. Evaluation of soil properties for use in earthquake response analysis. *Proc. Int. Symp. Num. Models in Geomech.*, Zurich, 237–259.

Ishihara, K. 1993. Liquefaction and flow failure during earthquakes. *Géotechnique*, 43(3):351–415.

Ishihara, K. 1996. *Soil Behavior in Earthquake Geotechnics*. Clarendon Press, Oxford.

Iwasaki, K., Tsuchiya, H., Sakai, Y. and Yamamoto, Y. 1991. Applicability of the Marchetti dilatometer test to soft ground in Japan. *Proc. GEOCOAST '91*, Yokohama, September, 1/6.

Jackson, A.B. 1969. *Undrained shear strength of a marine sediment*. Monash University.

Jaky, J. 1944. The coefficient of earth pressure at rest. *J. Soc. Hungarian Architects and Engineers*, Budapest, 355–358.

Jamiolkowski, M. 2001. Where are we going? *Proc. Pre-failure Deformation Characteristics of Geomaterials*, Torino. AA Balkema Publishers, The Netherlands, 2:1251–1262.

Jamiolkowski, M. and Lo Presti, D.C.F. 2003. Geotechnical characterization of Holocene and Pleistocene Messina sand and gravel deposits. *Charact. and Engng Properties of Natural Soils*, Singapore, AA Balkema Publishers, The Netherlands, 2:1087–1120.

Jamiolkowski, M., Ladd, C.C., Germaine, J.T. and Lancellotta, R. 1985. New developments in field and laboratory testing of soils. *11th Int. Conf. Soil Mech. Found. Engng*, San Francisco, 1:57–153.

Jamiolkowski, M., Ghionna, V., Lancellotta, R. and Paqualini, E. 1988. New correlations of penetration tests for design practice. *Proc. Int. Symp. of Penetration Testing, ISOPT-1*, Orlando, AA Balkema Publishers, The Netherlands, 1:263–296.

Jamiolkowski, M., Leroueil, S. and Lo Presti, D.C.F. 1991. Theme lecture: design parameters, from theory to practice. *Proc. Geo-Coast '91*, Yokohama, 2:877–917.

Jamiolkowski, M., Lancellotta, R. and Lo Presti, D.C.F. 1995a. Remarks on the stiffness at small strains of six Italian clays. Keynote Lecture 3, *Proc. Int. Symp. on Pre-failure Deformation Charact. of Geomaterials*, Sapporo, 2:817–836.

Jamiolkowski, M., Lo Presti, D.C.F. and Pallara, O. 1995b. Role of in-situ testing in geotechnical earthquake engineering. *3rd Int. Conf. on Recent Advances in Geotech. Earthquake Engng and Soil Dynamics*. State-of-the-Art Report 7, 3:1523–1546.

Jamiolkowski, M., Lo Presti, D.C.F. and Manassero, M. 2003. Evaluation of relative density in shear strength of sands from cone penetration tests (CPT) and flat dilatometer (DMT). *Soil Behaviour and Soft Ground Construction*, ASCE, GSP119, 201–238.

Janbu, N. 1985. Discussion. *Proc. ASCE Specialty Conf. on In Situ Measurement of Soil Properties*, North Canadian University, Raleigh, NC, 2:150–152.

Janbu, N. and Senneset, K. 1974. Effective stress interpretation of in situ static penetration tests. *Proc., 1st Eur. Symp. on Penetration Testing*, Nat. Swedish Build. Res., Stockholm, 2(2):189–193.

Jardine, R.J. 1985. *Investigations of pile–soil behaviour with special reference to the foundations of offshore structures*. PhD thesis, University of London, London.

Jardine, R.J. 1992. Some observations on the kinematic nature of soil stiffness. *Soils and Found.*, 32(2):111–124.

Jardine, R.J., Zdravkovic, L. and Porovic, E. 1977. Panel contribution: anisotropic consolidation including principal stress axis rotation: experimental, results and practical implications. *Proc. 14th Int. Conf. Soil Mech. Found. Engng*, Hamburg, 4:2165–2168.

Jardine, R.J., Symes, M.J. and Burland, J.B. 1984. The measurement of soil stiffness in the triaxial apparatus. *Géotechnique*, 34(3):323–340.

Jardine, R.J., St. John, H.D., Hight, D.W. and Potts, D.M. 1991. Some practical applications of a non-linear ground model. *10th European Conf. Soil Mech. Found. Engng*, Florence, 1:223–228.

Jardine, R.J., Lehane, B.M., Smith, P.R. and Gildea, P.A. 1995. Vertical loading experiments on rigid pad foundations at Bothkennar. *Geotéchnique*, 45(4): 573–597.

Jardine, R.J., Kuwano, R., Zdravkovic, L. and Thornton, C. 1999. Some fundamental aspects of the pre-failure behaviour of granular soils. *Proc. Pre-failure Charact. of Geomaterials*, AA Balkema Publishers, Toronto, 2:1077–1112.

Jefferies, M.G. 1988. Determination of horizontal geostatic stress in clay with self-bored pressuremeter. *Can. Geotech. J.*, 25:559–573.

Jefferies, M.G. and Davies, M.P. 1991. Use of the CPTu to estimate equivalent SPT N60, *Geotech. Testing J.*, 16(4):458–468.

Jefferies, M.G. and Been, K. 2000. Implications for critical state theory from isotropic compression of sand. *Géotechnique*, 50(4):419–429.

Jones, D.L. and Rust, E. 1989. Foundations on residual soil using a pressuremeter moduli. *Proc. 12th Int. Conf. Soil Mech. Found. Engng*, Rio de Janeiro, 1: 519–524.

Josa, A., Balmaceda, A., Gens, A. and Alonso, E.E. 1992. An elastoplastic model for partially saturated soils exhibiting a maximum of collapse. *Proc. 3rd Int. Conf. Comput. Plasticity*, Barcelona, 1:815–826.

Kaito, T., Sakaguchi, S., Nishigaki, Y., Miki, K. and Yukami, H. 1971. Large penetration test. *Tsuchi-to-Kiso*, Japan, 629:15–21.

Kamey, T. and Iwasaki, K. 1995. Evaluation of undrained shear strength of cohesive soil using a flat dilatometer. *Soils and Found.*, 35(2):111–116.

Karlsrud, K., Lunne, T. and Brattlieu, K. 1996. Improved CPTU correlations based on block samples. *Nordisk Geoteknikermote*, Reykjavik.

KarWinn, H., Rahardjo, H. and She, C.P. 2001. Chracaterisation of residual soils in Singapore. *Geotechnical Engineering*, 23(1):1–14.

Kavvadas, M. and Amorosi, A. 1998. A plasticity approach for the mechanical behaviour of structured soils. *Proc. 2nd Int. Symp. on the Geotech. of Hard Soils–Soft Rocks*, Napoli, 2:603–613.

Kavvadas, M., Anagnostopoulos, A. and Kalteziontis, N. 1993. A framework for the mechanical behaviour of the cemented Corinth marl. *Int. Symp. on Geotech. Engng of Hard Soils–Soft Rocks*, Athens, 1:577–583.

Kilbourn, N.S., Treharne, G. and Zarifian, V. 1988. The use of the Standard Penetration Test for the design of bored piles in the Keuper Marl of Cardiff. *Proc. Conf. on Penetration Testing in the UK*, Birmingham, Thomas Telford, London, 128–132.

Kim, Y.-S. and Paik, S. 2006. DMT dissipation analysis using equivalent radius and optimization. *Proc. from the 2nd Int. Flat Dilatometer Conf.*, Washington, DC, 313–318.

Koester, J.P., Daniel, C.R. and Anderson, M.L. 2000. In situ investigation of liquefiable gravels. *Research Record 1714*, Transportation Research. Board, Washington, 75–82.

Kokusho, K. 1980. Cyclic triaxial test of dynamic soil properties for wide strain range. *Soil. Found.*, 20:45–60.

Konrad, J.M. 1987. Piezo-friction-cone penetrometer testing in soft clays. *Can. Geotech. J.*, 24:645–652.

Konrad, J.M. 1997. In situ sand state from CPT: evaluation of a unified approach at two Canlex sites. *Can. Geotech. J.*, 34(1):120–130.

Konrad, J.M. 1998. Sand state from cone penetrometer tests: a framework considering grain crushing stress. *Géotechnique*, 48(2):201–216.

Konrad, J.M. and Law, K. 1987. Preconsolidation pressure from piezocone tests in marine clays. *Géotechnique*, 37(2):177–190.

Koskinen, M., Karstunen, M. and Wheeler, S.J. 2002. Modelling destructuration and anisotropy of a natural soft clay. *5th European Conf. Num. Methods in Geotech. Engng*, Paris, 11–20.

Kovacs, W.D. and Salomone, A. 1982. SPT hammer energy measurement. *J. Geotech. Engng Div.*, ASCE, 108(4):599–620.

Kratz de Oliveira, L.A., Schnaid, F. and Gehling, W.Y.Y. 2001. On the dilatancy of unsaturated residual soils, *Brazilian Symp. Unsaturated Soils*, Porto Alegre, 1:218–234.

Kulhawy, F.H. and Mayne, P.W. 1990. *Manual of estimating soil properties for foundation design.* Geotech. Engng Group, Cornell University, Ithaca.

Kulhawy, F.H., Birgisson, B. and Grigoriu, M.D. 1992. Reliability-based foundation design for transmission line structures: transformation models for in-situ tests. Report No. EL-5507(4), Electric Power Research Institute, Palo Alto, CA.

La Rochelle, P., Roy, M. and Tavernas, F. 1973. Field measurements of cohesion in Champlain clays. *Proc. 8th Int. Conf. Soil Mech. and Found. Engng*, Moscow, 1:229–236.

Lacasse, S. 1986. *In situ site investigation techniques and interpretation for offshore practice.* Norwegian Geotechnical Inst., Report 40019-28, September.

Lacasse, S. and Lunne, T. 1986. Dilatometer tests in sand. *Proc. In Situ '86, ASCE Spec. Conf. on Use of In Situ Tests in Geotechn. Engineering*, Virginia Tech, Blacksburg, VA, June, ASCE Geotechn. Special Publ. 6:686–699.

Lacasse, S. and Lunne, T. 1988. Calibration of dilatometer correlations. *Proc. Int. Symp. of Penetration Testing, ISOPT-1*, Orlando, AA Balkema Publishers, The Netherlands, 1:539–548.

Lacasse, S. and Lunne, T. 1998. Geologial and geotechnical challenges at deepwater sites. *Proc. Int. Cong. International Association of Engineering Geologists, 8.* Vancouver, BC, Canada, 5:3801–3818.

Lacerda, W.A. and Almeida, M.S.S. 1995. Engineering properties of regional soils: residual soils and soft clays. *Proc. 10th Pan. Am. Conf. Soil Mech. Found. Engng*, 1:31–39.

Ladanyi, B. 1972. In situ determination of undrained stress-strain behaviour of sensitive clays with the pressuremeter. *Can. Geotech. J.*, 9(3):313–319.

Ladanyi, B. and Johnston, G.H. 1974. Behaviour of circular footings and plate anchors embedded in permafrost. *Can. Geotech. J.*, 11:531–553.

Ladd, C.C. 1969. The prediction of in situ stress-strain behaviour of soft saturated clay during undrained shear. *Proc. Bolkesjö Symp.*, Norway, Norwegian Geotechnical Inst., 14–20.

Ladd, C.C. 1991. Stability evaluation during staged construction. *J. Geotech. Engng Div.*, ASCE, 117(4):540–615.

Ladd, C.C., Foott, R., Ishihara, K., Schlosser, F. and Poulos, H.G. 1977. Stress-deformation and strength characteristics. State-of-the-Art Report. *9th Int. Conf. Soil Mech. Found. Engng*, Tokyo, 2:421–494.

Lagoia, R. and Nova, R. 1995. An experimental and theoretical study of the behaviour of a calcarenite in triaxial compression. *Géotechnique*, 45(4):633–648.

Lambe, T.W. 1967. Stress path method. *J. Soil Mech.*, ASCE, 93(SM6):309–331.

Lambe, T.W. and Withman, R.V. 1979. *Soil Mechanics*, SI version, John Wiley and Sons, New York.

Larrson, R. 1980. Undrained shear strength in stability calculation of embankments and foundations on soft clays. *Canadian Geotech. J.*, 17(4):591–602.

Larsson, R. 1992. *CPT-sondering. Spetstrycksondering med och utan portryksmätning; en in situ method för bestämning av lagerföljd och egenskaper I jord; utforande och utvärdering.* Swedish Geotechnical Institute. Linköping, Information, 15, in Swedish.

Larsson, R. 1995. Use of a thin slot as filter in piezocone tests. *Int. Symp. Cone Penetration Tests, CPT '95*, Linkoping, Sweden, 2:35–40.

Larsson, R. and Mulabdic, M. 1991. *Piezocone tests in clay.* Swedish Geotechnical Institute, Linkoping Report, 42.

LCPC – SETRA 1985. *Pile foundation design rules from pressuremeter results,* Laboratoires des Ponts et Chaussées, Service d'Etudes Techniques des Routes et Autoroutes, Paris, in French.

Leach, B.A. and Thompson, R.P. 1979. The design and performance of large diameter bored piles in weak mudstone rocks. *Proc. 7th Conf. Soil Mech. Found. Engng*, Brighton, British Geotechnical Society, 3:101–108.

Lee, K.L. and Seed, H.B. 1967. Drained strength characteristics of sand. *J. Soil Mech. and Found. Div.*, ASCE, 93(6):117–141.

Leon, E., Gassman, S.L. and Talwani, P. 2006. Accounting for soil aging when assessing liquefaction potential. *J. Geotech. Geoenvironm. Engng*, ASCE, 132(3):363–377.

Leong, E.C., Rahardjo, H. and Tang, S.K. 2003. Characterisation and engineering properties of Singapore residual soils. *Charact. and Engng Properties of Natural Soils*, 2, AA Balkema Publishers, The Netherlands, 1279–1304.

Leroueil, S. and Vaughan, P.R. 1990. The general and congruent effects of structure in natural soils and weak rocks. *Géotechnique*, 40(3):467–488.

Leroueil, S. and Hight, D.W. 2003. Behaviour and properties of natural and soft rocks. *Charact. and Engng Properties of Natural Soils*, AA Balkema Publishers, The Netherlands, 1:29–254.

Leroueil, S., Tavenas, F., Samson, L. and Morin, P. 1983. Preconsolidation pressure of Champlain clays – Part II: laboratory determination. *Canadian Geotech. J.*, 20(4):803–816.

Leroueil, S., Demers, D., La Rochelle, P., Martel, G. and Virely, D. 1995. Practical applications of the piezocone in Champlain sea clays. *Int. Symp. on Cone Penetration Testing, CPT-95*, Linköping, 515–522.

Leroueil, S., Hamouche, K., Tavenas, F., Boudali, M., Locat, J., Virely, D., Roy, M., La Rochelle, P. and Leblond, P. 2002. Geotechnical characterization and properties of a sensitive clay from Québec. *Proc. Int. Workshop on Charact. and Engng Properties of Natural Soils*, Singapore, 2:363–394.

Levadoux, J.-N. and Baligh, M.M. 1986. Consolidation after undrained piezocone penetration. *J. Geotech. Engng*, ASCE, 112(7):707–726.

Li, S.S. and Dafalias, Y.F. 2000. Dilatancy for cohesionless soils. *Géotechnique*, 52(3):449–460.

Liao, S.S.C. and Whitman, R.V. 1985. Overburden correction factors for SPT in sand. *J. Geotech. Engng*, 112(3):373–377.

Lim, T.T. 1995. *Shear strength characteristics and rainfall-induced matric suction changes in a residual soil slope*. MEng thesis, School of Civil and Structural Engineering, Nanyang Technological University, Singapore.

Lo, K.W., Leung, C.F., Hayata, K. and Lee, S.L. 1988. Stability of excavated slopes in weathered Jurong Formation of Singapore. *Proc. of 2nd Int. Conf. on Geomech. in Tropical Soils*, Singapore, 1:277–284, AA Balkema Publishers, The Netherlands.

Lo Presti, D.C.F. 1989. Proprietà dinamiche dei terreni. *Atti delle Conferenze di Geotecnica del Politecnico di Torino*.

Lo Presti, D.C.F., Jamiolkowski, M., Pallara, O., Cavallaro, A. and Pedroni, S. 1997. Shear modulus and damping of soils. *Géotechnique*, 47(3):603–617.

Lo Presti, D.C.F., Pallara, O., Jamiolkowski, M. and Cavallaro, A. 1999. Anisotropy of small strain stiffness of undisturbed and reconstituted clays. *Proc. 2nd Int. Symp. on Prefailure Deformation Charact. of Geomaterials*, IS-Torino 99, Torino, 1:11–18.

Lo Presti, D.C.F., Shibuya, S. and Rix, G.J. 2001. Innovation in soil testing. *Proc. Symp. on Pre-failure Charact. of Geomaterials*, Torino, 2:1027–1076.

Lo Presti, D.C.F., Jamiolkowski, M. and Pepe, M. 2003. Geotechnical characterization of the subsoil of Pisa Tower. *Charact. and Engng Properties of Natural Soils*, AA Balkema Publishers, The Netherlands, 2:909–946.

Long, M. and Phoon, K.K. 2004. General report: innovative technologies and equipment. *Proc. 2nd Int. Conf. on Site Charact.*, Milpress, Porto, 1:625–634.

Lund, S.A., Soares, J.M.D. and Schnaid, F., 1996. *Vane Test and its Applicability in Soft Clay Deposits*. Federal University of Rio Grande do Sul, CE-51/95, Porto Alegre, in Portuguese.

Lunne, T. and Christoffersen, H.P. 1983. *Interpretation of cone penetrometer data for offshore sands*. Norwegian Geotechnical Institute, Oslo, Report 52108–52115.

Lunne, T. and Powell, J.J.M. 1992. Recent developments in in situ testing in offshore soil investigation. *SUT Conf.: Offshore Site Investigation Found. Behaviour*, Kluwer, Dordrecht, 147–180.

Lunne, T., Christoffersen, H.P. and Tjelta, T.I. 1985. Engineering use of piezocone data in North Sea clays. *11th Int. Conf. Soil Mech. Found. Engng*, San Francisco, 2:907–912.

Lunne, T., Eidsmoen, T., Powell, J.J.M. and Quartermann, R.S.T. 1986. Piezocone testing in overconsolidated clays. *Proc. 39th Canadian Geotech. Conf.*, Canadian Geotechnical Society, Ottawa, preprint volume, 209–218.

Lunne, T., Lacasse, S. and Rad, N.S. 1989. SPT, CPT, pressuremeter testing and recent developments on in-situ testing of soils. *12th Int. Conf. Soil Mech. Found. Engng*, Rio de Janeiro.

Lunne, T., Lacasse, S. and Rad, N.S. 1994. General report: CPT, PMT, and recent developments in in-situ testing. *Proc., 12th ICSMFE*, Vol. 4, Rio de Janeiro, 2339–2403.

Lunne, T., Robertson, P.K. and Powell, J.J.M. 1997. *Cone penetration testing in geotechnical practice*, Blackie Academic and Professional, 312pp.

Lutenegger, A.J. 1988. Current status of the Marchetti dilatometer test. *Proc. Int. Symp. on Penetration Testing, ISOPT-1*, Orlando, AA Balkema Publishers, The Netherlands, 1:137–155.

Lutenegger, A.J. 2006. Cavity expansion model to estimate undrained shear strength in soft clay. *Proc. 2nd Int. Flat Dilatometer Conf.*, Washington DC, 319–328.

Lutenegger, A.J. and Adams, M.T. 2006. Flat dilatometer method for estimating bearing capacity of shallow foundations. *Proc. 2nd Int. Flat Dilatometer Conf.*, Washington DC, 334–340.

Maccarini, M., Teixeira, V.H., Santos, G.T. and Ferreira, R.S. 1988. Quaternarium sediments from Santa Catarina Coastal Area. *Symp. on Quaternarium Brazilian Clay Deposits*, Rio de Janeiro, 2:62–93, in Portuguese.

Machado, O.V.B. 1988. Experimental testing field using stone piles. *Symp. on Quaternarium Brazilian Clay Deposits*, Rio de Janeiro, 2:37–61, in Portuguese.

McNeilan, T.W. and Bugno, W.T. 1985. Cone penetration test results in offshore California silts. *Strength testing of marine sediments: laboratory and in situ test measurements, ASTM STP 833*, ASTM, Philadelphia, 55–71.

Magnan, J.P. and Tavenas, F. 1985. Embankment on soft clay. *Technique Documentation, Lavoisier*, 342 pp., in French.

Mair, R.J. and Wood, D.E. 1987. *Pressuremeter testing: methods and interpretation*. Ciria Report, Butterworths, UK, 160pp.

Manassero, M. 1989. Stress–strain relationships from drained self-boring pressuremeter test in sand. *Géotechnique*, 39(2):293–308.

Manassero, M. 1994. Hydraulic conductivity assessment of slurry wall using piezocone test. *J. Geotech. Engng*, ASCE, 120(10):1725–1746.

Mantaras, F.M. 2000. *Cavity expansion theory in cohesive-frictional materials*. PhD thesis, Federal University of Rio Grande do Sul, Brazil, in Portuguese.

Mantaras, F.M. and Schnaid, F. 2002. Cavity expansion in dilatant cohesive-frictional soils. *Géotechnique*, 52(5):337–348.

Marchetti, D., Marchetti, S., Monaco, P. and Totani, G. 2007. Risultato di prove in sito mediante Dilatometro Sismico (SDMT). *Proc. 23rd Italian Conf. SMGE*, Padova, May 2007, in Italian.

Marchetti, S. 1980. In situ tests by flat dilatometer. *J. Geotech. Engng Div.*, ASCE, 106(3):299–321.

Marchetti, S. 1982. Detection of liquefiable sand layers by means of quasi static penetration tests. *Proc. 2nd European Symp. on Penetration Testing, ESOPT-II*, Amsterdam, May, 2:689–695.

Marchetti, S. 1985. On the field determination of $K0$ in sand. Discussion session no. 2A, *Proc. XIth Int. Conf. Soil Mech. Found. Engng*, San Francisco, 5: 2667–2673.

Marchetti, S. 1997. The flat dilatometer: design applications. Keynote lecture. *Proc. 3rd Int. Geotech. Engng Conf.*, Cairo, 421–448.

Marchetti, S. 1999. *On the calibration of the DMT membrane*. L'Aquila University, unpublished report, March.

Marchetti, S. 2001. The Flat Dilatometer Test (DMT) and its applications. *18th Conferenze Geotecnica Torino*, 56pp.

Marchetti, S. 2006. Origin of the flat dilatometer. *Proc. 2nd Int. Flat Dilatometer Conf.*, Washington DC, 2–4.

Marchetti, S. and Crapps, D.K. 1981. *DMT Operating Manual*. Prepared by GPE Inc., Gainesville, FL.

Marchetti, S. and Totani, G. 1989. C_h evaluations from DMTA dissipation curves. *Proc. 12th Int. Conf. Soil Mech. Found. Engng*, Rio de Janeiro, 1:281–286.

Marchetti, S., Totani, G., Calabrese, M. and Monaco, P. 1991. p–y curves from DMT data for piles driven in clay. *Proc. 4th Int. Conf. on Piling and Deep Foundations*, DFI, Stresa, 1:263–272.

Marchetti, S., Monaco, P., Totani, G. and Calabrese, M. 2001. The flat dilatometer test (DMT) in soil investigation. *Report by the ISSMGE Committee TC 16. Proc. Int. Conf. on In Situ Measurement of Soil Properties*, Bali, 95:132.

Marchetti, S., Monaco, P., Calabrese, M. and Totani, G. 2004. DMT – predicted vs measured settlements under a full-scale instrumented embankment at Treporti (Venice, Italy). *Proc. 2nd Int. Conf. on Site Charact.*, Milpress, Porto, 2: 1511–1518.

Marchetti, S., Monaco, P., Totani, G. and Marchetti, D. 2007. In situ tests by seismic dilatometer. ASCE Geotechnical Special Publication honoring Dr John H. Schmertmann. *From Research to Practice in Geotechnical Engineering GSP 170*, Geo-Institute Meeting, New Orleans.

Marcuson, W.F. and Bieganousky, W.A. 1977. Laboratory standard penetration test on fine sands. *J. Geotech. Engng*, 103(6):565–588.

Marsland, A. and Randolph, M.F. 1977. Comparisons of the results from pressuremeter tests and large in-situ plate tests in London Clay. *Géotechnique*, June, 27(2):217–243.

Martin, G.K. and Mayne, P.W. 1997. Seismic flat dilatometer tests in Connecticut Valley varved clay. *Geotech. Test. J.*, 20(3):357–361.

Martin, R.E., Seli, J.J., Powell, G.W. and Bertoulin, M. 1987. Concrete pile design in tidewater. *J. Geotech. Engng*, ASCE, 113(6):568–585.

Masood, T. and Mitchell, J.K. 1993. Estimation of in situ lateral stress in soils by cone penetration tests. *J. Geotech. Engng*, ASCE, 110(10):1624–1639.

Massad, F. 1985. *Quaternary clays from Baixada Santista: geotechnical characteristics and properties*. São Paulo Polytechnic, USP, São Paulo, in Portuguese.

Massad, F. 1988. Geological history and properties of clays from lower lands: Brazilian experience. *Proc. Symp. on Brazilian Quaternary Near-shore Deposits*. Rio de Janeiro, 3:1–34, in Portuguese.

Massarsch, K.R. 2004. Deformation properties of fine-grained soils from seismic testing. *Proc. 2nd Int. Conf. on Site Charact.*, Milpress, Porto, 1:133–146.

Matlock, H. 1970. Correlation for design of laterally loaded piles in soft clay. *Proc. 2nd Offshore Technical Conf.*, Houston, 1, 577–594.

Matlock, H. and Reese, L.C. 1960. Generalised solutions for laterally loaded piles, *J. Soil Mech. Found. Div.*, 86(5):63–91.

Matthews, M.C., Clayton, C.R.I. and Own, Y. 2000. The use of field geophysical techniques to determine geotechnical stiffness parameters. *J. Geotech. Engng*, 143(1):31–42.

Maugeri, M. and Monaco, P. 2006. Liquefaction potential evaluation by SMDT. *Proc. 2nd Int. Flat Dilatometer Conf.*, Washington DC, 296–311.

Mayne, P.W. 1987. Determining preconsolidation pressures from DMT contact pressures. *Geotech. Testing J.*, 10:146–150.

Mayne, P.W. 1991. Determination of OCR in clays by piezocone tests using cavity expansion and critical state concepts. *Soil Found.*, 31(1):65–76.

Mayne, P.W. 1992. Tentative method for estimating σ'_{h0} from Q_c data in sands. *Int. Symp. Calibration Chamber Testing*, Potsdam, New York, 249–256.

Mayne, P.W. 1997. Enhancements in Geotechnical Site Characterization. *IX Jornada Geotecnicas de Ingenieria Colombiana*, Santafe de Bogota, 15–17 October 1997, 26pp.

Mayne, P.W. 2001. Stress-strain-strength and flow parameters from enhanced in-situ tests. *Int. Conf. In-Situ Measurement of Soil Properties and Case Histories*, Bali, 27–48.

Mayne, P.W. 2005. Unexpected but foreseeable mat settlements of Piedmont residuum. *Int. Conf. Geoengng Case Histories*, 1:5–17.

Mayne, P.W. 2006. Undisturbed sand strength from seismic cone tests. The 2nd James K. Mitchell Lecture. *J. Geomech. and Geoengng*, 1(4):239–258.

Mayne, P.W. and Kulhawy, F.H. 1982. K_0–OCR relationships in soil. *J. Geotech. Engng*, 108(6):851–872.

Mayne, P.W. and Bachus, R.C. 1988. Profiling OCR in clays by piezocone soundings. *Proc. 1st Int. Symp. on Penetration Testing, ISOPT-1*, Orlando, AA Balkema Publishers, The Netherlands.

Mayne, P.W. and Mitchell, J.K. 1988. Profiling of OCR in clays by field vane. *Can. Geotech. J.*, 25(1):150–157.

Mayne, P.W. and Kulhawy, F.H. 1991. Calbration chamber database and boundary effects correction for CPT data. *Proc 1st Int. Symp. Calibration Chamber Testing*, Potsdam, New York, 257–264.

Mayne, P.W. and Rix, G.J. 1993. G_{max}–q_c relationships for clays. *Geotech. Testing J.*, ASTM, 16(1):54–60.

Mayne, P.W. and Martin, G.K. 1998. Seismic flat dilatometer test in Piedmont residual soils. *Proc. 1st Int. Conf. on Site Charact. ISC '98*, Atlanta, GA, 2:837–843.

Mayne, P.W., Schneider, J.A. and Martin, G.K. 1999. Small- and large-strain soil properties from seismic flat dilatometer tests. *Proc. 2nd Int. Symp. on Pre-failure Deformation Characteristics of Geomaterials*, Torino, 1:419–427.

Mayne, P.W., Brown, D., Vilson, J., Schneider, J.A. and Finke, K.A. 2000. *Site characterisation of Piedmont residual soil at the NGES, Opelika, Alabama, National Geotechnical Experimental Sites*, GPS 93, ASCE, Reston, VA, 160–185.

McClelland, B. 1974. Design and performance of deep foundation. *Special Conf. on Perf. of Earth and Earth-supp. Structs*, ASCE, 2:111–117.

McNeilan, T.W. and Bugno, W.T. 1980. Cone penetration test results in offshore California soils. *Strength Testing of Marine Sediments: Laboratory and In Situ Test Measurements*, ASTM, SPT 833, 55–71.

Meigh, A.C. 1987. *Cone penetration testing – methods and interpretation*. CIRIA, London, 141pp.

Ménard, L. 1957. *Measurement of the in situ physical properties of soils*. Annales des Ponts et Chaussées, Paris, in French.

Ménard, L. 1963. Calculation of the bearing capacity of foundations based on the results of pressuremeter tests. *Sols Soils*, 5:9–28, text in French, summary in English, see Figs. 2, 3a, 36.

Ménard, L. 1975. Interpretation and application of pressuremeter results. *Sols–Soils*, 26, in French.

Ménard, L. and Rousseau, J. 1962. Evaluation of settlements: tendencies. *Sols–Soils*, 1, in French.

Menzies, B.K. and Merrifield, C.M. 1980. Measurements of shear distribution at the edges of a shear vane blade. *Géotechnique*, 30(3):314–318.

Mesri, G. 1975. Discussion: new design procedure for stability on soft clays by Ladd and Foott. *J. Geotech. Engng Div.*, ASCE, 101(4):409–412.

Mesri, G. 1989. A re-evaluation of $s_{u(mob)} = 0.22\ \sigma N_p$. *Can. Geotech. J.*, 26(1): 162–164.

Mesri, G. 2001. Primary compression and secondary compression. *Proc. Ladd Symposium*, Massachusetts Institute of Technology, 5–6 October.

Meyerhof, G.G. 1951. The ultimate bearing capacity of foundations. *Géotechnique*, 2(1):301–332.

Meyerhof, G.G. 1956. Penetration tests and bearing capacity of cohesionless soils. *J. Soil Mech. Found. Div.*, ASCE, 82(1):1–19.

Meyerhof, G.G. 1957. Discussion on research on determining the density of sands by spoon penetration testing. *Proc. 4th Int. Conf. Soil Mech. Found. Engng*, London, 3:110–112.

Meyerhof, G.G. 1957. The ultimate bearing capacity of foundations. *Géotechnique*, 2(4):301–302.

Meyerhof, G.G. 1963. Some recent research on bearing capacity of foundation. *Can. Geotech. J.*, 1:16–26.

Migliori, H.J. and Lee, H.J. 1971. *Seafloor Penetration Tests: Presentation and Analysis of Results*. US Naval Civil Engineering Laboratory, Technical Note N-1178, 60pp.

Milititsky, J., Clayton, C.R.I., Talbot, J.C.S. and Dikran, S. 1982. Settlement predictions in granular soils from SPT data: critical review. *Proc. 7th Brazilian Conf. Soil Mech. Found. Engng*, 1:133–150.

Mimura, M. 2003. Characteristics of some Japanese natural sands – data from undisturbed frozen samples. *Charact. and Engng Prop. of Natural Soils*, AA Balkema Publishers, The Netherlands, 2:1149–1168.

Mitchell, J.K. 1976. *Fundamentals of Soil Behaviour*. John Wiley, 422pp.

Mitchell, J.K. (2001) Selected geotechnical papers – civil engng. Classics, ASCE, 934 pp.

Mitchell, J.K. and Lunne, T. 1978. Cone resistance as a measurement of sand strength. *J. Geotech. Engng*, 104(7):995–1012.

Mitchell, J.K. and Solymar, Z.V. 1984. Time-dependent strength gain in freshly deposited or densified sand. *J. Geotech. Engng*, ASCE, 110(11):1559–1576.

Monaco, P. and Marchetti, S. 2007. Evaluating liquefaction potential by seismic dilatometer (SDMT) accounting for aging. *Proc. 4th Int. Conf. on Earthquake Geotech. Engng*, Thessaloniki.

Monaco, P. and Schmertmann, J.H. 2007. *Accounting for soil aging when assessing liquefaction potential*. Discussion by Leon, E. *et al.* (in J. *Geotech. Geoenv. Engrg*, ASCE, 2006, 132(3):363–377). *J. Geotech. Geoenv. Engrg*, ASCE, 133(9):1177–1179.

Monaco, P., Marchetti, S., Calabrese, M. and Totani, G. 1999. *The flat dilatometer test*. Draft of the Report to the ISSMGE Committee TC 16.

Monaco, P., Marchetti, S., Totani, G. and Calabrese, M. 2005. Sand liquefiability assessment by flat dilatometer test (DMT). *Proc. 16th Int. Conf. Soil Mech. Geotech. Engng (ICSMGE)*, Osaka, 4:2693–2697.

Monaco, P., Totani, G. and Calabrese, M. 2006. DMT-predicted vs. observed settlements: a review of the available experience. *Proc. 2nd Int. Conf. on the Flat Dilatometer*, Washington DC, 244–252.

Moss, R. 2003. *CPT-based probabilistic assessment of seismic soil liquefaction initiation*. PhD thesis, University of California, Berkeley, CA.

Muir Wood, D. 1990. Strain dependent moduli and pressuremeter tests. *Géotechnique*, 40(3):509–512.

Murchison, J.M. and O'Neill, M.W. 1984. Evaluation of *p–y* relationships in cohesionless soils. *Analysis and Design of Pile Foundations*, J.R. Meyer (ed.), ASCE, New York, 174–191.

Nash, D.F.T., Powell, J.J.M. and Lloyd, I.M. 1992. Initial investigations of the soft clay test-bed site at Bothkennar. *Géotechnique*, 42(2):163–182.

NBR 6484, 2001. *Standard penetration test: testing procedure*. Brazilian Standard, in Portuguese.

NCEER.1997. *Proc. NCEER Workshop on Evaluation of Liquefaction Resistance of Soils*. Edited by Youd, T.L. and Idriss, I.M., Technical Report No. NCEER-97-0022, 31 December.

NEN 5140, 1996. Nederlands Normalisatie Institute (NNI). *Geotechnics: Determination of the Cone Resistance and Sleeve Friction of Soil. Electric Cone Penetration Test*, Dutch Standard, The Netherlands.

NFP 94-110-1, 2000. *Soils: Characterization Tests – Ménard Pressuremeter Tests. Part 1: Tests without cycles*. French Standard, in French.

NFP 94-113, 1989. *Soils: Characterization and Testing: Static Penetration Testing*. French Bureau of Normalization of Soils, 16pp, in French.

Nixon, I.K. 1982. Standard penetration test, state-of-the-art report. *Proc. European Symp. on Penetration Testing, ESOPT-2*, Stockholm, AA Balkema Publishers, The Netherlands, 1:3–24.

Nova, R. and Wood, D.M. 1979. Constitutive model for sand in triaxial compression. *Int. J. Num. and Anal. Method. in Geomech.*, 3(3):255–298.

Novais Ferreira, H. 1985. Characterisation, identification and classification of tropical lateritic and saprolitic soils for geotechnical purposes. *General Report, Int. Conf. Geomech. in Tropical Lateritic and Saprolitic Soils*, Brasília, 3:139–170.

Nutt, N.R.F. and Houlsby, G.T. 1992. Calibration tests on the cone pressuremeter in carbonate sand. *Int. Symp. Calib. Chamber Testing*, Potsdam, New York, 265–276.

Nyirenda, Z.M. 1989. *Piezocone studies in lightly overconsolidated clay*. DPhil. thesis, Department of Engineering Science, University of Oxford.

Oda, M., Koishikawa, I. and Higuchi, T. 1978. Experimental study of anisotropic shear strength of sand by plane strain test. *Soils and Found.*, 18(1):25–38.

Odebrecht, E. 2003. *Energy measurements in SPT test*. PhD thesis, Porto Alegre Federal University of Rio Grande do Sul, in Portuguese, 203pp.

Odebrecht, E., Schnaid, F., Rocha, M.M. and Bernardes, G.P. 2004. Energy measurements for standard penetration tests and the effects of the length of rods. *Proc. 2nd Int. Conf. on Site Charact.*, Milpress, Porto, 1:351–358.

Odebrecht, E., Schnaid, F., Rocha, M.M. and Bernardes, G.P. 2005. Energy efficiency for Standard Penetration Tests. *J. Geotech. Geoenvironm. Engng*, 131(10):1252–1263.

Ohta, Y. and Goto, N. 1978. Empirical shear wave velocity equations in terms of characteristic soil indexes. *Earth Engng and Struct. Dynamics*, 6:61–73.

Olsen, R.S. 1997. Cyclic liquefaction based on the cone penetrometer test. *Proc. NCEER Workshop on Evaluation of Liquefaction Resistance of Soils*, National Center for Earthquake Engineering Research, State University of New York at Buffalo, Report No. NCEER-97-0022:225–276.

Olson, S.M. and Stark, T.D. 1998. CPT-based liquefaction resistance of sandy soils. *Proc. Geotechnical Earthquake Engineering and Soil Dynamics III*, GTP No. 75 ASCE, Seattle, WA, 325–336.

Orihara, K., Chan, M.L., Chabayashi, K., Okamoto, S., Teo, P.P.T. and Tan, C.G. 2001. Excavation of new Dhoby Ghaut Station for MRT northest line. *Proc. of Underground Singapore 2001*, Singapore, 183–192.

Ortigão, J.A.R. 1980. *Failure of a trial embankment on Rio de Janeiro gray clay.* DSc. thesis, COPPE/UFRJ, Rio de Janeiro, Brazil, in Portuguese.

Ortigão, J.A.R. 1988. Onshore and offshore vane test experiences. *Symp. on New Concepts in Lab. and In Situ Tests.* Universidade Federal do Rio de Janeiro, Rio de Janeiro, 94pp, in Portuguese.

Ortigão, J.A.R. 1995. *Soil Mechanics in the Light of Critical State Theories – An Introduction.* AA Balkema Publishers, The Netherlands, 299pp.

Ortigão, J.A.R. and Collet, H.B. 1987. Errors caused by friction in field vane testing. *ASTM Symp. on Laboratory and Field Vane Shear Strength Testing*, STP 1014, Tampa, 104–116.

Palmer, A.C. 1972. Undrained plane-strain expansion of a cylindrical cavity in clay: a simple interpretation of the pressuremeter test. *Géotechnique*, 22(3): 451–457.

Parkin, A.K., Holden, K., Aamot, K., Last, N. and Lunne, T. 1980. *Laboratory of CPTs in sand.* Norwegian Geotechnical Institute, Oslo, Report 52108-9, p. 45.

Parkin, A.K. and Lunne, T. 1982. Boundary effects in the laboratory calibration of a cone penetrometer for sand. *Proc. 2nd European Symp. on Penetration Testing, ESOPT-II*, Amsterdam, 2:761–768.

Part 1: Electric friction cone and piezocone tests

Part 1: General rules

Part 11: Flat plate dilatometer.

Part 2: Ground investigation and testing

Part 3: Design assisted by field testing.

Part 3: Standard penetration test

Part 4: Ménard pressuremeter test

Part 6: Self-boring pressuremeter test

Part 9: Field vane test

Peck, R.B. 1969. Advantages and limitations of the observational method in applied soil mechanics. 9th Rankine Lecture. *Géotechnique*, 19(2):171–187.

Peck, R.B., Hanson, W.E. and Thornburn, T.H. 1974. *Foundation Engineering.* 2 edn. New York: John Wiley and Sons.

Pelnik, T.W., Fromme, C.L., Gibbons, Y.R. and Failmezger, R.A. 1999. Foundation design applications of CPTU and DMT tests in Atlantic Coastal Plain Virginia. *Transp. Res. Board, 78th Annual Meeting*, January, Washington, DC.

Perlow, M. and Richards, A.F. 1977. Influence of shear velocity on vane shear strength. *J. Geotech. Engng Div.*, ASCE, 103 (GT1):19–32.

Pestana, J.M. and Whittle, A.J. 1999. Formulation of a unified constitutive model for clays and sands. *Int. J. Num. and Anal. Methods in Geomech.*, 23:2115–2143.

Pinto, C.S. and Abramento, M. 1997. Pressuremeter tests on gneissic residual soil in São Paulo, Brazil. *Proc. XIV Int. Conf. on Soil Mech. and Found. Engng*, Hamburg, 1:1154–1158.

Poh, K.B., Chua, N.L. and Tan, S.B. 1985. Residual granite soil of Singapore. *Proc. 8th Southeast Asian Geotec. Conf.*, Kuala Lumpur, 3:1–39.

Poulos, H.G. 1989. Pile behaviour – theory and application. 29th Rankine Lecture. *Géotechnique*, 39(3):363–416.

Powell, J.J.M. and Butcher, A.P. 2004. Small strain stiffness assessments from in situ tests. *Proc. 2nd Int. Conf. on Site Charact.*, Milpress, Porto, 2:1717–1729.

Powell, J.J.M. and Quarterman, R.S.T. 1988. The interpretation of cone penetration tests in clays, with particular reference to rate effects. *Proc. Int. Symp. on Penetration Testing, ISOPT-1*, Orlando, AA Balkema Publishers, The Netherlands, 2:903–910.

Powell, J.J.M. and Uglow, I.M. 1988. Interpretation of the Marchetti dilatometer test in UK clays. *Proc. Conf. on Penetration Testing in the UK*, Birmingham, 121–125.

Powell, J.J.M., Quarterman, R.S.T. and Lunne, T. 1988. Interpretation and use of the piezocone test in UK clays. *Proc. Geotech. Conf.: Penetration Testing in the UK*, Birmingham, Swedish Geotechnical Society, 151–156.

Rahardjo, H. 2000. *Rainfall-induced slope failures*. NSTB 17/6/16 Main Report, Nanyang Technological University, Singapore.

Rampello, S. and Viggiani, G.M.B. 2001. Pre-failure deformation characteristics of geomaterials. *Proc. Symp. on Pre-failure Charact. of Geomaterials*, Torino, 2:1279–1289.

Randolph, M.F. 2004. Characterisation of soft sediments for offshore applications. *Proc. 2nd Int. Conf. on Site Charact.*, Milpress, Porto, 1:209–232.

Randolph, M.F. and Wroth, C.P. 1979. An analytical solution for the consolidation around a driven pile. *Proc. Int. J. Num. and Anal. Methods in Geomech.*, 3(3):217–229.

Randolph, M.F. and Houlsby, G.T. 1984. The limiting pressure on a circular pile loaded laterally in a cohesive soil. *Géotechnique*, 34(4):613–623.

Randolph, M.F. and Hope, S. 2004. Effect of cone velocity on cone resistance and excess pore pressures. *Proc. Int. Symp. on Engng Practice and Performance of Soft Deposits*, Osaka.

Randolph, M., Hefer, P.A., Geise, J.M. and Watson, P.G. 1998. Improved seabed strength using T-bar penetrometer. *Int. Conf. Offshore Site Investigation and Found. Behaviour – New Frontiers*, Society for Underground Technology, London, 221–235.

Randolph, M., Cassidy, M., Gourvenec, S. and Erbrich, C. 2005. Challenges of offshore geotechnical engineering. State-of-the-art report. *16th Int. Conf. Soil Mech. Geotech. Engng (ICSMGE)*, Osaka, 1:123–176.

Rangeard, D., Hicher, P.Y. and Zentar, R. 2003. Determining soil permeability from pressuremeter tests. *Int. J. Num. and Anal. Methods in Geomech.*, 27(1):1–24.

Rangeard, D., Zentar, R., Moulin, R. and Hicher, P.Y. 2001. Strain rate effect on pressuremeter test in soft clay. *Proc. Int. Conf. Soil Properties and Case Histories*, Bali, 379–384.

Ranzini, S. 1988. SPTF, Technical Note, *Soils and Rocks*, 11:29–30, in Portuguese.

Reese, L.C. 1977. Laterally loaded piles – program documentation. *J. Geotech. Engng*, 103(4):287–305.

Reese, L.C. and Desai, C.S. 1977. *Laterally loaded piles*. In: Desai, C.S. and Christian, J.T. (eds), *Numerical Methods in Geotechnical Engineering*, McGraw-Hill.

Reese, L.C., Cox, W.R. and Koop, F.D. 1974. Analysis of laterally loaded piles in sand. *Proc. 5th Annual Offshore Technology Conf.*, Paper No. OTC 2080, 473–485.

Reyna, F. and Chameau, J.L. 1991. Dilatometer based liquefaction potential of sites in the Imperial Valley. *Proc. 2nd Int. Conf. on Recent Advances in Geotech. Earthquake Engrg and Soil Dynamics*, St Louis, MO, May.

Richard, F.E. Jr, Hall, J.R. and Woods, R.B. 1970. *Vibrations of Soils and Foundations*. Prentice Hall: New Jersey, 414pp.

Richardson, A.M. and Whitman, R.V. 1963. Effect of strain-rate upon undrained shear resistance of a saturated remoulded fat clay. *Géotechnique*, 13(4):310–324.

Rix, G.J. and Stokes, K.H. 1992. Correlation of initial tangent modulus and cone resistance. *Int. Symp. Calib. Chamber Testing*, Potsdam, New York, 351–362.

Robertson, P.K. 1990. Soil classification using the cone penetration test. *Can. Geotech. J.*, 27(1):151–158.

Robertson, P.K. 2004. Evaluating soil liquefaction and post-earthquake deformations using the CPT. *Proc. 2nd Geotech. Geophys. Site Charact., ISC '3*, Millpress, Rotterdam, 1:233–252.

Robertson, P.K. and Campanella, R.G. 1983. Intepretation of cone penetration tests. *Can. Geotech. J.*, 20(4):734–745.

Robertson, P.K. and Campanella, R.G. 1985. Liquefaction potential of sands using the CPT. *J. Geotech Engng*, 111(3):384–403.

Robertson, P.K. and Campanella, R.G. 1986. Estimating liquefaction potential of sands using the flat plate dilatometer. *Geotech. Test. J.*, ASTM, 38–40.

Robertson, P.K. and Hughes, J.M.O. 1986. Determination of properties of sand from self-boring pressuremeter tests. *Proc. 2nd Symp. on Pressuremeter and its Marine Applications*, ASTM SPT 950, 443–457.

Robertson, P.K. and Campanella, R.G. 1988. *Guidelines for geotechnical design using CPT and CPTU*. University of British Columbia, Vancouver, Department of Civil Engineering, Soil Mechanics Series 120.

Robertson, P.K. and Campanella, R.G. 1989. *Design manual for use of CPT and CPTU*. University of British Columbia, Vancouver, BC.

Robertson, P.K. and Fear, C.E. 1995. Liquefaction of sands and its evaluation. *1st Int. Conf. on Earthquake Geotech. Engng*, Tokyo '95, Keynote lecture.

Robertson, P. K. and Wride, C.E. 1997. Cyclic liquefaction and its evaluation based on the SPT and CPT, NCEER-97-0022. *Proc. NCEER Workshop on Evaluation of Liquefaction Resistance of Soils*, 41–87.

Robertson, P.K., Campanella, R.G., Gillespie, D.G. and Greig, J. 1986. Use of piezocone data. *ASCE Specialty Conf. In Situ '86: Use of In Situ Tests in Geotech. Engng*, Blacksburg, VA, 1263–1280.

Robertson, P.K., Davies, M.P. and Campanella, R.G. 1987. Design of laterally loaded driven piles using the flat plate dilatometer. *Geotech. Testing J.*, 12(1):30–38.

Robertson, P.K., Campanella, R.G., Gillespie, D.G. and By, T. 1988. Excess pore pressure and the flat dilatometer test. *Proc. Int. Symp. Penetration Testing, ISPOT-1*, Orlando, AA Balkema Publishers, The Netherlands, 1:567–576.

Robertson, P.K., Sully, J.P., Woeller, D.J., Lunne, T., Powell, J.J.M. and Gillespie, D.G. 1992. Estimating coefficient of consolidation from piezocone tests. *Can. Geotech. J.*, 29(4):551–557.

Robertson, P.K., Woeller, D.J. and Addo, K.O. 1992. Standard penetration test energy using a system based on the personal computer. *Can. Geotech. J.*, 29: 551–557.

Robertson, P.K., Wride, C.E., List, B.R., Atukorala, U., Biggar, K., Byrne, W., Campanella, R.G., Finn, W.D., Howie, J.A., Hughes, J., Lord, E.R.F., Morgenstern, N.R., Watts, B.D. and Zavodni, Z. 1998. The Canadian liquefaction experiment: an overview. *Can. Geotech. J.*, 37(3):499–504.

Robertson, P.K., Wride, C.E. (Fear), List, B.R. , Atukorala, U. , Biggar, K.W., Byrne, P.M. *et al.* 2000. *Can. Geotech. J.* 37(3):563–591.

Rocha Filho, P. 1987. Determination of the undrained shear strength of two soft clay deposits using piezocone tests. *Int. Symp. Geotech. Engng Soft Soil*, Mexico, 1:201–211.

Rodrigues, C.M.G. and Lemos, L.J.L. 2004. SPT, CPT and CH tests results on saprolitic soils from Guarda, Portugal. *Proc. 2nd Int. Conf. on Site Charact.*, Milpress, Porto, 2:1345–1352.

Roque, R., Janbu, N. and Senneset, K. 1988. Basic interpretation procedures of flat dilatometer tests. *Proc. Int. Symp. on Penetration Testing, ISOPT-1*, Orlando, AA Balkema Publishers, The Netherlands, 1:577–587.

Roscoe, K.H. and Burland, J.B. 1968. On the generalized stress-strain behaviour of 'wet' clay. *Proc. Symp. on Plasticity*, Cambridge, 535–610.

Rotta, G.V., Consoli, N.C., Pritto, P.D.M., Coop, M.R. and Graham, J. 2003. Isotropic yielding in an artificially cemented soil cured under stress. *Géotechnique*, 53(5):493–501.

Rouainia, M. and Muir Wood, D. 2000. A kinematic hardening constitutive model for natural clays with loss of structure. *Géotechnique*, 50(2):153–164.

Rowe, P.W. 1962. The stress-dilatancy relation for static equilibrium of an assembly of particles in contact. *Proc. Royal Soc.*, London, A269, 500–527.

Rowe, P.W. 1963. Stress-dilatancy, earth pressure and slopes. *J. Soil Mech. Found. Div.*, ASCE, 89(3):37–61.

Rowe, P.W. 2001. *Geotechnical and Geo-environmental Engineering Handbook*. Kluwer Academic Publishing, Norwell, MA.

Roy, M. and LeBlanc, A. 1988. The in-situ measurement of the undrained shear strength of clays using the field vane. *Vane Shear Strength Testing in Soils: Field and Laboratory Studies*, ASTM STP 1014, A.F. Richards (ed.), Philadelphia, 117–128.

Salgado, R., Mitchell, J.K. and Jamiolkowski, M. 1997. Cavity expansion and penetration resistance in sand. *J. Geotech. Geoenvironm. Engng*, 123(4):344–354.

Samara, V., Barros, J.M.C., Marco, L.A.A., Belicanta, A. and Wolle, C.M. 1982. Some properties of marine clays from Santos, Brazil. *Proc. 7th Brazilian Cong. Soil. Mech. Geotech. Engng*, Recife, 4:301–318, in Portuguese.

Sandroni, S.S. 1991. Young metamorphic residual soils. *9th Pan Am. Conf. Soil Mech. Found. Engng*, Viña del Mar, Argentina.

Sandroni, S.S. and Maccarini, M. 1981. Triaxial and direct shear test in a Gneiss residual soil. *Brazilian Symp. on Tropical Soils*, COPPE, Rio de Janeiro, 324–339.

Sandven, R. 1990. *Strength and deformation properties of fine grained soils obtained from piezocone tests*. PhD thesis, Norwegian Institute of Technology, Trondheim, Norway.

Sandven, R., Sennesset, K. and Janbu, N. 1988. Interpretation of piezocone test in cohesive soils. *Proc. 1st Int. Symp. on Penetration Testing, ISOPT-1*, Orlando, AA Balkema Publishers, The Netherlands, 2:939–953.

Santamarina, J.C., Klein, K.A. and Fam, M.A. 2001. *Soils and Waves*. John Wiley and Sons Ltd, Ontario, Canada, 488pp.

Sayão, A.S.F.J. 1980. *Laboratory tests on the soft clay at excavation site of Sarapuí*. MSc thesis, PUC-RIO, Rio de Janeiro, Brazil, in Portuguese.

Schapery, R.A. and Dunlap, W.A. 1978. Prediction of storm-induced sea bottom movement and platform forces. *Proc. 10th Offshore Tech. Conf.*, Houston, OTC No. 3259.

Schmertmann, J.H. 1970. Static cone to compute static settlement over sand. *J. Soil Mech. Found. Div.*, ASCE, New York, 96(SM3):1011–1043.

Schmertmann, J.H. 1978. *Guidelines for cone penetration test, performance and design*. US Federal Highway Administration, Washington, DC, Report FHW-TS-78-209.

Schmertmann, J.H. 1982. A method for determining the friction angle in sands from the Marchetti dilatometer test (DMT). *Proc. 2nd European Symp. on Penetration Testing, ESOPT-II*, Amsterdam, 2:853–861.

Schmertmann, J.H. 1986. Suggested method for performing Flat Dilatometer Test. *Geotech. Testing J.*, 9(2):93–101.

Schmertmann, J.H. 1988. *Guidelines for using CPT, CPTU and Marchetti DMT for geotechnical design.* Report FHWA-PA-87-022+84-24, Office of Research and Special Studies, USA.

Schmertmann, J.H. 1991. The mechanical aging of soils. *J. Geotech. Engng*, ASCE, 117(9):1288–1330.

Schmertmann, J.H. and Palacios, A. 1979. Energy dynamics of SPT. *J. Soil Mech. Found. Div.*, ASCE, 105(8):909–926.

Schnaid, F. 1990. *A study of the cone pressuremeter test in sand.* DPhil thesis, Oxford University.

Schnaid, F. 1997. Panel discussion: evaluation of in situ tests in cohesive frictional materials. *Proc. 14th Int. Conf. Soil Mech. Found. Engng*, Hamburg, 4:2189–2190.

Schnaid, F. 1999. On the interpretation of in situ tests in unusual soil conditions. *Proc. Symp. on Pre-failure Charact. of Geomaterials*, Torino, 2:1339–1348.

Schnaid, F. 2005. Geocharacterisation and properties of natural soils by *in situ* tests. State-of-the-art report. *Proc. 16th International Conference on Soil Mechanics and Geotechnical Engineering (ICSMGE)*, Osaka, 1:3–46.

Schnaid, F. and Houlsby, G.T. 1991. An assessment of chamber size effects in the calibration of in situ tests in sand. *Géotechnique*, 41(4):437–445.

Schnaid, F. and Houlsby, G.T. 1992. Measurement of the properties of sand in a calibration chamber by cone pressuremeter test. *Géotechnique*, 42(4):578–601.

Schnaid, F. and Houlsby, G.T. 1994. Interpretation of shear moduli from cone-pressuremeter tests in sand. *Géotechnique*, 44(1):147–164.

Schnaid, F. and Mantaras, F.M. 2003. Cavity expansion in cemented materials: structure degradation effects. *Géotechnique*, 53(9):797–807.

Schnaid, F. and Mantaras, F.M. 2004. Interpretation of pressuremeter tests in a gneiss residual soil from Sao Paulo, Brazil. *Proc. 2nd Int. Conf. on Site Charact.*, Milpress, Porto, 2:1353–1360.

Schnaid, F. and Yu, H.S. 2005. Theoretical interpretation of the seismic cone test in granular soils. *Géotechnique*, 57(3):265–272.

Schnaid, F. and Yu, H.S. 2007. Interpretation on the seismic cone test in granular soils. *Géotechnique*, 57:265–272.

Schnaid, F., Soares, J.M. and Bica, A.V.D. 1990. Properties of the Porto Alegre soft clay deposit. *Brazilian Cong. Soil Mech. Geotech. Engng*, Foz do Iguaçu, 2:127–134, in Portuguese.

Schnaid, F., Bica, A.D.V. and Soares, J.M. 1997a. Determination of the characteristics of a soft clay deposit in southern Brazil. *Symp. on Recent Developments on Soil and Pavement Mechanics*, Rio de Janeiro, 1:296–302.

Schnaid, F., Sills, G.C., Soares, J.M. and Nyirenda, Z. 1997b. Predictions of the coefficient of consolidation from piezocone tests. *Can. Geotech. J.*, 34(2):143–159.

Schnaid, F., Kratz de Oliveira, L.A. and Gehling, W.Y.Y. 2000. Use of pressuremeter tests in the assessment of soil collapse. *Soils & Rocks*, 3(1):12, Brazil, in Portuguese.

Schnaid, F., Prietto, P.D.M. and Consoli, N.C. 2001. Characterization of cemented sand in triaxial compression. *J. Geotech. Geoenvironm. Engng*, ASCE, 127(10): 857–868.

Schnaid, F., Lehane, B.M. and Fahey, M. 2004. *In situ* test characterisation of unusual geomaterials. *Proc. 2nd Int. Conf. on Site Charact.*, Milpress, Porto, 1:49–74.

Schnaid, F., Gaudin, C. and Garnier, J. 2005. Sand characterisation by combined centrifuge and laboratory tests. *Int. J. Physical Modelling in Geotechnics*, 5(4):98–112.

Schnaid, F., Odebrecht, E. and Rocha, M. 2007. On the mechanics of dynamic penetration tests. *Geomech. and Geoengng: an Int. J.*, 3:12–16.

Schnaid, F., Odebrecht, E., Rocha, M.M. and Bernandes, G.P. 2008. Prediction of soil properties from the concepts of energy transfer in dynamic penetration tests. *J. Geotech. Geoenvironm. Engng*, ASCE, in publication.

Schneider, J.A., Lehane, B. and Schnaid, F. 2007. Velocity effects on piezocone measurements in normally and overconsolidated clays. *Int. J. Physical Modelling in Geotechnics*, 7:23–34.

Schofield, A. and Wroth, P. 1968. *Critical State Soil Mechanics*. McGraw-Hill, London, 310pp.

Schofield, A.N. and Steedman, R.S. 1988. State-of-the-art report: recent developments of dynamic model testing in geotechnical engineering. *Proc. 9th World. Conf. on Earthquake Engng*, Tokyo/Kyoto, Japan, 8:813–824.

Schultze, E. and Sherif, G. 1973. Prediction of settlements from evaluated settlement observations for sand. *Proc. 8th Int. Conf. on Soil Mech. Found. Engng*, Moscow, 1(3):225–230.

Seah, T.P., Ranjith, P.G., Zhao, J., Henfy, A.M. and Williams, I.O. 2001. Effects of ground conditions on performance of pressurized soft-ground tunnel boring machines. *Proc. of Underground Singapore 2001*, Singapore, 148–158.

Seed, H.B. 1979. Soil liquefaction and cyclic mobility evaluation for level ground during earthquakes. *J. Geotech. Engng*, ASCE, 105(2):201–255.

Seed, H.B. and Idriss, I.M. 1971. Simplified procedure for evaluating soil liquefaction potential. *J. Geotech. Engng*, ASCE, 97(9):1249–1273.

Seed, H.B. and Idriss, I.M. 1981. Evaluation of liquefaction potential of sand deposits based on observations of performance in previous earthquakes. *Proc. Session on In-situ Testing to Evaluate Liquefaction Susceptibility*, ASCE National Convention, St. Louis, MO, October.

Seed, H.B. and Idriss, I.M. 1982. *Ground motion and soil liquefaction during earthquakes*. Monograph, Earthquake Engineering Research Institute, Oakland, CA.

Seed, H.B., Idriss, I.M. and Arango, I. 1983. Evaluation of liquefaction potential using field performance data. *J. Geotech. Engng*, ASCE, 109(3):458–482.

Seed, H.B., Tokimatsu, K., Harder, L.F. and Chung, R.M. 1984. The influence of SPT procedures in soil liquefaction resistance evaluations. *Earthquake Engineering Research Center Rep. No. UCB/EERC-84/15*, University of California, Berkeley, CA.

Seed, H.B., Tokimatsu, K., Harder, L.F. and Chung, R. 1985. Influence of SPT procedures in soil liquefaction resistance evaluations. *J. Geotech. Engng*, ASCE, 111(12):1425–1445.

Seed, R.B., Cetin, K.O., Moss, R.E.S., Kammerer, A.M., Wu, J., Pestana, J.M. and Riemer, M.F. 2001. Recent advances in soil liquefaction engineering and seismic site response evaluation. *4th Int. Conf. Recent Advances in Geotechnical Earthquake Engineering and Soil Dynamics*, San Diego, CA, March.

Senneset, K. and Janbu, N. 1985. Shear strength parameters obtained from static penetration tests. *Symp. Shear Strength of Marine Sediments*, San Diego, ASTM Special Technical Publication, STP 883, 41–54.

Senneset, K., Janbu, N. and Svano, G. 1982. Strength and deformation parameters from cone penetrometer tests. *Proc. 2nd European Symp. on Penetration Testing, ESOPT-II*, Amsterdam, 2:863–870.

Senneset, K., Sandven, R., Lunne, T., By, T. and Amundesen, T. 1988. Piezocone testing in silty soil. *Penetration Testing '88*, AA Balkema Publishers, The Netherlands, 955–966.

Sharifounnasab, M. and Ullrich, R.C. 1985. Rate of shear effects on vane shear strength. *J. Geotech. Engng*, ASCE, 111(1):135–139.

Sharma, P.V. 1997. *Environmental and engineering geophysics*, Cambridge University Press, Cambridge, 475pp.

Sheahan, T.C., Ladd, C.C. and Germaine, J.T. 1996. Rate dependent undrained behavior of saturated clay. *J. Geotech. Engng*, ASCE, 122(2):99–108.

Shibata, T. and Teparaksa, W. 1988. Evaluation of liquefaction potential of soils using cone penetration testing. *Soils and Foundations, J. Japanese Soc. of Soil Mech. and Found. Engng*, 28(2):49–60.

Shibuya, S., Mitashi, T., Yamashita, S. and Tanaka, H. 1996. Recent Japanese practice for investigating elastic stiffness of ground. *Advances in Site Investigation Practice*, Thomas Telford, London, 875–886.

Shibuya, S., Hwang, S.C. and Mitachi, T. 1997a. Elastic shear modulus of soft clays from shear wave velocity measurement. *Géotechnique*, 47(3):593–601.

Shibuya, S., Mitachi, T., Fukuda, F. and Hosomi, A. 1997b. Modelling of strain-rate dependent deformation of clay at small strains. *Proc. 12th Int. Conf. on Soil Mech. and Geotech. Engng*, Hamburg, 1:409–412.

Shibuya, S., Yamashita, S., Watabe, Y. and Lo Presti, D.C.F. 2004. In situ seismic survey in characterising engineering properties of natural ground. *Proc. 2nd Int. Conf. on Site Charact.*, Milpress, Porto, 1:167–185.

Shioi, Y. and Fukui, J. 1982. Application of N-value to design of foundations in Japan. *Proc. 2nd European Symp. on Penetration Testing, ESOPT-II*, Amsterdam, 1:159–164.

Simonini, P. and Cola, S. 2000. Use of piezocone to predict maximum stiffness of Venetian soils. *J. Geotech. Geoenvironm. Engng*, ASCE, 126(4):378–381.

Simons, N.E. 1975. General report 'Normally consolidated and highly over-consolidated cohesive materials'. *Proc. Conf. of the British Geotech. Society on Settlements of Structures*, Cambridge, 500–530.

Simons, N.E. and Menzies, B.K. 1977. *A Short Course in Foundation Engineering*. London: Newnes-Butterworths.

Skempton, A.W. 1944. Notes on the compressibility of clays. *Q. J. Geol. Soc.*, London, 100:119–135.

Skempton, A.W. 1948. The $\phi = 0$ analysis of stability and its theoretical basis. *Proc. 2nd ICSMFE, Int. Conf. on Soil Mechanics and Found. Engng*, Rotterdam, 1:72–78.

Skempton, A.W. 1951. The bearing capacity of clays. *Proc. Building Reseach Congress*, 180–189.

Skempton, A.W. 1954. The pore-pressure coefficients A and B. *Géotechnique*, 4:143–147.

Skempton, A.W. 1957. Discussion of 'Planning and design of New Hong Kong Airport'. *Proc. Instn Civ. Engrs*, 7:305–307.

Skempton, A.W. 1986. Standard penetration test procedures and effects in sands of overburden pressure, relative density, particle size, aging and over consolidation. *Géotechnique*, 36(3):425–447.

Skempton, A.W. and Northey, R.D.T. 1952. The sensitivity of clays. *Géotechnique*, 3(1):72–78.

Skiles, D.L. and Townsend, F.C. '1994. Predicting shallow foundation settlement in sands from DMT. *Proc. Settlement '94 ASCE Spec. Conf.*, Texas A&M University Geotechnical Special Publication, 40(1):132–142.

Skov, R. 1982. Evaluation of stress wave measurements. DMT Grümdungstechnik, Hamburg, Germany.

Sladen, J.A., D'Hollander, R.D. and Krahn, J. 1985. The liquefaction of sands, a collapse surface approach. *Can. Geotech. J.*, 22(4):564–578.

Smith, A.D. and Richards, A.F. 1975. Vane shear strength at two high rotation rates. *Proc. Civil Engineering in the Oceans III*, ASCE, Delaware, I:421–433.

Smith, E.A.L. 1960. Pile driving analysis by the wave equation. *J. Soil Mech. Found. Div.*, ASCE, 86(4):35–61.

Smith, M.G. and Houlsby, G.T. 1995. Interpretation of the Marchetti dilatometer in clay. *Proc. 11th Int. Conf. Soil Mech. Found. Engng*, 1:247–252.

Soares, J.M.D. 1997. *Characterization of the Porto Alegre soft clay deposits*. PhD thesis, UFRGS, Porto Alegre, in Portuguese.

Sousa Coutinho, A.G.F. 1990. Radial expansion of cylindrical cavities in sandy soils: application to pressuremeter tests. *Can. Geotech. J.*, 27:737–748.

Souza Pinto, C. and Massad, F. 1978. Coefficient of consolidation of soils from Baixada Santista. *Proc. 6th Brazilian Conf. Soil Mech. Found. Engng*, Rio de Janeiro, 4:358–389.

Stark, T.D. and Olson, S.M. 1995. Liquefaction resistance using CPT and field case histories. *J. Geotech. Engng*, 121(12):856–869.

Stewart, D.P. and Randolph, M. 1991. A new site investigation tool for the centrifuge. *Int. Conf. Centrifuge Modelling – Centrifuge '91*, Boulder, Colorado, 531–538.

Stokoe II, K.H. and Nazarian, S. 1985. Use of Raleigh waves velocity by cross-hole method. *Proc. Measurement and Use of Shear Wave Velocity for Evaluating Dynamic Soil Properties*, ASCE, 1–17.

Stokoe II, K.H. and Santamarina, J.C. 2000. Seismic-wave-based testing in geotechnical engineering. *Proc. Int. Conf. on Geotech. and Geol. Engng*, Melbourne, 1:1490–1536.

Stokoe II, K.H., Lee, J.N.K. and Lee, S.H.H. 1991. Characterization of soil in calibration chambers with seismic waves. *Proc. 1st Int. Symp. on Calibration Chamber Testing*, Potsdam, New York.

Stokoe II, K.H., Hwang, S.K., Lee, J.N.K. and Andrus, R.D. 1994. Effects of various parameters on the stiffness and damping of soils at small to medium strains. *Proc. 1st Int. Conf. on Pre-failure Deformation Charact. of Geomaterials*, Sapporo, 2:785–816.

Stokoe II, K.H., Joh, S.H. and Woods, R.D. 2004. The contributions of in situ geophysical measurements to solving geotechnical engineering problems. *Proc. 2nd Int. Conf. on Site Charact.*, Milpress, Porto, 1:97–132.

Stroud, M.A. 1974. The standard penetration testing in insensitive clays and soft rocks. *Proc. 2nd European Symp. on Penetration Testing, ESOPT-II*, Amsterdam, AA Balkema Publishers, The Netherlands, 1:367–375.

Stroud, M.A. 1988. The standard penetration test – its application and interpretation. *Proc. Geotech. Conf. on Penetration Testing in the UK*, Thomas Telford, London, 89–95.

Stroud, M.A. and Butler, F.G. 1975. The standard penetration test and the engineering properties of glacial materials, *Symp. Engng Properties Glacial Materials*, Midlands Geotechnical Society, Birmingham, 117–128.

Strutymky, A.L., Sandiford, R.E. and Cavliere, D. 1991. Use of piezometric cone penetration testing with electrical conductivity measurements for the detection of hydrocarbon contamination in saturated granular soil. *Current Practices in Groundwater and Vadozone Investigations*, ASTM Special Technical Publication, STP 1118.

Su, S.F. and Liao, H.J. 2002. Influence of strength anisotropy on piezocone resistance in clay. *J. Geotech. Geoenvironm. Engng*, ASCE, 128(2):166–173.

Sully, J.P., Campanella, R.G. and Robertson, P.K. 1988. Overconsolidation ratio of clays from penetration pore pressure. *J. Geotech. Engng*, ASCE, 114(GT2):209–216.

Suzuki, Y., Tokimatsu, K., Taya, Y. and Kubota, Y. 1995. Correlation between CPTU data and dynamic properties of in situ frozen samples. *Proc. Int. Conf. on Recent Advances in Geotech. Earthquake Engng and Soil Dynamics*, St Louis, University of Missouri Rolla.

Suzuki, Y., Tokimatsu, K. and Koyamada, K. 1997. Prediction of liquefaction resistance based on CPT tip resistance and sleeve friction. *Proc. 14th Int. Conf. of Soil Mech. and Found. Engng*, Hamburg, Germany, 603–606.

Sykora, D.W. and Stoke, K.H. 1983. *Correlations of in situ measurements of shear wave velocity, soil characteristics and site conditions.* Geotech. Report GR83-33, Dept. Div. Engng, University of Texas, Austin, TX.

Tan, S.B., Tan, S.L., Lim, T.L. and Yang, K.S. 1987. Landslide problems and their control in Singapore. *Proc. 9th Southeast Asian Geotech. Conf.*, Bangkok, Thailand, 1:25–36.

Tanaka, H. 2002. A comparative study on geotechnical characteristics of marine soil deposits worldwide. *Proc. Int. J. Offshore and Polar Engng*, 12(2):81–89.

Tanaka, H. and Tanaka, M. 1998. Characterization of sandy soils using CPT and DMT. *Soils and Foundations*, Japanese Geotechnical Society, 38(3):55–65.

Tanaka, H., Tanaka, M. and Iguchi, H. 1994. Shear modulus of soft clay measured by various kinds of tests. *Proc. Symp. on Pre-failure Deformation of Geomaterials*, Sapporo, 1:235–240.

Tatsuoka, F. and Shibuya, S. 1991. Deformation characteristics of soils and rocks from field and laboratory tests. *Proc. 9th Asian Regional Conf. on Soil Mech. and Found. Engng*, Bangkok, 2:101–170.

Tatsuoka, F., Lo Presti, D.C.F. and Kohata, Y. 1995. Deformation characteristics of soils and soft rocks under monotonic and cyclic loads and their relationships. *Proc. 3rd Int. Conf. on Recent Advances in Geomech. Earthquake Engng and Soil Dynamics*, St Louis, MO, 2:851–879.

Tatsuoka, F., Jardine, R.J., Lo Presti, D., Di Benedetto, H. and Kodaka, T. 1997. Theme lecture: characterising the pre-failure deformation properties of geomaterials. *Proc. 14th Int. Conf. Soil Mech. Found. Engng*, Hamburg, 4:2129–2164.

Tavenas, F. and Leroueil, S. 1977. Effects of stresses and time on yielding of clays. *Proc. 9th Int. Conf. on Soil Mech. and Found. Engng*, Tokyo, 1:319–326.

Tavenas, F. and Leroueil, S. 1987. State-of-the-art on 'Laboratory and in situ stress-strain-time behavior of soft clays'. *Proc. Int. Symp. on Geotech. Engng of Soft Soils*, Mexico City, 2:1–46.

Tavenas, F., Jean, P., Leblond, P. and Leroueil, S. 1983. The permeability of natural soft clays. Part II: Permeability characteristics. *Can. Geotech. J.*, 20(4):645–660.

Taylor, R.N. 1995. *Geotechnical Centrifuge Technology*. Blackie Academic and Professional, London.

Teh, C.I. and Houlsby, G.T. 1989. An analytical study of the cone penetration test in clay. *OUEL Report No. 1800/89 (Soil Mechanics Report No. SM099/89)*, Department of Engineering Science, University of Oxford.

Teh, C.I. and Houlsby, G.T. 1991. An analytical study of the cone penetration test in clay. *Géotechnique*, 41(1):17–34.

Teixeira, A.H. 1988. Pile bearing capacity of driven piles in Quaternarium sediments. *Symp. on Quaternarium Brazilian Clay Deposits*, Rio de Janeiro, 2:1–25, in Portuguese.

Teixeira, A.H. 1996. *Design and construction of foundations*. Seminar on Special Found., São Paulo, Brazil, in Portuguese.

Terzaghi, K. 1943. *Theoretical Soil Mechanics*. John Wiley and Sons, New York.

Terzaghi, K. and Peck, R.B. 1948. *Soil Mechanics in Engineering Practice*. John Wiley and Sons, New York.

Terzaghi, K., Peck, R.B. and Mesri, G. 1996. *Soil Mechanics in Engineering Practice*. 3rd edn, John Wiley and Sons, New York.

Thompson, R.P. and Leach, B.A. 1989. The application of the SPT in weak sandstone and mudstone rocks. *Proc. ICE Conf. on Penetration Testing in the UK*, Birmingham, Thomas Telford, London, 21–24.

Thorburn, S. and Mac Vicar, S.L. 1971. Pile load tests to failure in the Clyde alluvium. *Behaviour of Piles*, ICE, London, 1(7):53–54.

Tice, J.A. and Knott, R.A. 2000. Geotechnical Planning, Design, and Construction for the Cape Hatteras Light Station Relocation. *Geo-Strata-Geo Institute of ASCE*, Vol. 3, No. 4, 18–23.

Timoshenko, S.P. and Goodier, J.N. 1970. *Theory of Elasticity*. McGraw-Hill Book Co., New York, NY.

Tokimatsu, K. 1988. Penetration tests for dynamic problems. *Proc. Int. Symp. on Penetration Testing, ISOPT-1*, Orlando, 1:117–136, AA Balkema Publishers, The Netherlands.

Tokimatsu, K. and Yoshimi, Y. 1983. Empirical correlation of soil liquefaction based on SPT N-value and fines content. *Soils and Foundations, J. Japanese Soc. of Soil Mech. and Found. Engng*, 23(4):56–74.

Tomlinson, M.J. 1969. *Foundation Design and Construction*. Pitman Publishing, 2:785.

Torstensson, B.A. 1977a. The pore pressure probe. *Norsk Jord-Og Fjellteknisk Forbund*. Oslo, Foredrag 34.1–34.15, Trondheim, Norway.

Torstensson, B.A. 1977b. Time-dependent effects in the field vane test. *Proc. Int. Symposium on Soft Clay*, Bangkok, 387–397.

Uriel, S. and Serrano, A.A. 1973. Geotechnical properties of two collapsible volcanic soils of low bulk density at the site of two dams in Canary Islands (Spain). *Proc. 8th Int. Conf. on Soil Mech. and Found. Engng*, Moscow, 2(2): 257–264.

US Army Corps of Engineers 2001. *Engineering Design: Geotechnical Investigations*. Manual 1110-1-1804.

Vaid, Y.P., Robertson, P.K. and Campanella, R.G. 1979. Strain rate behaviour of Saint-Jean-Vianney clay. *Can. Geotech. J.*, 16(1):34–42.

Van Impe, W.F. and Van der Broeck, M. 2001. Geotechnical characterization in offshore conditions. *Conf. Di Geotecnica di Torino*, Italy, 1–15.

Vargas, M. 1974. Engineering properties of residual soils from southern-central region of Brazil. *Proc. 2nd Int. Cong., IAEG*, São Paulo, 1:5.1–5.26.

Vaughan, P.R. 1985. Mechanical and hydraulic properties of tropical lateritic and saprolitic soil, General Report. *Int. Conf. Geomech. in Tropical Lateritic and Saprolitic Soils*, Brasilia, 3:231–263.

Vaughan, P.R. 1997. Engineering behaviour of weak rock: some answers and some questions. *Proc. 1st Int. Conf. on Hard Soils and Soft Rocks*, Athens, 3:1741–1765.

Veismanis, A. 1974. Laboratory investigation of electrical friction cone penetrometers in sands. *Proc. Europ. Symp. on Penetration Testing*, Stockholm, 2:407–419.

Vesic, A.S. 1972. Expansion of cavities in infinite soil mass. *J. Geotech. Engng Div.*, ASCE, 98(3):265–290.

Vesic, A.S. 1975. Principles of pile foundation design. *Soil Mech. Series*, 38, Durham, NC.

Vesic, A.S. 1977. *Design of pile foundations: synthesis of Highway Practice 42*, Transportation Research Board, National Research Council, Washington DC, 68 pp.

Viana da Fonseca, A. 1996. *Geomechanics of residual soils from Porto Granite: design criteria for shallow foundations*. PhD thesis, Porto University, Porto, in Portuguese.

Viana da Fonseca, A. 2003. Characterization and deriving engineering properties of a saprolitic soil from granite in Porto. *Charact. and Engng Properties of Natural Soils*, AA Balkema Publishing, The Netherlands, 2:1341–1378.

Villet, W.C.B. and Mitchell, J.K. 1981. Cone resistance, relative density and friction angle. *Cone Penetration Testing and Experience; Session at the ASCE National Convention*, St. Louis, American Society of Engineers, 178–207.

Walker, R.F. 1983. Vane shear strength testing. *In-situ Testing for Geotechnical Investigation*. AA Balkema Publishers, The Netherlands.

Watabe, Y., Tanaka, M. and Takemura, J. 2004. Evaluation of in-situ K_0 for Ariake, Bangkok and Hai-Phongs clays. *Proc. 2nd Int. Conf. on Site Charact.*, Milpress, Porto, 2:1765–1774.

Weltman, A.J. and Head, J.M. 1983. *Site Investigation Manual*. Construction Industry Research and Information Association, CIRIA Special Publication 25/ PSA Civil Engineering Technical Guide 35, London, 144pp.

Wheeler, S.J. and Sivakumar, V. 1995. An elasto-plastic critical state framework for unsaturated soil. *Géotechnique*, 45(1):35–53.

Wheeler, S.J., Cudny, M., Neher, H.P. and Wiltafsky, C. 2003. Some developments in constitutive modelling of soft clays. *Proc. Int. Workshop on Geotech. Soft Soils Theory and Practice*, 17–19 September, Noordwijkerhout, The Netherlands, Verlag Gluckauf GmbH, pp. 3–22.

Whittle, A.J. and Aubeny, C.P. 1993. The effects of installation disturbance on interpretation of in situ tests in clays. *Predictive Soil Mechanics*, Thomas Telford, London, 742–767.

Whittle, A.J. and Kavvadas, M.J. 1994. Formulation of MIT E3 constitutive model for overconsolidated clays. *J. Geotech. Cngng*, 120(1):173–198.

Whittle, R.W. 1999. Using non-linear elasticity to obtain the engineering properties of clay – a new solution for the self-boring pressuremeter test. *Ground Engng*, 30–34.

Wiesel, C.E. 1973. Some factors influencing in-situ vane tests results. *Proc. 8th Int. Conf. Soil Mech. Found. Engng*, Moscow, 1(2):475–479.

Windle, D. 1976. *In situ testing of soils with a self-boring pressuremeter*. PhD thesis, Cambridge University.

Windle, D. and Wroth, C.P. 1977. The use of a self-boring pressuremeter to determine the undrained properties of clays. *Ground Engng*, Sept., 10(6): 37–46.

Winn, K., Rahardjo, H. and She, C.P. 2001. Characterisation of residual soil in Singapore. *Geotech. Engng*, 23(1):1–14.

Withers, N.J., Schaap, L.H.J. and Dalton, C.P. 1986. The development of a full displacement pressuremeter. *Proc. 2nd Int. Symp. on the Pressuremeter and its Marine Applications*, College Station, Texas, ASTM Special Technical Publication, STP 950, 38–56.

Withers, N.J., Howie, J., Hughes, J.M.O. and Robertson, P.K. 1989. Performance and analysis of cone pressuremeter tests in sand. *Géotechnique*, 39:433–454.

Woeller, D.J., Weemees, I., Kohan, M., Jolly, G. and Robertson, P.K. 1991. Penetration testing for ground water contaminants. *Geotech. Engng Conf.*, ASCE, Boulder, Colorado, 1:76–83.

Woodward, M.B. and McIntosh, K.A. 1993. Case history: shallow foundation settlement prediction using the Marchetti dilatometer. *ASCE Annual Florida Sec. meeting* – abstract and conclusions.

Wride, C.E., Robertson, P.K., Biggar, K.W., Campanella, R.G., Hofmann, B.A., Hughes, J.M.O., Kupper, A. and Woeller, D.J. 2000. Interpretation of *in situ* test results from the Canlex sites. *Can. Geotech. J.*, 37(3):505–529.

Wright, S.J. and Reese, L.C. 1979. Design of large diameter bored piles. *Ground Engng*, 17–50.

Wroth, C.P. 1982. British experience with the self-boring pressuremeter. *Proc. Symp. on the Pressuremeter and its Marine Applications*. Institut du Petrole, Laboratoires des Ponts et Chaussées, Paris. Editions Technip, Collections Colloques et Seminaires, 37:143–164.

Wroth, C.P. 1984. The interpretation of in situ soil test. 24th Rankine Lecture. *Géotechnique*, 34(4):449–489.

Wroth, C.P. 1988. Penetration testing – a more rigorous approach to interpretation. *Proc. Int. Symp. on Penetration Testing, ISOPT-1*, Orlando, AA Balkema Publishers, The Netherlands, 1:421–432.

Wroth, C.P. and Basset, N. 1965. A stress-strain relationship for the shearing behaviour of sand. *Géotechnique*, 15(1):32–56.

Wroth, C.P. and Hughes, J.M.O. 1973. An instrument for the in situ measurement of the properties of soft clays. *Proc. 8th Int. Conf. on Soil Mech. and Found. Engng*, Moscow, 1(2):487–494.

Wroth, C.P. and Wood, D.M. 1978. The correlation of index properties of soils. *Can. Geotech. J.*, 15(2):137–145.

Wroth, C.P. and Houlsby, G.T. 1985. Soil mechanics – property characterization and analysis procedures. Theme lecture no. 1. *Proc. 11th Int. Conf. Soil Mech. Found. Engng*, 1:1–56.

Wroth, C.P., Randolph, M.F., Houlsby, G.T. and Fahey, M. 1979. *A Review of the Engineering Properties of Soils with Particular Reference to Shear Modulus*. International Report CUED/D Soils TR75, Cambridge University.

Wu, B.L., King, M.S. and Hudson, J.A. 1991. Stress induced anisotropy in rock and its influence on wellbore stability. *Rock Mechanics as a Multidisciplinary Science*, AA Balkema Publishers, The Netherlands, 941–950.

Yamashita, K., Tomono, M. and Kakurai, M. 1987. A method for estimating immediate settlement of piles and pile groups. *Soils and Foundations, J. Japanese Soc. of Soil Mech. and Found. Engng*, 27(1):61–76.

Yang, K.S. and Tang, S.K. 1997. Stabilising the slope of Bukit Gombak. *Proc. 3rd Youth Geotech. Engng*, Singapore, 589–605.

Yoshida, N., Tokimatsu, K., Yasuda, S., Kokusho, T. and Okimura, T. 2001. Geotechnical aspects of damage in Adapazari City during 1999 Kocaeli, Turkey earthquake. *Soils and Foundations*, 41:25–45.

Yoshida, Y., Kokusho, T. and Montonori, I. 1988. Empirical formulas of SPT blow-counts for gravelly soils. *Proc. Int. Symp. Penetration Testing, ISOPT-1*, Orlando, AA Balkema Publishers, The Netherlands, 1:381–391.

Yoshimi, Y., Tokimatsu, K. and Hosaka, Y. 1989. Evaluation of liquefaction resistance of clean sands based on high-quality undisturbed samples. *Soils and Foundations*, 29(1):93–104.

Youd, T.L. and Idriss, I.M. 2001. *Liquefaction Resistance of Soils: Summary Report from the 1996 NCEER and 1998 NCEER/NSF Workshops on Evaluation of Liquefaction Resistance of Soils*. ASCE, 127(4):297–313.

Young, R.N., Chen, C.K., Sellappah, J. and Chong, T.S. 1985. The characterization of residual soils in Singapore. *Proc. 8th Southeast Asian Geotech. Conf.*, Kuala Lumpur, Malaysia, 1:3-19–3-26.

Young, S.K. and Carter, J.P. 1990. Interpretation of the pressuremeter test in clay allowing for membrane end effects and material non-homogeneity. *Proc. 3rd Int. Symp. on Pressuremeters*, Oxford, 199–208.

Yu, H.S. 1990. *Cavity expansion theory and its application to the analysis of pressuremeters*. DPhil. thesis, Oxford University, England.

Yu, H.S. 1993. A new procedure for obtaining design parameters from pressuremeter tests. *Australian Civil Engng Trans.*, 35(4):353–359.

Yu, H.S. 1994. State parameter from self-boring pressuremeter tests in sand. *J. Geotech. Engng*, ASCE, 120(12):2118–2135.

Yu, H.S. 1996. Interpretation of pressuremeter unloading tests in sands. *Géotechnique*, 46(1):17–31.

Yu, H.S. 1998. CASM: a unified state parameter model for clay and sand. *Int. J. Num. Analy. Meth. Geomech.*, 22:621–653.

Yu, H.S. 2000. *Cavity Expansion Methods in Geomechanics*. Kluwer Academic Publishers, UK, 385pp.

Yu, H.S. 2004. The James K. Mitchell Lecture. In situ testing: from mechanics to prediction. *Proc. 2nd Int. Conf. on Site Charact.*, Milpress, Porto, 1:3–38.

Yu, H.S. and Houlsby, G. 1991. Finite cavity expansion in dilatant soils: loading analysis. *Géotechnique*, 41(2):173–183.

Yu, H.S. and Houlsby, G.T. 1995. A large strain analytical solution for cavity contraction in dilatant soils. *Int. J. for Numerical and Analytical Methods in Geomechanics*, 19(4):793–811.

Yu, H.S. and Collins, I.F. 1998. Analysis of self-boring pressuremeter tests in overconsolidated clays. *Géotechnique*, 48(5):689–693.

Yu, H.S. and Mitchell, J.K. 1998. Analysis of cone resistance: a brief review of methods. *J. Geotech. Geoenvironm. Engng*, ASCE, 124(2):140–149.

Yu, H.S. and Whittle, A.J. 1999. Combining strain path analysis and cavity expansion theory to estimate cone resistance in clay. Unpublished notes.

Yu, H.S., Carter, J.P. and Booker, J.R. 1993. Analysis of the dilatometer test in undrained clay. *Predictive Soil Mechanics*, London, 783–795.

Yu, H.S., Schnaid, F. and Collins, I.F. 1996. Analysis of cone pressuremeter tests in sands. *J. Geotech. Engng*, 122(8):623–632.

Yu, H.S., Hernann, L.R. and Boulanger, R.W. 2000. Analysis of steady cone penetration in clay. *Int. J. Num. Analyt. Methods in Geomech.*, 126(7):594–609.

Zhang, Z. and Tumay, M.T. 1999. Statistical to fuzzy approach toward CPT soil classification. *J. Geotech. Geoenvironm. Engng*, 125(3):179–186.

Zhou, Y. 2001. Engineering geology and rock mass properties of the Bukit Timah Granite. *Proc. of Underground Singapore 2001*, Singapore, 308–314.

Zhu, H. 2000. *Evaluation of load transfer behaviour of bored piles in residual soils incorporating construction effects.* PhD thesis, School of Civil and Structural Engineering, Nanyang Technological University, Singapore.

Zuidberg, H.M. 1988. Piezocone penetration testing – probe development. *Proc. Int. Symp. on Penetration Testing, ISOPT-1*, Orlando, Specialty Session no. 13, 24 March. AA Balkema Publishers, The Netherlands.

Zuidberg, H.M. and Post, M.L. 1995. The cone pressuremeter: an efficient way of pressuremeter testing. *Proc. Conf. on the Pressuremeter and its New Avenues*, Sherbrooke, Canada, AA Balkema Publishers, The Netherlands, 387–394.

Index

Milton Keynes UK
Ingram Content Group UK Ltd.
UKHW021630071024
449327UK00020BA/1262

9 780415 433860